THE
PEOPLE
SHAPERS

BOOKS BY VANCE PACKARD

The Hidden Persuaders
The Status Seekers
The Waste Makers
The Pyramid Climbers
The Naked Society
The Sexual Wilderness
A Nation of Strangers
The People Shapers

THE

PEOPLE SHAPERS

VANCE PACKARD

LITTLE, BROWN AND COMPANY　　BOSTON • TORONTO

Third Printing

T 10/77

LIBRARY OF CONGRESS CATALOGING IN PUBLICATION DATA

Packard, Vance Oakley, 1914–
  The people shapers.

  Includes bibliographical references and index.
  1. Eugenics. 2. Behavior modification. 3. Human
genetics — Social aspects. 4. Human engineering.
I. Title.
HQ751.P28   301.42'6   77–8973
ISBN 0–316–68750–2

Designed by Janis Capone

Published simultaneously in Canada
by Little, Brown & Company (Canada) Limited

PRINTED IN THE UNITED STATES OF AMERICA

To Jane Eager

We can choose to use our growing knowledge to enslave people in ways never dreamed of before, depersonalizing them, controlling them by means so carefully selected that they will perhaps never be aware of their loss of personhood.

— Carl Rogers, humanistic psychologist

Contents

III. Concerns and Countermeasures

THE
PEOPLE
SHAPERS

The Emerging Plastic Image of Man

1

We have not yet seen what man can make of man.
— B. F. Skinner, behavioral psychologist

B. F. Skinner's ringing pronouncement reflects ambition as much as fact. But dramatic efforts are indeed under way to reshape people and their behavior. These efforts obviously have profound implications. In quite a few instances the implications are disquieting.

Human engineers are at work in a variety of fields. They are increasing the capacity of a relatively small number of people to control, modify, manipulate, reshape the lives of a great number of other people. And they are functioning in many countries, especially in the United States, Great Britain, Germany, France, Japan, Canada, Israel, Russia, Australia, the Netherlands, and Scandinavia.

These new technologists draw primarily upon discoveries in the behavioral, biological, and computer sciences. Control is being achieved over human actions, moods, wishes, thoughts. As the psychologist Perry London put it: "Never in human history has this occurred before, except in fantasy."

The results may force us to alter our concepts of the nature of humanness.

Many of the revolutionists are caught up in a fervor not normally seen in scientists, who ordinarily speak cautiously. Columbus wasn't cautious and in substantial numbers they are not. They are now certain that new worlds can be discovered inside the human being. And they are eager to get on with the discovering. They see themselves out on the cutting edge of science.

One of the geneticists, Nobel Laureate Joshua Lederberg, called on the U.S. Congress to appropriate at least $10,000,000 to set up a national genetic task force. This force would make a crash effort to broaden knowledge of the genetic code, which would simplify the biological engineering of people.

Robert Sinsheimer has exulted that for the first time since the Creation a living creature can undertake to redesign its future. He has been serving as the head of the Biology Division at the California Institute of Technology and is a driving force in the current biological revolution. Sinsheimer asserted: "We can be the agent of transition to a wholly new path of evolution."

Among behavioral psychologists there are a host of restless revolutionaries. That most famed of the behaviorists, B. F. Skinner of Harvard University, has called for "a technology of behavior" because "we need to make vast changes in human behavior."

A few years ago a group of his disciples, in trying to describe "What Behavioral Engineering Is," explained: "For openers, we can develop a technology for routinely producing superior human beings. . . . We have the technology for installing any behavior we want."[1]

James V. McConnell, a wide-ranging psychological explorer from the University of Michigan, was quoted in 1974 as proposing: "We should reshape our society so that we all would be trained from birth to do what society wants us to do." (Knowing him, I suspect he was carried away by his natural exuberance and did not intend to make any ominous call for total conformity.)

Some of the revolutionists seem to relish the powers being achieved. José M. R. Delgado, a pioneer in brain probing and a most civil man, has called for physical control of the mind in order to develop a "psycho-civilized society." He has urged the U.S. government to make "conquering the mind" a national goal.

Other noted scientists, I should add, are cautioning the more outspoken activists. Leon Kass, a molecular biologist and ethicist, calls it an arrogant presumption that scientists think they are wise enough to remake Man.*

Some of the projects to reshape or control Man are simply intriguing. Many are disturbing. Some may make your skin crawl. By the latter I mean such plans as keeping people under surveillance by locking transmitters to their bodies, creating subhumans for menial work and as a source of spare parts for human bodies, transplanting heads, creating

---

* Throughout this book I capitalize the word *man* when using it in the dictionary sense of the human race, in order to avoid the awkward his-or-her construction. Unfortunately, males seem to have controlled the forming of English words describing the earth's foremost primate. Note the common singular forms hu*man* and *man*kind. Even words referring to persons of the feminine gender have a masculine base, as in wo*man* and fe*male*. There is, of course, *Homo sapiens,* but *Homo* means "man" in Latin. Modernization obviously is in order.

humans with four or more parents, and pacifying troublesome people, including children, by cutting into the brain.

The strategies being pressed by the human engineers have caused some observers to suggest that we are hurling toward the fictional worlds envisioned by two Englishmen. I refer to George Orwell's *1984* and Aldous Huxley's *Brave New World,* projected for six centuries from now. Actually, Orwell's Big Brother was pretty heavy-handed and simplistic compared with Huxley's World Controller, Mustapha Mond.

Orwell, writing in the period of Stalin and Hitler, primarily extrapolated totalitarian policing techniques. His Big Brother controlled by intrusion, such as installing a watching TV eye in every home and by setting up Thought Police. He ruled by exquisite forms of coercion, and by massive indoctrination.

Huxley's Controller, conceived earlier, in 1932, developed his controls by far more sophisticated scientific techniques. His Controller saw that total control should start at conception. In hatcheries made possible by reproductive biology, embryos were molded to order by genetic means to become humans of certain types. The level of intelligence was controlled in part by manipulating the amount of oxygen given the fetuses. Future sewer workers, who needed few brains, were mass-produced on low levels of oxygen.

Once new humans were born, a variety of controls continued, from infancy on. Persons were induced to love their assigned status and the regime by the use of "neo-Pavlovian conditioning" techniques, by sleep teaching, and by a wondrous "soma" drug. There is no single soma drug as such today but the states it produced — tranquillity, stimulation, colorful visions, and apparently high suggestibility — all can now be produced by specific drugs or precise techniques.

Most of the techniques Huxley fantasized for the distant future are already becoming available or are at least being forecast by reputable scientists. One of them is sleep teaching. Modern experiments have shown that a person can learn from messages whispered in his ear as he is falling asleep. Huxley (erroneously as it turned out) had the person deeply asleep when learning from the whispered messages, but his idea was valid in the main.

Today there is no all-powerful Controller in sight in the Western world. The coming crunch on natural resources combined with rampant overbreeding in many areas of the world make it likely that we will be hearing calls for more authoritarian governments within the coming quarter century. At any rate there are already a host of technologists in a variety of fields who qualify as people controllers or people shapers. They are becoming, willingly or not, a new elite. Many work for institutions, including governments, to help those institutions increase their power to control us and to impose their values and views on others. Meanwhile, the ordinary person's sense of power over his life is threatened.

A few years ago I participated in a conference in Wyoming on the topic "Captive Man in a Free Society." One of the other speakers was the late Sidney Jourard, a noted humanistic psychologist. In talking of the technologists of control — and some of them were present — he made a comment that remains etched in my mind:

"The worst thing that can happen is that a person doesn't count."

We have always had masters, not only Presidents but tax assessors, school boards, draft boards, zoning commissions, and a hierarchy of on-the-job bosses. Now in a sense we are confronted with a new set of masters to add to those we already have. Many are earnest benefactors out to improve the well-being of Man and are achieving commendable results. But the potential is there for influencing our lives, our society, and future generations, sometimes in adverse ways the benefactors have not thought about.

We need to get to know these people who possess a new ability to shape our behavior and life development. And often it is not easy to find them. Some flourish behind a scientific verbiage that discourages understanding by anyone outside their particular discipline. They toss around such words and phrases as *neuropsychopharmacology, heterozygotes, contingencies of reinforcement, psychoneurobiochemistry.*

But the activists have not hedged in talking about what they aim to achieve. I will be using phrases encountered in their own literature when I refer to "behavior control," "human engineering," "genetic tailoring," "biomedical engineers," "programming people," "manipulating human behavior," "shaping behavior." In fact, behavior control was the theme of the Master Lecture Series at the 1976 convention of the American Psychological Association in Washington. The eight speakers all gave talks with "behavior control" in the title.

The techniques scientists are using or plan to use for reshaping people and their behavior include the following:

> Stimulating and modifying the brain
> Programming behavior
> Manipulating genes
> Controlling people by radio
> Creating refined techniques for managing large groups of people
> Marrying man to animal and man to machine
> Creating new and startling forms of surveillance
> Manufacturing hearts and other organs
> Altering dramatically the beginning and end of life

IMAGES OF MAN

Such capabilities make it timely that we think anew about a number of fundamental questions:

What is Man?
How should humans be treated by their fellow humans?
Is Man something special, a being set apart from other creatures, or not?
Are humans essentially self-directed or is their behavior determined essentially by other forces?
What paths should future human development ideally follow?
What is a reasonable model of Man for tomorrow?

The image of Man we hold today and the model of Man we want to realize in the future are important. They substantially influence how people perceive themselves. And they help determine what goals are desirable.

Man's nature has been a source of argument for at least thirty centuries, and there have recently been some significant shifts in emphasis.[2]

Judeo-Christian doctrine generated the widespread belief that God created Man in his own image. That was comforting, if ambiguous, and implied that unspoiled Man had only good qualities. Judeo-Christian theologians also had great success in promoting the idea that humans, and only humans, had a touch of divinity. The popularity of this image largely explains why the first reports of Darwin's theory that humans evolved from animals created such widespread consternation.

In the seventeenth century the philosophers Thomas Hobbes and John Locke took up radically different positions on the nature of Man. In Hobbes's view Man was an elaborate machine controlled by outside stimuli. Individual humans were so irresponsible, aggressive, and ego-driven that they needed an absolute monarch to control their lives in detail.

Locke believed that each individual deserved to be accepted as a rational, responsible, tolerant, self-directed being until proved otherwise. An individual grew from experience and awareness rather than being in the grip of instincts, and should concede only limited regulation of his affairs to government.

In the eighteenth century two influential shapers of the new nation in America — Alexander Hamilton and Thomas Jefferson — paired off in much the same way as Hobbes and Locke did, except that Hamilton was not as arrogant or pessimistic as Hobbes. Jefferson's view was that people normally will be sensible and fair in dealing with each other and with any government, and deserve to be trusted. This view became accepted by most officials and laymen alike during the early decades of the American Republic.

The life roles people are expected to play have had their impact on some latter-day images of Man. For example:

• With industrialization, people often became valued only to the extent that they were efficient units of production.
• More recently, with the superabundance of goods made possible by mass production in many countries, families have become esteemed by business and government leaders largely to the extent that they are ardent consumers.
• In those societies experimenting with Marxism or Maoism, the individual is valued only to the extent that he selflessly contributes to the group effort.

In recent decades, scientists have come up with a number of novel images of Man. Six of them are described in highly abbreviated form in the next paragraphs. Most have substantial contemporary support. (For a more extensive analysis of "Man's nature" arising from psychological and philosophical thought, see the Appendix.)

Man is a bad animal. This image, which got some support from Freudian views about the often ugly unconscious side of Man, was recently revived. Archaeologists claimed to have uncovered evidence that very early Man began as a meat-eating hunter. From this it was extrapolated that aggression was a part of Man's biological nature. Experts in animal behavior (ethologists) pointed out that almost all species except Man have a built-in prohibition against killing their own kind if they can possibly help it. In quarrels over mates and territory they try to keep contention as bloodless as possible. Man stands almost alone as a systematic murderer of his own kind. In wars he often kills wantonly and with seeming enthusiasm.

The best-known ethologist, Konrad Lorenz, has argued that the discovery of weapons made it easy for men to kill each other. The adoption of weapons was so rapid that inhibition against their use did not have time to evolve. Lorenz's theory may account for the beginning of Man's homicidal habits. But since we have had effective weapons for hundreds of thousands of years now, we have certainly had time to evolve an inhibition if we really wanted to. It is more plausible to suggest that weapons have made people-killing more tempting.

On the other side of the argument, a number of critics consider the view of Man as a bad animal overblown. They have observed, for example, that Man has never been just a carnivore. He has eaten everything he could get. Time and time again men have demonstrated that they can be altruistic and cooperative, that they abhor violence. According to one hypothesis about Man the Hunter, altruism and cooperation had survival value. And if Man were restricted to killing the hard way, by tooth and claw, without weapons, as animals must do, we surely would see a spectacular decline in intentional homicide.

The important point is that scientists in many fields are coming up with strategies to remove the alleged inborn aggressiveness of Man.

Man is controlled by his genes. This tendency to physiological predestination has been known to college freshmen for decades. Recent argument has centered around the contention of two educational psychologists that scores on intelligence tests are primarily determined by genes. They have asserted further that consistent differences in the scores were related to race or social class. Their conclusions upset those who believe that basically we start off with a fairly clean slate and that major differences arise from the way we are reared. (As of late 1976, the Genetics Society of America was contending that it had no convincing evidence of any appreciable genetic difference in intelligence among races.)

One of the educational psychologists cited above, Arthur Jensen, has maintained that 80 percent of IQ differences are based on genes, not on the effects of environment. His claims have been denounced. Several leading geneticists have advanced rebuttals or strong reservations. One, Theodosius Dobzhansky, pointed out that you don't find an 80-percent degree of heritability in far simpler matters, such as chicken-egg production.

In late 1976 the genetic dogmatists were subdued by a scientific scandal. It appeared that an early English pioneer in intelligence testing, who had tied intelligence to genes quite impressively from his study of twins, may have invented some of his data.

Whatever the facts, the big new thing is that genes are now considered manipulatable, not immutable. The genetic engineers have a host of plans, some already tested, for modifying Man by modifying his genetic patterns.

Man is shaped by primal instincts and early childhood experiences. This view, pioneered by Sigmund Freud, remains the cornerstone of one of modern psychology's three major schools of thought about the nature of Man, the other two being the humanistic and the behaviorist.

The instincts that concerned Freud were those of sexuality, aggression, pleasure-seeking, and so on. They may be masked by people, but they remain in force and influence behavior. The challenge is to cope with these debilitating urges so that civilization can stay on the track.

Modern neo-Freudians continue to use Freud's terms and some of his concepts, but most do not see Man as essentially base. And they give credit to the way we live as a shaping influence.

Man is captain of his fate. Humanists such as Abraham Maslow, Carl Rogers, and Rollo May have seen Man as possessing a considerable amount of free will despite genes, instincts, and the environment. The individual can size up his situation, weigh alternatives, and decide upon courses of action. Some humanists from the existential wing, notably Jean-Paul Sartre, have insisted that Man can be a *totally* free agent. Those from the vitalist wing believe there is no physical accounting for the inspired behavior of some people. The historian-philosopher Arnold Toynbee, for example, has contended that there is a spark of "creative spiritual power in every human being."

Humanists in general hold that life provides the greatest satisfactions if people largely manage their own affairs. They work to enhance our capacity for self-fulfillment, creativity, open communication, and psychological growth.

Man is a mere reactor to prods from the environment. This proposition is assumed by most, but not all, behavioral psychologists. For some years behaviorism has been in the ascendancy. Shortly we will be looking at the activities of certain behaviorists in detail.

Behaviorists tend to see themselves as far more scientific than the humanists or psychoanalysts. They deal only with observable behavior and are wary of the "mind" since what goes on in it is not observable. They measure behavior, sometimes with stopwatches, and design precise testing strategies. Often they aim for, and get, quick, visible results. The more ardent behaviorists argue that real progress for Man can come only if people rid themselves of prescientific ideas about freedom, will, consciousness, and dignity.

As a general rule behaviorists are fascinated with techniques of control. Many favor using conditioning techniques. They modify the behavior of humans by shifting factors in the environment and manipulating incentives. Many believe that Man can progress fastest by permitting his environment to be systematically managed.

In short, they tend strongly to the view that Man is highly malleable and needs molding.

Man is an adjustable, chemically controlled machine. The most forthright spokesman of this still-marginal view is a neo-Hobbesian and a radical behaviorist, H. L. Newbold. He is a New York orthomolecular psychiatrist, and thus is a believer in using chemicals to enhance the success of behavioral therapy.

Behaviorists have often been described by critics as being mechanistic. Newbold goes further and calls Man a machine. In his book *The Psychiatric Programming of People,* which is addressed to psychiatrists, psychologists, and those preparing to be psychiatrists, he says of Man's nature: "Man is, for psychological purposes, a computer. . . . This computer can fail to function properly if the hardware (central nervous system) is physically or chemically damaged or if the software (Biologically Programmed Computer or Socially Programmed Computer) is abnormal. . . . It follows then that, if damaged or abnormal, the computer must be, in so far as is possible, normalized . . . with psychochemicals" to get it to function more effectively.[3]

Newbold is not alone in his opinions. One of the speakers of the Master Lecture Series at the 1976 convention of the American Psychological Association alluded to the Man-is-a-machine concept. He said that the flowering of behavior modification has forced people to take the concept seriously. The approach is to modify Man in a desired direction by pressing social and biological levers.

So where do we stand? If we ponder the underlying assumptions of most of the scientific images of Man cited, a common thread emerges. The same thread appears when we consider assumptions of pioneering activists from such diverse fields as reproductive biology, psychosurgery, and molecular biology.

What emerges is a pervading assumption that humans are creatures of almost limitless plasticity.

People are raw material that needs perfecting, modifying, or at least improving, either for their own good or to suit the wishes of others. Malleable people are more likely to be controllable people.

Whereas the old believers in the perfectibility of people thought primarily in moral terms, the new revolutionists want to change people physically, emotionally, mentally. Often their efforts are underwritten by the government.

If their view prevails people will indeed become different. The revolutionary brain prober José Delgado suggests that the main question is no longer "What is Man?" but rather, "What kind of man are we going to construct?"[4]

Under expert management, some scientists feel, people will benefit by becoming more efficient and predictable. Quite probably as a side effect people will also lose some of what we think of as humanness.

In each of my books of social comment I have sought to examine what is happening to the individual in the face of rapid social and technological changes in a specific area of life. This book will seek to bring into convergence developments from several fields of scientific exploration in which investigators are testing the plasticity of Man. The investigators' goal in all cases is to modify or control people or shape the way they develop. Implications for both the present and the future will be discussed.

The ascent of Man as described by Jacob Bronowski covered tens of thousands of years. The reshaping of Man now under way can occur within a few decades.

TECHNIQUES FOR CONTROLLING BEHAVIOR

I

Pioneers in Programming Pigeons and People 2

The technology of behavior control makes it possible today to exact individual conformity with greater reliability and less resistance than ever before.

— Perry London, psychologist

Programming people means getting others to act consistently as you want them to act. Stern parents or employers often are pretty good at this, at least while the subjects are under observation. Hypnotists can obtain excellent results in achieving desired behavior from suggestible subjects for short periods.

What interests us here are precise techniques for altering long-term behavior patterns in predictable ways. These new patterns may be desired by the subject or by the programmer or by the organization employing him.

For achieving certain kinds of long-term programmed behavior the programmer need not be a scientifically trained technologist. Consider how the intense and unattractive Charles Manson horrified and fascinated millions of people a few years ago by his control methods. He had an ability to induce sustained zombie-like behavior in his followers, mostly girls. They committed random murders in the Los Angeles area. When a number of his "slaves" faced trial they vigorously asserted that the murders were their own idea. They wanted to protect Charlie, who was always somewhere else when the butcheries occurred.

In order to prove his theory that Manson had masterminded the killings the prosecutor, Vincent Bugliosi, had to spend months uncovering and analyzing the sources of Manson's control over the presumably free and footloose young people.[1] His most important findings were these:

- Manson was gifted at perceiving the psychological needs of others. He assured runaway girls needing a father that he would be their father. He assured plain-looking girls that they were beautiful.
- He was careful to destroy preexisting identities. All the members of his clan had to take on new names.
- He systematically destroyed inhibitions as a part of his obedience training. He supervised naked orgies. When he sensed that a recruit was reluctant to engage in an unusual sexual act he ran the recruit through the whole repertoire of homosexuality, cunnilingus, fellatio, sodomy.
- He offered these insecure youngsters a bizarre religion, in which he was the Infinite Being who would lead them to a world of milk and honey.
- He was careful to identify and probe what each recruit was most afraid of, and to play on it.
- Finally, Manson apparently had some hypnotic powers.

Bugliosi succeeded in convincing the jury that Manson was, indeed, responsible for the murders.

THE THREE D'S OF CONVERSION UNDER COERCION

It is possible to reprogram people's personalities and beliefs even when they resist if coercive techniques can be employed. A number of studies have been made of the scientific brainwashing techniques used by Asian Communists on captured American military personnel. Studies also have been made of techniques used by the Russians in their staged trials during the 1930's, and by the Nazis on early political dissidents. Investigators have searched for the common strategies used in shattering and remolding personalities.[2]

Brainwashing is not as mysterious as it once seemed. It involves very little reliance on drugs. The so-called truth drugs used even for interrogation have been overrated. They bring out fantasies as well as fact. Drugs have a modest success rate in ferreting out facts the subject does not wish to reveal, especially if he is stable and highly motivated. In true brainwashing there is also relatively little use of excruciating torture, such as is still used in several dozen countries in the routine interrogation of suspects, dissidents, and the like. The brainwashers were less interested in obtaining information than in obtaining converts. The latter could be used for propaganda purposes in broadcasting and for writing denunciations of their homeland.

For the personality-transformers a major early objective is to destroy the prisoner's identity, disorganize his self concept. That was also the goal of Charlie Manson. The Nazis, for example, called everyone by number. The use of names was forbidden.

Three states of mind or body, sometimes called the Three D's, contribute to the destruction of identity and a readiness for conversion:

Debilitation. The prisoner is made to stand still for many exhausting hours. He is put in a sweatbox. He is interrogated night and day, with only a few hours for sleep. He is rationed small amounts of water. All this unsettles the normal internal stability of the body and mind. The mind becomes further debilitated as the prisoner is isolated in dark places. Tests at McGill University in Canada have shown that such isolation can dramatically alter mental functioning within a few days.

Dread. The prisoner is constantly threatened with torture or death. He is treated roughly, sometimes whipped. He is encouraged to doubt that he will ever see his homeland again, or even survive. He comes to dread humiliation or degradation. Asians sometimes tied prisoners' arms behind their backs and forced them to eat like dogs.

Dependency. The prisoner is led to realize that his fate is at the whim of his captors. He is helplessly dependent on them for sleep or food. As the prisoner shows signs that his identity is in disarray and may be approaching a crisis point the captors will suddenly, briefly, become kindly. They will become apologetic, offer cigarettes, chat. Then, just as suddenly, they go back to the strategies of dread and debilitation. The prisoner is given a glimpse that his captors can be welcome friends if only he accepts their viewpoint.

When a prisoner's personality starts collapsing he is welcomed as a convert and is immediately sent to all-day classes on thought reform. Fellow prisoners, already transformed, join in pressing the new way of thinking. Some prisoners, it should be added, have been able to resist the full Three-D strategy.

More recently, brainwashing figured in the trial of Patty Hearst. She had been converted to being an active member of the gang that had kidnapped her. After her arrest she was charged with participating in a holdup. At the trial one psychiatrist contended that she had been brainwashed and reprogrammed by the Three-D method of "debility, dependency and dread." Another psychiatrist, however, cited evidence that she had been a rebel before the kidnapping and was "ripe for the plucking."

APPROACHES OF PSYCHOLOGISTS IN CHANGING MENTAL STATES

Consider the reprogramming of people as practiced by psychoanalytically oriented therapists. The goal usually is to lead the client to new insights about himself by a dredging of formative events in his life. In the process the therapist moves into a position of ascendancy. Often the client becomes extremely eager for the therapist's approval, and as the therapy continues, may seek to take on the therapist's values or those of society as interpreted to him by the therapist. In the latter situation the therapist can be a kind of double agent.

Humanistically oriented therapists use many of the traditional insight techniques but make much more point of leveling with the people they work with. They typically encourage the client to set the goals he wants to work toward, such as greater self-reliance.

It is the conditioners in the behaviorist school of psychology who have been openly fascinated with the control, manipulation, and programming of people. Some are psychiatrists, some are not. The people to be controlled need not necessarily have severe emotional problems. The pioneering by behaviorists in conditioning humans started in mental hospitals. However, as time went on, the techniques were applied to people of all ages without emotional problems. Now, many of the practitioners want to transform society for the greater happiness of all.

Behaviorists tend to reject the Freudian concept that you change a person's behavior by getting inside his mind and helping him gain insight. They feel you don't need to help him understand his levels of consciousness, his id, his superego, and so on. Most behaviorists want to change observable behavior in the subjects they are working with. Above all they are pragmatists. Members of a major and fast-growing wing of behavioral psychology seek behavioral change by conditioning (or programming) subjects. When these conditioners leave the laboratory and move out into society they call themselves behavior shapers or behavioral engineers. They are confident enough to take on any project for changing people's behavior that challenges them.

THE ORIGINS OF SCIENTIFIC CONDITIONING

What the conditioners call the *science* of behavior control started in a Russian laboratory, with Ivan Pavlov's discovery of the conditioned reflex. This was at the beginning of the century. As most readers know, Pavlov rigged up a tubing arrangement to measure how much a dog salivated. He put ground meat on the tongue of a harnessed dog and in the same instant rang a bell. After numerous repetitions and measurement of saliva, he began ringing the bell but omitting the meat. The outpouring of saliva was the same as when the meat was offered. The result has become known as "classical conditioning." It is conditioning based on a simple *reaction* to stimuli.

At about the same time the educational psychologist Edward Thorndike in the United States was tempting cats with food at Columbia University. The cats were inside cages. As they thrashed around trying to reach the food in full sight, their claws finally hit a loop of string and opened the cage. The cats didn't mentally make the cage-string connection right away. But by trial and error they did in subsequent tests claw closer and closer to the string. Soon they learned to go to it directly upon the sight of food. The cats were not just *reacting* by salivating like Pavlov's dogs. They were conditioned by the reward of food to take affirmative *action:* looking for the string each time they saw the food.

Thorndike assumed from this that human learning might proceed faster if students were rewarded for their correct answers.

About one and a half decades later an idea-popping psychologist at Johns Hopkins University, John B. Watson, heard of Pavlov's experiments. He let his imagination take flight and mapped out a whole new science, which he called behaviorism. As noted, its central doctrine is that observed behavior offers the only acceptable psychological data. What goes on inside the mind is irrelevant. Watson proposed that you could, by conditioning, program an animal or person into performing quite complex patterns of behavior. The strategy was to analyze all the actions that had to be performed and condition the subject one step at a time. His extravagant imagination later led him to become a very successful advertising expert. Still he had announced a formal science. And he had correctly forecast how future behavioral engineers could operate by fragmenting behavior patterns into bits to be mastered one at a time, in sequence. The arranged "environment" was all-important in shaping behavior. He even claimed he could select at random a healthy infant and train it to become any kind of specialist you might suggest, whether doctor, lawyer or safecracker.

A couple of decades later another pioneer in conditioning, the South Africa-born psychiatrist Joseph Wolpe, emerged and still has wide influence. He experimented with desensitizing people who were unreasonably scared of something. His best-known study was of people with snake phobias. First he induced the subjects into deep relaxation. Then he introduced the least threatening stimulus possible, a passing reference to snakes. Step by step he went from getting the phobic patient to handle toy snakes, to watching a small, harmless snake sleeping thirty feet away. In a few weeks he had the former snake phobic playing with snakes.

ENTER SKINNER AND HIS DISCIPLES

It was a Harvard psychologist named Burrhus Frederic Skinner who is generally recognized as the founder of modern behaviorism, especially that aspect involving what is called operant conditioning. A lean, impatient, cantankerous, often beguiling man, Skinner has become a dominant figure in the world of psychology.

Almost all of his early research before 1950 was with pigeons and rats. He has in the past two decades done some research involving people, especially when he was perfecting his concept of the teaching machine. But many readers of his books find his pronouncements about people exasperatingly short on documentation.

At any rate, Skinner has developed a great array of ideas about changing the world of people. Many of his ideas about people are unflattering. He sees broad-scale control of people and the reconditioning of them as our best hope of saving the Western world. Otherwise, he warns, "some other group" may become more proficient at controlling behavior and directing it "into paths we consider undesirable." That group, we are left to assume, might be a Communist country or some nasty Big Brother. When he was asked if a totalitarian regime could use his concepts of behavior modification to set up or maintain a repressive state he acknowledged that it probably could.

Skinner until recently has rarely been modest. He states that behaviorism (as he sees it) "calls for probably the most drastic change ever proposed in our way of thinking about man."

So we had better get to know him and his disciples a little better.

Skinner set out in life from Susquehanna, Pennsylvania, to become a writer. He soon decided he had nothing important to say. Because of his boyhood fascination with animal behavior he turned to psychology.

His longtime interest in mechanics may also have influenced the type of research he did. (And it may have contributed to his mechanistic view of human behavior.) As a boy, he relates, he was always inventing things. He converted a discarded water boiler into a steam cannon. While selling elderberries door to door he invented a flotation system that separated ripe from green berries. He confesses that he worked for years, in vain, on designs for a perpetual-motion machine.

At any rate he won his first real recognition as a psychologist largely because of his mechanical ingenuity. He developed cages that automatically fed rats on a given schedule if the rats behaved as desired. A machine automatically counted the rat's rate of eating. This evolved into the now widely used "Skinner box" for conditioning animals.

Skinner began conditioning pigeons to perform fairly complex feats by breaking each feat down into its component parts (as Watson had proposed). The pigeon was rewarded with a food pellet for each correct move, no matter how slight. This gave the bird an incentive to repeat the move. To get a pigeon to walk in a circle Skinner rewarded its first slight clockwise movement. Soon he had pigeons doing figure eights, dancing, and playing a kind of table tennis. Later he got animals to perform on cue even though they had to make a certain number of moves before they got their reward. The precise spacing and timing of reinforcements at each step became an important part of Skinner's contribution to conditioning.

Skinner and others also conditioned animals to avoid certain prohibited behavior by setting up automatic punishments such as electric stings.

By the 1950's, after visiting his daughter's class in arithmetic, Skinner began to devise teaching machines. He also programmed texts for a variety of school subjects. His concept again was the same as teaching a pigeon to walk in circles. Each bit of the whole problem or lesson was learned a step at a time. The reward was a signal that notified the learner that he had been successful and provided him with immediate reinforcement.

Readers may wonder what is so revolutionary about Skinner's feats of conditioning. Most of them had been anticipated by Thorndike or Watson. In fact, the step-at-a-time conditioning of animals to perform impressive feats had been perfected by Russian bear trainers and Austrian horse trainers hundreds of years ago. And the phrase "the carrot and the stick," meaning two alternative ways for getting behavioral results, has been in the English language for decades. Nearly two hundred years ago Jeremy Bentham gained wide publicity from his concept that Man seeks pleasure, avoids pain. What is brand-new about Skinner's achievements is the use of precise, scientific conditioning techniques on humans.

After Skinner had published one of his articles on what he was doing with his animals, two European psychologists picked an argument with him. They claimed to have earlier published quite similar findings. In order to defend himself he began trying to think through what kind of conditioning had been going on in his experiments.[3]

Clearly he had gone beyond Pavlov, whose animals had simply reacted. Animals under Skinner acted. They had acted repeatedly in predictable patterns (just as Thorndike's cats had). Why had they acted? They repeated their actions on cue because they remembered the *consequences* that had occurred in the same situation. They had either been rewarded afterward or they had escaped pain afterward by performing correctly. This memory developed a strong incentive always to perform as Skinner, the controller, desired. That was it! Behavior is controlled by the *consequences* of previous behavior in the same situation. More recently Israel

Goldiamond, also a specialist in conditioning, put it this way: "The consequences are contingent on the behavior and maintain it."

Skinner put the formidable label *operant conditioning* on what happened under his method. He felt that *operant* was a good word to use because "behavior which operates upon the environment to produce consequences [operant behavior] can be studied by arranging environments in which specific consequences are contingent upon it." Get it? You put Thorndike's cat in an environment (cage with a string) in which specific consequences (getting the food) are contingent upon hitting the string.

Skinner has since acknowledged that he and his followers have simply taken Thorndike's law of effect seriously.[4] Seriously indeed! They have constructed a precise methodology and describe it in terms incomprehensible to laymen — or to a Russian bear trainer. For example:

> "Stimulus-response-reinforcement paradigm"
> "The technology of contingency management"
> "Aversive stimuli"
> "A schedule of intermittent reinforcement"

The word *environment* is central to any operant conditioner's vocabulary. Environment (not heredity or passion or thought) is what makes us behave the way we do. And you control the behavior of animals and people by manipulating their environment to suit your purposes.

Trying to grasp the meaning of "environment" as conditioners use it, however, can be as difficult as trying to grasp a greased pig. Environment can be a physical setup. It can be a prevailing social pattern. It can be a person's life history. I like the word *situation* better.

A *stimulus* can be a promise of M & M candy drops that makes you act as desired. And the enjoyment of the candy *reinforces* the probability that you will repeat the performance on cue. You can even be induced to accept a *token economy,* another favorite concept. You can be trained to work hard for plastic chips because you know that when you acquire the specified number you will be given a whole bag of M & M's.

To sum up this technology of operant conditioning that attracts so many practitioners: We fall into patterns of behavior desired by the controller because he gives us incentives (consequences) to repeat the pattern again and again. The incentive can be a *positive reinforcer* like candy, or a *negative reinforcer* like a whack from a stick. Negative reinforcers are also known as aversive stimuli.

And the whole process is often called behavior shaping.

What today's behavior shapers have added to what the Russian bear trainers know is this: The conditioning method can be developed into a precise technology that can be applied, under the right conditions, to pigeons and people, and in a variety of situations.

This technology of behavior shaping was being worked out by many besides Skinner. Some of these fellow pioneers were flamboyant plungers, some very cautious experimenters. Two of Skinner's disciples, Nathan H. Azrin and Ogden R. Lindsley, moved his concepts from pigeons to people by working with mental patients. Lindsley later reported in an advertising journal on ways to test the effectiveness of TV commercials by utilizing insights gained from operant conditioning. His technique measured how hard a viewer will work to see the commercial clearly. And still later he set up the Behavior Research Company in Kansas City with a computerized "behavior bank" containing all kinds of facts and programs for modifying behavior.

Charles B. Ferster collaborated with Skinner in working out precise schedules for reinforcing behavior. Lloyd E. Homme worked with him in programming teaching machines. Fred S. Keller sought to analyze how humans learn. David Premack, now famous for teaching chimpanzees to communicate with him, hammered out the Premack Principle, which broadened understanding of the various kinds of reinforcements that can motivate desired behaviors. Leonard Krasner, who, at a chance encounter, first opened my eyes to the scope of the people-shaping movement, pioneered the use of reinforcement in psychotherapy. In the United States, enthusiasts of behavioral engineering have tended to congregate at Harvard, Southern Illinois University, the University of Kansas, Western Michigan University, the University of Washington, and Arizona State University.[5]

At some universities considerable elbowing has been going on between behaviorists and nonbehaviorists. At the University of Washington for example, the behaviorists lost out in a power play and many went to the University of Kansas, where behaviorists were more firmly entrenched. On the other hand, two scholars at another northwestern university reportedly failed to get tenure because behaviorists had the upper hand in the department.

While there were many pioneers, B. F. Skinner remains the guru. Partly this is because of the publicity given to his outspoken opinions about the nature of Man and how Man should be reshaped.

SKINNER'S VISIONS OF HAPPY, CONTROLLED SOCIETIES

In 1948, Skinner, the frustrated writer, drew upon his experiments with rats to write a novel, *Walden II,* about a human utopia. It was quite a leap. Skinner's Walden II is a commune where everything, including work, is shared. Everyone is contented. Everyone is free of jealousy. Everyone's behavior is substantially controlled by sound behavioral engineering principles. The founder of the commune, Frazier, is a man who talks a lot like Skinner.

Below Frazier there are six Planners. Below them are Managers for every aspect of community life. Members are sometimes called controllees. They follow the commune's rather austere "Code of Conduct." Midnight snacks are forbidden. Members are to be quite puritanical about sex. In regard to the outside world, the Political Manager develops a Walden Ticket for local, state, and federal elections. Everyone votes for it. Founder Frazier explains: "And why not? . . . Remember our interests are all alike and our Political Manager is in the best position to tell us what candidates will act in those interests. Why should our members take the time — and it does take time — to inform themselves on so complex a matter?"

Through positive and negative "reinforcement" all significant behavior that the Planners want to encourage or discourage is guided. Frazier adds: "If it's in our power to create any of the situations which a person likes or to remove any situation he doesn't like, we can control his behavior." People, in short, are seen as high on plasticity.

In Walden II, education is handled as an aspect of "human engineering" (that is, conditioning). Children are cared for primarily by the group and as little as possible by parents. Frazier: "Home is no place to raise children." Marriage bonds too are tenuous. The Planners hope that with the weakening of family structure, experimental breeding will become feasible.

Meanwhile husband and wife sleep in separate rooms.

Even in his utopian Walden II of 1948, Skinner suggested that while everyone was conditioned to feel "free," freedom itself is an illusion. His head man, Frazier, states: "I deny that freedom exists at all. I must deny it — or my program would be absurd."

By 1971 Skinner had pondered human behavior further, and had seen and encouraged a broad range of efforts to control it. Also he had developed further his thinking about freedom and so-called human dignity. His stark tract, the nonfictional *Beyond Freedom and Dignity,* created quite an uproar.

In it Skinner suggests that human survival depends upon deciding how people must behave and then using behavioral engineering techniques to see that they do. His basic argument is that our behavior is shaped by external influences — not by any conscious decision making inside our heads. He implies that if this affronts our concept of human dignity, so be it.

Man's struggle for freedom is not due to any inner will to be free, he argues, but simply to a reaction to adverse stimuli in the environment.

He suggests that we are so preoccupied with the illusion of freedom that we are unprepared for the next step, "which is not to free people from control but to analyze and change the kinds of controls to which they are exposed."

He writes, perhaps wistfully, of the fact that there is more "order" in the laboratory than in the "confused" world outside. But he expresses confidence that the science of behavior will soon be able to reduce these differences. And at another point he says, "What is needed is more control, not less." He dismisses the so-called inner man (who makes decisions by processes of thought) as an understandable human conceit.

In sum, Skinner views thoughts and emotions mainly as by-products of human behavior, which is largely shaped by external influences.

Thanks more to Skinner than anyone else the behavioral engineers are on the march, now literally in the thousands. They have perfected techniques and he has given them ambitious goals to become movers and changers out in the world. What they think they can do was perhaps best summed up by James McConnell. He is a broad-ranging behaviorist, a friend of Skinner's, but a man who speaks lightly of the limitations of the true believers. In addressing a group of lawyers he tried to shake them out of their frozen view that prisoners can't be changed, so punish them. To make the point that people can indeed be reshaped he matter-of-factly made this statement:

"I believe that the day has come when we can combine sensory deprivation with drugs, hypnosis and astute manipulation of reward and punishment to gain absolute control over an individual's behavior."

With that thought in mind let us proceed to look at some of the current efforts of the psychological behavior shapers.

The Behavior Shapers Take On the Public

3

Today's band of human-behavior controllers . . .
can be found . . . in classrooms, kitchens, mental
hospitals, rehabilitation wards, prisons, nursing
homes, day care centers, factories, movie theaters,
national parks, community mental health centers,
stores, recreation centers, and right next door.
— Kenneth Goodall, in *Psychology Today*

The thousands of experts at conditioning are now trying out their be-
havior-changing technology on tens of thousands of people. The demand
is there. Many kinds of institutions are eagerly looking for ways to change
the behavior of people they must deal with.

Even the federal government has developed an array of programs designed according to Skinnerian principles. This includes programs the
Department of Defense has initiated. One psychologist who took part in a
defense program recalled: "Actually the Department of Defense . . .
wanted to know how better to control human behavior, not how to understand behavior." The federal government has channeled many millions of
dollars into behavioral-change programs outside the government.

In 1975 the National Institute of Mental Health undertook to make an
evaluation of the proliferating behavior-modification programs. It came up
with a recommendation that more effort be made to try out behavior-
modification techniques on *larger* numbers of persons in the *general* population, outside institutions.

Some control programs involve elaborate equipment. At Patton State
Hospital in California, twenty-one television cameras were connected to a
computer console. A nurse monitoring the cameras pushed buttons to
record observations. When she felt that a patient's behavior warranted
reward she pushed a button. The computer thereupon dispensed the reward

through a slot in the wall by the patient and a recorded voice simultaneously congratulated the patient.[1]

Research Media, Incorporated, of Cambridge, Massachusetts, offers institutions a $450 package course complete with cassettes and slides on behavior-modification skills. Among those in the "target population" to be trained in behavior-modification techniques are nurses, social workers, teachers, students, managers, and correctional facility personnel. The firm advertises several dozen users. They include the Shelby County Penal Farm, the Container Corporation of America, the Montgomery County Public Schools, and the Department of the Army, Fort Bragg.

Here are some of the kinds of situations in which psychological conditioners are attempting to engineer people's behavior:

THE REPROGRAMMING OF DISAPPROVED BEHAVIOR

Conditioners have been seeking quick alteration in sexual deviants, whether they are homosexuals or adults attracted to children. One aversive-conditioning technique often used with male homosexuals is to show the subject photographs that arouse him sexually. At the same time, an unpleasant electric shock is applied to his genital area. Then he is shown pictures of sensuous-looking women but without any accompanying shock.

This technique was used at a state prison in Somers, Connecticut. The subjects, all volunteers, were serving sentences for molesting children. According to the program director, before-and-after psychological tests demonstrated that the desired personality change had occurred — at least temporarily — in two thirds of the subjects. Those who had improved were paroled. Subsequently, three prisoners who were also serving sentences for molesting children but who had not volunteered for the treatment sued the prison. They alleged that prisoners had been denied parole because they had not participated in the program. The American Civil Liberties Union supported their suit. Many civil libertarians are concerned that such programs are a form of governmental thought control. Thus they could lead to broader applications with political dissidents.

Some behavior shapers prefer to use a drug that induces vomiting when the pictures are shown. Or an image of vomiting may be presented verbally.

Three psychologists in Vermont have experimented with describing to deviants in obnoxious terms a scene ordinarily erotic to them.[2] The subjects were men. Their degree of deviance was first tested by recording the extent that penile erection occurred when they were shown scenes of sexually attractive males or children.

One male homosexual was induced to undergo the treatment because he had recently fallen so deeply in love with a boy friend that it was threatening his marriage. He was told to relax and listen carefully as a psychologist verbally took him to his boy friend's apartment. For the first ten seconds the trip was made to seem very erotic; then this narration followed:

"As you get closer to the door you notice a queasy feeling in the pit of the stomach. You open the door and see Bill lying naked and you can sense that puke is filling up your stomach and forcing its way to your throat. You walk over to Bill and see him clearly, as you reach out for him you can taste the puke, bitter and sticky and acidy on your tongue, you start gagging and retching and chunks of vomit are coming out of your mouth and nose, dropping on your shirt and all over Bill's skin."

Poor Bill.

Many similar nauseous episodes were related at subsequent sessions. The psychologists reported a decrease in deviance — at least a short-term decrease — in this and other cases.

A large number of conditioning programs have been undertaken to overcome objectionable behavior in children. Bed-wetting in children aged six to fourteen has been one target, particularly in Great Britain. The conditioning techniques used there seem to be more Pavlovian than Skinnerian since the goal is simply to create a reaction, not a pattern of action. One behavior shaper attached electrodes to the child's genitalia to administer a mild shock when the bed-wetting began. He claimed improvement in fifty-two out of fifty-eight cases. A more common approach has been to use a buzzer that goes off when as much as a teaspoon of urine penetrates a thin mat containing metallic gauze. In one such test success was claimed in sixty-five out of one hundred "hard cases." One question not answered is whether bed-wetting is merely a particularly bothersome symptom of emotional disturbance. In any case a number of commercial firms in England have marketed buzzer-mat equipment to parents without specifying any need for supervision by a physician.

Another kind of undesirable childish behavior the conditioners have been attacking is what is called the brat syndrome. Here the conditioners go into the child's home with tape recorders, cameras, and stopwatches to size up the problem in its natural setting. They itemize the episodes of alleged brattish behavior as they recur over a long period of filming and recording. Then they map a strategy of punishments and rewards for de-bratting the subject.

Three psychologists at the University of Oregon reported in 1976 that they captured the flavor of brattish behavior by "bugging" the children. Parents strapped radio transmitters to the children for most of the day. A device in a closet was set to record, periodically, fifteen-minute samples of dialogue. The child did not know when the recordings were being made, and often the parents did not know either. One child refused to wear the transmitter, so it was placed in his room.

Two of the pioneer efforts to de-brat children dealt with a boy named "Peter"[3] and with a boy named "Jeff."[4]

Peter's repertory of "O" (objectionable) behavior included nine offenses, among them sticking out his tongue, calling other people bad names, removing or threatening to remove his clothing, saying no loudly, threatening to break things, throwing objects, and pushing his little sister. After sixteen sessions the observers had established a statistical baseline of his "O" behaviors. Then they set out to treat the nine objectionable patterns. The mother was instructed to cope with Peter, a preschooler, by following signals from the psychologists. An "A" signal meant that she should tell Peter to stop whatever he was doing. "B" meant that he deserved a reward of affectionate contact, such as a pat or a hug. "Treatment" at each of the forty-five sessions lasted an hour.

The mother reported that after the first phase of a dozen sessions the possibly bewildered Peter had become less O-some. By thirty sessions, even when he turned O-ish there was less belligerence in his actions. The therapists working on Peter acknowledged that any success depended on a cooperative parent.

In de-bratting Jeff, a different team of behavior shapers concluded that the challenge of changing the mother was bigger than that of changing eight-year-old Jeff. He was enrolled in a school for the disturbed.

Jeff had temper tantrums and assaulted his mother, his teacher, and other children. He was also a bed wetter and had skin allergies and chronic asthma, all of which suggested that he was not just being objectionable for kicks. Possibly even more pertinent, his parents had fought a great deal in his presence. Both had undergone psychiatric treatment and were currently living apart much of the time.

The analysts soon decided to concentrate on changing the mother's behavior. She obviously had become intimidated by Jeff and had fallen into the habit of avoiding scenes at any cost. She constantly tried to pacify him. She kept up a murmur of mollifying patter, accompanied by threats of punishment she rarely carried out. When her annoyance reached a certain point she would explode by thrashing Jeff severely.

The controllers concentrated on getting her to reduce her verbal outpouring and to ignore abusive behavior for a while. If Jeff persisted, she was to express her anger and order him to stop. Then if he still persisted, she was to spank him. If he obeyed, she was to praise him warmly. Over a period of twenty-five weeks she developed more confidence in coping with him. On camera at least, he became less abusive. His bed-wetting stopped, but not his asthma or allergies.

Experimenters from the University of Kansas came to the rescue of a mother who was upset because her two daughters were so thoughtless. They didn't make their beds or pick up their clothes. They left the bathroom a mess. Et cetera.[5] The girls were asked to make a chart of eight habits which the mother and the psychologists agreed needed to be changed. They were given a nickel a day for perfect reports and a fifteen-cent bonus for a perfect week. Their habits changed almost miraculously. After two months they were told to continue charting but because they had done so well they would be paid in full regardless of what the chart showed. The girls reverted to their old sloppy ways abruptly. So they were told they would have to earn their nickels as before. Again, habits improved dramatically. The parents professed to be pleased.

Such experiments have caused some behavior shapers to call for broad-scale programs to teach parents how to be good behavior managers through operant conditioning. Thousands of parents already have taken courses on behavior management. R. P. Hawkins of Western Michigan University proposed that high school students, male and female, take a compulsory parent-training course that stresses behavior-modification techniques. Regarding such proposals, a New York psychologist expressed strong skepticism to me that behavior shapers had all the answers on child rearing. She added: "In fact some of the most difficult children I have ever met belong to behavior modifiers."

DISPELLING ADULT ANXIETIES AND MALFUNCTIONING

The desensitization approach that Wolpe used on snake phobics is now being used to attack a wide variety of neurotic anxieties.

Compulsive behavior is treated by the blast-out method (implosive therapy). A compulsive handwasher obsessed with dirt is repeatedly thrust into dirt-involved situations. He may, for example, be told to imagine himself strolling knee-deep in a cesspool. After a few such mental strolls his concern about a few germs on his hands often diminishes.

Yet another conditioning technique is used on an involutionary bodily function in order to bring it under "environmental" control. For instance, a woman complains of chronic constipation. She is persuaded to strap on a belt that presses two electrodes to her spine. The electrodes give off a pleasant tingle when activated. She is then trained to press the tingle button just before defecation is about to occur. She is also trained to concentrate on the good consequences that follow the tingle. One investigator reports that laxatives can "generally" be abandoned after twenty to thirty days of tingle conditioning.[6]

CONDITIONING MARRIED COUPLES TO COPE

An ardent people-programmer at the Camelot Behavior Systems in Kansas suggests that husbands and wives can "reinforce" each other. They begin by charting the number of times they smile at each other each day for a couple of weeks.

More ambitious programs aimed at lifting ailing marriages out of the doldrums by using operant-conditioning techniques have been tried out on couples by investigators in Texas and Michigan.[7] Both involved setting up token economies in which each partner acquired points for behavior desired by the other. Also, in both programs the partners were required to list the three or four behavior changes they most desired in the spouse. They were to keep careful charts of the occurrence of the changes. The wife might most want help with her burden of household chores. The husband might most want more explicit affection.

The University of Michigan psychologist handling the overhauling of four couples believed that a sound marriage boils down to a balancing of reinforcing rewards. He set up a year-long experiment. Each wife was instructed to purchase a kitchen timer and carry it about the house. Whenever her husband came home she was to start the timer and give him a chip or other token for each hour in which he conversed on what she considered to be a satisfactory level.

The wife likewise collected tokens for satisfying his three main desires. For example, wives won three tokens for kisses, five tokens for "heavy petting," and fifteen tokens for intercourse. Tokens could be redeemed from a list of what each partner wanted most. Tokens earned and spent were recorded on the Behavior Chart. All eight charts showed a rise in both conversation and sex during the first month, but from then on there was considerable see-sawing until the experiment ended. On questionnaires they completed the couples indicated that "marital satisfaction" had risen in the early months, then leveled off.

I know eminently happy partners who would rather quit marriage than spend a year keeping daily records of the rewards they give each other. On the other hand, the experiment was clearly of value to the participants in the initial phase, when each was frank about what his or her partner needed to change. But for that you don't need operant conditioning.

BEHAVIOR SHAPING IN THE CLASSROOM

In less than a decade, the behavioral technologists have moved in a major way into our compulsory, government-operated public schools. Just one ardent behavior shaper, Charles H. Madsen, Jr., of Florida State University, has, as a consultant, taught operant-conditioning techniques to many thousands of teachers. And in Kansas City, H & H Enterprises, Incorporated, operated by the behaviorist R. Vance Hall, has sold to schools many thousands of courses in managing behavior.[8] By 1976 the behaviorists Susan O'Leary and Daniel O'Leary found that somewhere between 10 and

20 percent of all teachers on the eastern seaboard of the United States were systematically using behavior-modification techniques in their classrooms.

One of the most enthusiastic of all behavior shapers has been Roger E. Ulrich, an editor of a two-volume report on behavior-shaping studies, *Control of Human Behavior.* He has urged that compulsory public school education begin with two-year-olds in order to get good behavior patterns started early.

While most of the experimenting and behavior shaping have been done in the lower grades, they are also carried on in public high schools. Psychologists and graduate students from Stanford University have been coaching local high school teachers on how to manipulate students into proper, obedient behavior patterns — without the students' awareness — by the systematic use of rewards and punishment.[9]

Two teachers in Butte, Montana, were so enthusiastic about behavior shaping that in 1975 they arranged for the construction of a "behavior modification box." The box was four feet high, had no interior illumination, little ventilation. There were two holes for observation. The relatives of a retarded child made such a fuss because their child was locked in the box that the teachers lost their jobs.

Perhaps typical of intervention in lower grade schoolrooms was that by a team of behavior shapers from the University of Kansas. They arranged for a teacher and her students to play a "game." The teacher divided the class into two teams. The game consisted in seeing which team was guilty of the fewest violations of eleven rules. Samples of the rules: no one may leave his seat without permission; no one may "scoot" a desk; no one may talk or whisper without permission; no one may sit on top of a desk. The winners got victory tags, the right to first place in lunch lines, free time, and early recess.

The investigators, after studying their charts, asserted that the project "significantly and reliably modified the disruptive 'out of seat' and 'talking out' behavior of students."

Whether actual learning — presumably the main goal of education — is enhanced by any of the projects just described was not indicated.

Behavior shapers often contend that praise is just as effective in producing desired behavior as material rewards like candy. One experiment to change the conspicuously disruptive behavior of seven students in a classroom by praise, however, didn't produce the expected results. Only one of the seven students seemed to moderate his disruptiveness as a result of praise. When tokens were introduced that could be exchanged for candy, comic books, and perfume, all but one holdout fell into acceptable behavior. The trend in rewards offered — at least as indicated in research programs being reported — has been away from costly tangibles like candy, trinkets, and money, and toward such "natural" reinforcers as extra recess, privileges, and special activities.

In the early 1970's an argument exploded in the behavior shapers' own *Journal of Applied Behavior Analysis* about conditioning in classrooms.[10] R. A. Winett and R. C. Winkler, then of the State University of New York at Stony Brook, reviewed all the reports the *Journal* had published over a three-year period on the use of behavioral engineering in classrooms. They concluded that the behavior shapers were almost totally absorbed with training students to "be still, be quiet, be docile." Among "inappropriate" behaviors that had been targets for change were walking around, laughing, carrying on a conversation with anyone but the teacher, showing objects to another child, doing something different from what had been ordered. Not one of the investigations, they charged, questioned whether silence and lack of movement are necessary for learning, or whether behavior shapers should be in the business of molding passive, obedient rule followers. They questioned whether the "model" child should stay glued to his seat all day, should continually look at his teacher or work, should not laugh, and should pass silently in the halls.

A whole school of educational thought contends that such a rigid structure in fact inhibits learning. Advocates of "open" classrooms encourage walking about as a part of fulfilling projects. They encourage normal chatting with the teacher and other students, and so on. Winett and Winkler urged that the shapers give more thought to working with children rather than with the teacher, and to find what would be most reinforcing to *each* child.

There was a rejoinder to their article by a Stony Brook colleague, Daniel O'Leary. He contended that orderliness in students is in fact a critical concern of teachers. And he pointed out that Winett and Winkler had ignored reports of behavioral programs that had helped mentally retarded or handicapped students make progress in school work. However, Winett and Winkler had emphasized that their criticism was of behavior-modification programs in a normal classroom.

Ironically, one of the most mechanized techniques the behavioral engineers have introduced into education has one aspect humanists would applaud. The teaching machine, which apparently has peaked in popularity, allows each student to set his own pace. The information to be imparted is broken down into bits or frames. The information is doled out in such simple bits that the average student's chance of answering almost all the questions correctly is high. This "instant reinforcement" for correct answers is believed by some to help speed up certain simple kinds of learning.

And some teaching machines are not completely dehumanized. The machines will greet the student with "Good morning, Charles," or whatever the student's name is. That is something live teachers often don't bother to say. (The same reinforcing principles are also used in programmed teaching guides. These are booklets designed to accompany textbooks and are used through the college years.)

In quite a few colleges experiments are going forward to convert the professor from being a mentor to being an education engineer or an expert in contingency management.

At Western Michigan University, a highly inventive behaviorist named Richard Malott conducts classes based on behaviorist principles. His approach involves systematic bit-by-bit learning, setting one's own pace, immediate reinforcement of anything learned, and so on. Members of his "Student-centered Education Project" live in the same dormitory.

Malott is also an entrepreneur. His company, Behaviordelia, publishes texts and reference manuals on behaviorism. One is his red-covered *CONtingency MANagement in Education and Other Equally Exciting Places*. It is written primarily for college use and has sold more than forty thousand copies. The manual has the format of a rather wild Superman comic book. One statement it contains is that students find the underground comic-strip style "very reinforcing."

Another of his manuals shows on its cover a comic drawing of a man in cap and gown holding up a book and saying, "Hey, kids! This is it. This is the Real Thing," followed by the large words "Drugs! Booze! Sex!" On the opening page of Chapter 1, "The Importance of Social Reinforcement," is this example: "Powerful experiences like sexual stimulation are usually reinforcing experiences. A *reinforcing experience* is one that you'll repeat whenever the opportunity (or whatever) arises."

A behavioral psychologist at Pennsylvania State University, who was seeking ways to reduce noise in dormitories, turned to bribes. He set up a $100 "reinforcer" fund, which twenty students could split if they cooperated in keeping the noise down. The noise level dropped by 50 percent. However, noise rose to its former level when the reward was removed. He found the promise of an A in an experimental psychology course even more effective as a reinforcer. The noise level dropped by 90 percent.

TAKING ON WHOLE COMMUNITIES

While few behavior shapers working out in the real world are yet ready to implement Skinner's dream of reshaping a whole society, starts are being made in taking on whole communities. Or, as the shapers prefer to put it, "whole systems."

The behaviorist A. Jack Turner makes no claim of taking on a whole community, but he and his crew of three dozen have succeeded in introducing Skinnerian techniques into many corners of Madison County,

Alabama (pop. 186,000). Madison County consists mainly of Huntsville, the aerospace center.

Turner has had the modest title of associate director of the Huntsville–Madison County Mental Health Center. The center is in touch with many thousands of residents who are mentally ailing or have behavior problems. But more important, in terms of a whole system approach, the center operates a very active and out-reaching "consultation and education service." It holds classes for the public on behavior management, and it has persuaded the leaders of a host of the county's institutions to adopt conditioning techniques. Classes on operant conditioning for parents are also held (five two-hour sessions) and classes on "Positive Marriage Management." The center has also taken its "education" skills into many county agencies.

Turner explains: "A multitude of agencies are potential consumers of mental health consultation services. We have had programs with twenty-six agencies in the first three years of this project, including public schools, retardation programs, adult court, juvenile court, rehabilitation center, welfare department, nursery homes, law enforcement groups, and so on. Behavior modification has been the intervention strategy with most of these agencies."

After an agreement is signed Turner's aides train selected agency officials in behavior modification. One upshot is that a school system there has hired its own behavioral technician. And the juvenile court has begun offering behavior-modification classes to parents. Turner's staff claims credit for the fact that the number of mental patients admitted to the state hospital was reduced during the program's first three years. Another explanation might be that much of the turmoil and anxiety created by cutting back the aerospace program had eased during that period.

A second whole-systems approach has been developed in Lawrence, Kansas. Here the shapers are moving out from the campus of the University of Kansas, a stronghold of behaviorism, and into the region's schools. In addition, they have set up family-type "homes" for pre-delinquents run strictly on token economies of rewards and punishments. The results certainly are superior to those achieved at large institutions. Also, they oversee a program for many thousands of deprived children in a number of states. This program claims to have raised the reading scores of the children to above the national average for middle-class children.

Perhaps the effort to create a "community" most like Skinner's fictional Walden II is a commune in a cluster of buildings on a 123-acre farm called Twin Oaks near Louisa, Virginia. It was established in the late 1960's.

Like Walden II, Twin Oaks was set up with a board of Planners who regulate a corps of Managers (labor manager, food manager, health manager, clothing manager). In the creation of Twin Oaks the nuclear family was scorned, as in Walden II. Any children would be brought up on Skinnerian conditioning principles, with punishments forbidden. Only rewards would be given.

Obnoxious behavior was to be eliminated by refusing to reinforce it; that is, it would be ignored. Appropriate behavior was to be applauded. People who had job assignments they disliked, such as scrubbing floors, got more labor credits than those doing jobs they liked, such as repairing the pump. The dozen or so members I met had unconventional anti-establishment views. They seemed genuinely idealistic and were thoughtful and probing in their questions about the square outside world.

Twin Oaks found it had to modify a number of assumptions built into Walden II. Not all the members happily performed as they ought to because they wanted to, as Skinner forecast. There have been explosions of hostility. Skinner's leader, Frazier, couldn't imagine anyone's not wanting to work. Twin Oaks has had to tell a number of shirkers to take off.

The hard core of Twin Oaks residents have had many reasons for satisfaction but placid contentment has not been the prevailing mood as it was at Walden II. About half the members have left each year since the commune was established. Some have left because, unlike the prosperous Walden II, Twin Oaks has had a hand-to-mouth existence. It is too poor to use the material reinforcers or rewards that Skinnerians commonly use.

Also of interest: Twin Oaks's young residents have found that humanistic motivations, such as the simple sense of fulfillment in accomplishing an important task, work as potent motivators. At this writing, the future of Twin Oaks is still clouded for economic reasons. But as communes go it has lasted an unusually long time.

We will be encountering the operant conditioners again, some of them in surprising fields like sports, prisons, and business. But now seems a good time to consider what to make of them, and the implications of what they are doing. Here are some questions that seem relevant:

DOES BEHAVIOR SHAPING WORK?

Well, there are hundreds of success stories. As we shall see throughout this book, there is no question that professionally trained people can under proper conditions control, mold, and restructure the behavior of other people by a variety of techniques. Operant conditioning is one.

When that is said it should be added that at this stage Skinnerian conditioning is a technology that works sometimes and sometimes not. Thus it is hardly a true science. The general proposition that by using conditioning techniques you can get a whole free society to behave as desired is still pretty much just an ambition.

Conditioning works best when the people to be shaped are in a controlled environment, such as a hospital, prison, or school. It works best when the desired change in behavior is specific (for instance, stay in your seat). And it works best with people who aren't very far advanced in learning.

The degree of permanence of behavior changes created by conditioning also varies, once the "reinforcements" are removed. In many of the experiments subjects have undergone dramatic relapses less than six months after leaving the controlled situation.[11]

Another quibble is whether the "successful" treatment may have removed a symptom without significantly affecting the underlying cause. As indicated, the operant conditioners are interested only in what they can see and measure before their eyes; and are inclined to be uninterested in what is going on inside the subject's head.

Skinner himself has led the way in this by saying in *Beyond Freedom and Dignity:* "The dimensions of the world of the mind and the transition from one world to another do raise embarrassing problems, but it is usually possible to ignore them, and this may be good strategy."

Karl H. Pribram, who has done much of his major work as a brain physiologist at the Stanford University Medical School, spent a decade working in operant conditioning. He has lost patience with most Skinnerians. He charges them with being "anti-mentalist," and therefore naïve, parochial, and stuck with obsolete concepts. Of Skinner he says: "Skinner is against freedom and against dignity and against feelings and against values. He is against anything that smacks of mind, because mind is soft and ghostly and gets in the way of clear thinking about the control of behavior."[12] Pribram's experiments with the brain have led him to conclude that the brain is physically modified by experience.

SHOULD WE WANT IT?

Do we want skilled operators playing God in setting our goals, establishing the behavior patterns we are to fit into, and molding our social development?

Certainly some of the techniques and their purposes are better than others. For example:

- Behavior shaping that is oriented to serving people is preferable to behavior shaping oriented to giving someone better control over people.
- Behavior shaping in which the goals or targets are chosen by the people who want their behavior modified is preferable to behavior shaping in which the goals are set by the shapers or their employers. At Huntsville,

Turner sought to lower public resistance to his team of behavior shapers by announcing that the public would have the final say on any efforts that would be undertaken. He said, "You tell us how you want to change." However, it seems to me that "you" becomes a little obscure when "you" is the head of a large agency eager to change the behavior of employees or persons he is in charge of controlling. Efforts should be made to keep participation voluntary at the individual level.

• Behavior shaping in which people are simply trained in how to manage their own efforts at behavior modification are preferable to those in which the people remain under the supervision of controllers.

• Behavior shaping programs that are out in the open are certainly preferable to those in which the controller disguises his own motives (a common experimental strategy).

• Behavior shaping that involves the use of aversive stimuli is less reprehensible if the person agreeing to submit to it is not within the confines of an institution, where "voluntary" participation can be manipulated.

Behavior shapers themselves argue about the relative effectiveness of punishment versus reward. James McConnell offered me an impressive illustration of results that can be expected: "If you want to stamp out bad behavior in a prisoner, punishment will work. But if you want the prisoner to learn complex tasks necessary to succeed in the outside world, then you have to use positive reinforcement."

So rewards are usually preferable to punishment. But that raises another question:

IS LIFE JUST A BOWL OF REWARDS?

Behaviorists believe that rewards, in whatever form, are a prime force in making any society operate. Others now are wondering what kind of generation will emerge if we hook the young on the expectation of being constantly rewarded for good behavior. Many if not most young schoolchildren are naturally inquisitive. Would inquisitiveness disappear if school performance is linked continually to rewards?

And an expectation of reward for anything that helps others is hardly good training for altruism or community service.

Yet another question:

IS BEHAVIOR CONDITIONING DEHUMANIZING?

Well, more or less. It would be unfair to pin a dehumanizing label on all operant-conditioning projects. Many behavior shapers are working in humane ways to help people rid themselves of ailments, handicaps, or low self-esteem, and to help pupils learn at their own pace.

Still, the fundamental assumption of most conditioners is that you can control all "organisms," whether people or animals, by pushing buttons to manipulate reinforcements. There is a reliance on tools, such as stop-watches, cameras, shock machines, kitchen timers.

Certainly the generalization can be made that any person who senses that he is being manipulated will also sense that his personal integrity is being undermined.

And what about the basic philosophic position of Skinner? I think he has done us a service by stressing his startling assumption about the plasticity of man. It is making us think, and making us notice that this same assumption of plasticity is emerging in many other areas of scientific innovation.

However, the fact that Skinner has forced us to think doesn't mean his own assumptions are right, or even 60 percent right.

For many people today the assumptions of the Skinnerians are enticing. The Skinnerians assure them they are not responsible for their own acts. So why worry?

Finally it appears that the conditioners are starting to achieve some quite powerful tools for behavior modification, especially if used with large groups. And, historically, they are just getting started. They are still learning. Let us keep an appraising eye on them.

Mood Management | 4

*Predictable behavioral and mental responses may
be induced by direct manipulation of the brain.*
— José M. R. Delgado, brain prober

A host of scientists in several countries have been exploring that "soft
and ghostly," grayish blob containing the mind, which Skinner preferred
to ignore. And some are becoming more precise behavior shapers than the
psychological conditioners.

The brilliant, flamboyant, gracious José Delgado, a Spaniard, has been
in the forefront of those shaping behavior by brain stimulation. He spent
many years at Yale University and now directs research at a new institute
in Madrid.

His strategy of exploration has been to sink long needlelike shafts into
every corner of the brain. (He investigates the brain of both animals and
Man.) Then he applies electrical or chemical stimulations from the tip of
the probing metal and watches what happens. Sometimes what happens is
wondrous to behold.

Delgado has been particularly ingenious at developing instruments for
influencing the brain. Thanks partly to his inventions, there has been an
explosive growth of scientific interest in how to modify people by brain
stimulation. And independently there has been an explosive growth in
interest about what really goes on inside the brain.

The human brain is one of the prime marvels of Nature. That three-
pound ball inside your head is vastly more intricate and sophisticated —
if less quantitatively precise — than a room-sized electronic memory bank.
The actual working cells of the brain, the 10 billion or so nerve cells or
neurons, weigh only about six ounces. Most of the brain consists of about
90 billion glial cells that comprise a jellylike mortar holding the working
nerve cells in place.

Some of this human brain is very old, evolved to meet simple animal survival needs. Other portions are quite new and distinctively human. Sometimes the two pull us in conflicting directions: one highly emotional, the other more rational.

In the beginning of brain evolution hundreds of millions of years ago there was simply a brain stem, a knot of nerve tissue sitting atop the spinal cord. Reptiles have it and so do we. It mainly handles such basic functions as breathing and pulse rate. As animals became more complex, extra parts were elaborated to handle the increased demands, such as appetite control, sex drive control, aggression control. Most of these additions are called the limbic system. This old brain is pretty much the same in all mammals, Man included. The limbic system's parts contain such picturesque names as amygdala (Greek for almond), hypothalamus, and fornix. This old brain has a lot to do with our emotions. Also in the old part is the cerebellum, which controls muscular coordination in Man and mouse.

One brain specialist has good-naturedly referred to the kinds of behavior controlled by the limbic system as "the Four F's — fighting, feeding, fleeing and sex."

Most of this old brain is surrounded by the relatively new brain, the cerebrum. This is our thinking cap, particularly its outer cover, the cerebral cortex, and most particularly the front part of the cortex, our frontal lobes. All animals from the fish up have the beginnings of the cerebrum. In primates it becomes quite substantial. And in man it is a thick creviced mass that helps us read, write, make space ships, and crack jokes.

Another interesting point: The cerebrum, which is just under the skull, is divided down the middle into two hemispheres. The two are connected by a bundle of nerves. When exposed, the whole looks like a bushy wig combed down the middle. The left hemisphere controls in general the right side of the body, and the right hemisphere controls the left side.

Scientists have accumulated substantial evidence that dominance by one hemisphere over the other has a lot to do with our talents and personality. Roger Sperry of the California Institute of Technology was a pioneer in this research. He and other scientists have observed the behavior of people who have had their two hemispheres severed for medical reasons.

A strongly left-hemisphere person tends to be high on logic and verbal skills, whereas a person dominated by the right hemisphere tends to be more creative, intuitive. The right hemisphere of a typical adult has the sentence-forming ability of a five-year-old child. When the hemispheres are severed the person often seems to have two different personalities.[1]

The above pattern of contrasting hemispheric roles applies to right-handers (that is, to most of us) and to some of the one-in-ten who are left-handed. But for those who are biologically left-handed, the pattern seems to be reversed.

Scientists are now looking into the possibility of shifting the dominance from one hemisphere to the other. If they are eventually successful, we will conceivably be able to change our personalities on request.

One problem facing potential behavior shapers is that a single area of the brain may play a part in controlling several types of behavior. And few of our functions are totally controlled by any one brain area. E. Roy John of the New York Medical College has been studying brain activity by recording the electrical rhythms in many parts of the brain while learning is taking place. He recently asserted: "My research leads me to believe that vast regions of the brain are involved in every thought process, although some parts are more involved than others."

Added to all this is the fact that human head shapes vary considerably, which somewhat complicates precision in probing. However, as Delgado explained to me, scientists have found landmarks that show up in X rays as "reference points," regardless of head shape. He states: "The brain is like an ocean through which, by relying on instrument guidance, we can navigate without visibility and reach a specific destination."

The first notable brain stimulator, Walter Hess, a Swiss, won a Nobel Prize largely for inserting an electrode by hand into an unconscious cat's brain. When the cat awakened it wasn't particularly aware of the electrode, but when there was a mild electric stimulation the cat went into a rage.

The modern instrument guidance Delgado referred to is provided by a stereotaxic device that locks the head into place in relation to a metal arm holding the long needle to be inserted. Vertical and lateral controls permit adjustment until the needle is aimed on target for its gentle thrust through a hole bored in the skull. The patient may be wholly conscious and does not feel pain as the shaft slowly moves through his brain. (Brain cells have no nerve endings except for some at the base of the brain.)

If the stimulation Delgado plans to administer is electric, the shaft is an exceedingly thin steel-wire electrode coated with insulation except at the tip. Dozens of such needlelike wires may be inserted from one opening and can be attached to the same socket on top of the skull, or eventually inside it. If the stimulation to be administered is chemical, the shaft is a kind of oversized hypodermic needle.

Delgado has pioneered in the remote control of electrical stimulation. He began shaping the behavior of subjects while he was in a nearby room manning a push-button radio device. Now he can do this from thousands of feet away.

At first the sockets he was using to receive radio messages were outside the scalp. Now the equipment, built under a microscope, is the size of a coin and can be planted under the scalp and so is unnoticeable in a free-moving subject. Also, the device not only receives instructions but broadcasts back the subject's reactions. Delgado calls it a transdermal stimoceiver.

A very recent refinement, still being perfected, is for the information being received back from inside the brain to go to a tiny computer. This computer is being programmed to recognize abnormal brain-wave activity. There is now the possibility of forestalling epileptic seizures. The epileptic would "wear" one of these computers either inside or outside his brain. The computer would recognize the beginning of epileptic seizure in time to trigger a counterattack. This could be the electrical or chemical stimulation of a control organ in some other part of the brain.

Scientists have long assumed that the brain can learn only from input from the five senses: sight, sound, smell, taste, touch. Delgado with his devices has produced learned behavior changes completely independent of the five senses. An obstreperous chimpanzee became placid.

Increasingly Delgado has come to favor chemical stimulation over electrical stimulation in treating neurological problems. Electricity is a single tool, whereas chemicals come in thousands of varieties. His tiny electrochemical pumps, when activated by a radio signal, deliver minute quantities of chemical to the targeted site. And he now has a twin-shaft arrangement that enables him to withdraw fluid compounds deep inside the brain for analysis of changes created by the chemical stimulation.

Meanwhile, scientists are trying to learn how information gets transmitted and stored in the brain. Others are trying to find out how behavior can be shaped by drugs that reach the brain via the blood stream, whether they are taken orally or by injection.

Your brain is literally awash with blood. It requires ten times as much blood per square inch as the rest of your body. In the past twenty years hundreds of drugs that can be used to alter mood or shape behavior have become available and many more are on the way. These "psychotropic" drugs can steam you up or calm you down . . . make you gloomy or gay . . . make you weep . . . throw you into a talking spree . . . make you docile . . . make you promiscuous . . . befuddle or immobilize you . . . make you think you can walk through roaring traffic without getting hurt.

Most of the chemicals used to influence the brain are aimed at affecting brain substances believed to be critical as transmitters of information and sensations. Prominent among these transmitters are acetylcholine, serotonin, and norepinephrine.

Some scientists are so confident of the manipulative powers of the new drugs that they are making startling statements. When Kenneth B. Clark gave his presidential address to the American Psychological Association in 1971 he called for research on how, chemically, to control the behavior of powerful political leaders. The goal would be to curb dangerously aggressive behavior. His suggestion jolted even some of his colleagues. Powerful political figures, such as a chief of state and military chiefs, might be happy to improve their control over the behavior of subordinates or the masses. But they might resist the use of controls on themselves. Consent to being controlled could, of course, be made a part of their oaths of office.

Clark later explained that what inspired his proposal were the findings from recent neurophysiological, biochemical, psychopharmacological, and psychological research. And he told an American Psychological Association official: "I really was surprised at the number of my colleagues who still held firmly to a concept of man as having an untouchable spirit, an area of his being that should not be discussed in terms of scientific control. . . . That disappointed me."[2] Apparently they hadn't fully grasped the message about the new plasticity of Man.

Here are some of the kinds of manipulation of mood or behavior that scientists hope to achieve by stimulating the brain:

TURNING ON AGGRESSIVENESS OR PACIFISM

A number of years ago Delgado began mapping brain areas which if electrically stimulated cause quarrelsome animals to turn into placid ones. Normally ferocious rhesus monkeys would let him fondle their mouths without biting him when a certain area of the caudate nucleus near the midbrain was stimulated. And under stimulation a cat was induced to cringe from a mouse.

Delgado's most famous experiments involved inhibiting aggression in brave bulls. One of the experiments was filmed. A brave bull's head was rigged for radio-triggered electrical stimulation of the caudate nucleus. Delgado stood facing the animal. When the bull went into its charge he pushed a button. The bull skidded to a stop and walked away. This response was repeated a number of times. After several stimulations the bull permitted investigators into the ring without charging at them.

What had happened? According to Delgado, it was a "combination of motor effect, forcing the bull to turn to one side, plus behavioral inhibition of the aggressive drive." Elliot S. Valenstein, a psychologist at the University of Michigan who saw the film, suggested that the stimulation may not have really tamed the bull, but may have merely inhibited his motor activity by "activating a neural pathway controlling movement."

In any case, a dramatic behavioral change occurred after Delgado pushed the button. In alluding to Valenstein's speculation, Delgado told me that in more than twenty years of experimenting he found that caudate stimulation has inhibited aggression in all species studied. Included were not only bulls but cats, monkeys, gibbons, and chimpanzees.

He said that one magnificent old bull at a Spanish breeding farm was going to be killed because he was so dangerous to the herd. "After chronic caudate stimulation," he said, the dangerous bull "was tamed and could be maintained." He added that caudate stimulation is used routinely at his laboratory to inhibit monkeys and gibbons that need to be caught.

The reduction of aggressive behavior in humans by electrical brain stimulation has also been reported. Delgado has described a case involving a mentally disturbed person who habitually launched physical attacks upon members of his own family. After repeated stimulation of his amygdaloid nucleus the attacks were "considerably diminished."

Robert Heath of Tulane University, who pioneered the electrical stimulation of human brains, has equipped dangerously aggressive mental patients with self-stimulators. A film shows a patient working himself out of a violent mood by pushing his stimulator button.

The mechanisms that initiate violence are believed to be located in or near the old brain stem. Delgado has induced ferocious behavior in a peaceful colony of caged gibbons by stimulating their central gray matter near the brain stem. And Wilhelm Umbach, a neurosurgeon at the Free University in Berlin, has reported that aggressiveness in already aggressive patients was increased by stimulating the nucleus of the amygdala.

Delgado makes important qualifications about his ability to *incite* aggressiveness. He has found that the particular situation can make a big difference. And Umbach's findings would seem to support this because the same type of stimulation did not provoke aggression in nonaggressive patients.

One of the events that caused Delgado to be cautious was an experiment involving a male monkey. Under stimulation the monkey would lash out at a stranger or a pen rival, but he did not show any hostility to his female friend.[3]

Such findings were one reason Delgado set up a testing area on a one-and-a-half acre island in the Bermudas. Here he had a natural environment. His gibbons, with electrode under-the-skin implants are no longer "jailed" animals in laboratory cages. They have the full run of the island. In some tests, such as those in which movement is modified, the gibbons react exactly the same way as they did when tested back in "jail." But when their "central gray" is stimulated, they respond quite differently from the way they used to when they were jailed. Instead of becoming aggressively abusive toward the nearest gibbon they can find, they react by showing a considerable increase in running around and hollering.

Delgado told me: "We must conclude that social responses of the brain are powerfully influenced by the setting and we must be very careful about generalizing from laboratory experiments." This differing reaction of his gibbons, in his view, could lead to a "far more optimistic concept of man."

As noted, we have recently been hearing a great deal from ethologists about the inherent predatory nature of Man. Of all animal brains, Man's is the most immature at birth. Patterns are only lightly set. A human baby is helpless and for months can perform few functions beyond sucking, moving its limbs, vocalizing, and eliminating. Delgado believes that in Man it is not genes so much as accumulated sensory information that shapes future behavior patterns.

We have been talking only about the effect of electrical stimulation on aggression. What about chemicals? Are there actually potential "peace" pills we could give leaders to keep them from provoking dangerous confrontations? Or pills we could give to anyone judged to be unduly aggressive? Scientists have already developed what amounts to a "chemical straitjacket."

A drug called diphenylhydantoin has been under investigation as a specific relief for electrical brain disorders that can produce psychotic aggressiveness.[4] One problem in trying to find any single anti-aggression pill is that aggressiveness can develop for several different physiological reasons.

One provocative finding, however, was reported in *Science* by a team of psychologists at Princeton University. They found that aggression, at least in rats, could be both turned on and turned off chemically.[5] Some rats are naturally pacifistic as far as mice are concerned. Others are natural, instant killers. The scientists chose for the experiment rats who were either extremely aggressive or extremely pacifistic. Each chosen killer had, after spotting a mouse introduced into its cage, taken less than two minutes to dispatch it, and had done so on three successive days. Conversely, each chosen nonkiller had left its mouse alone, even though the mouse had stayed in the cage for several hours each day for seventeen days.

When a drug called carbachol was injected into the lateral part of the hypothalamus of the twelve pacifistic rats they all became mouse killers. Carbachol is believed to activate acetylcholine, the brain-transmitter substance. Another drug, atropine methyl nitrate, which inhibits acetylcholine, had the opposite effect. In five out of six cases, when it was injected into the same part of the hypothalamus of the killer rats, they let the mouse scurry about unharmed for as long as an hour.

The investigators caution against assuming that you could get the same turn-on, turn-off of aggression in Man with the drugs. But they did note that "a similar system may exist in other species. This raises the practical possibility that pharmacological manipulation of such a system could be used in the treatment of pathological aggressive behavior."

Findings such as theirs have caused scientists to contemplate the manipulative possibilities of chemicals that would incite or inhibit aggression in Man. Military people quite likely would be more intrigued by aggression inciters, and internal police in aggression inhibitors.

Heinz E. Lehmann, a psychiatrist at the Royal College of Physicians and Surgeons in Montreal, has speculated whether anti-aggression pills might have to be designed that would do the job of inhibiting aggression without decreasing achievement drive. And the psychologist Perry London has speculated about situations in which massive drug controls might be used to inhibit aggression. A subjugated population might be kept docile by having its drinking water dosed with a tranquilizing chemical. The citizens might be happy just to be hewers of wood. And they might become so docile that they might not care whether they obeyed or not. He adds, however, that "there are certainly some societies where the ruling powers would be glad to have a large portion of the population subjugated in just such a way."

On the other hand the ruling powers might be interested in mass arousal of aggressiveness. During this century most Americans were uninterested in becoming involved in foreign wars until heavy propaganda whipped up a militaristic mood. Some future leader might simply try adding aggression-inciting chemicals to drinking water, table salt, or the air. And then there are the men who must do the actual fighting. Military leaders would prefer to create gung-ho fighting men rather than wavering ones. And they are also understandably curious about the cause and prevention of anxiety under stress. In the United States, the navy and air force have funded research on mood modification by brain stimulation.

TURNING ANXIETY UP OR DOWN

Delgado has generated fear in a gibbon by stimulating its reticular formation. But he has found that reducing a generalized mood of anxiety by electrical stimulation of specific brain areas still poses very difficult problems.

Others, meanwhile, are claiming some success through a generalized electrical tickling of the whole brain.[6] In Russia it is called electro-sleep, in the United States, cerebral electrotherapy.

Usually, four electrodes are clamped to the head of the patient, who is in a resting position. Two are above the eyebrows. Two are behind the ears. At most, the patient feels a slight tingle while being treated for about a half hour. Often he seems to doze. Afterward, if the treatment has been successful, he is likely to exclaim how relaxed and refreshed he feels. The technique was developed by the Russians as a treatment for insomnia. But anxiety is a common cause of insomnia. The treatment spread to other countries and recently to the United States. Saul R. Rosenthal, psychiatrist at the University of Texas Medical School, San Antonio, did a control study with anxiety-prone patients for five days. One group got the current, the other group was told it was getting the current, but did not. Most of the stimulated patients showed marked improvement after five days; most of the control group did not. There remain questions about how long the relief lasts. And there are still skeptics who call for more careful testing.

Meanwhile we have seen in two decades a fantastic growth in the use of anti-anxiety drugs. They go under the seductive name of tranquilizer. Tens of millions of people in the Western world use them every day. Of the three drug prescriptions most prescribed in America, two are tranquilizers. Most of these pills sedate in much the same way that alcohol does. The strong anti-anxiety drugs like the ataractics have a good record in helping get mental patients out of strait jackets. Often the patients become calm enough to be sent home. The drug companies — ardent enthusiasts of the plasticity of Man concept — have been making much of their fortunes from the so-called minor or weak tranquilizers. These have become the pablum that doctors feed to unhappy, lonely, anxious patients who show up at their offices for largely nonmedical reasons. These people constitute more than half of all patients. Many of the drug companies, in their ads aimed at doctors, have been giving doctors clues on how to get these patients out of the office quickly and, perhaps, still collect a fee. The *Journal of the American Medical Association* carried an ad showing a woman sadly pondering her face and asking, "Am I old?" The message to the doctors was: "In the face of obvious decline, anxiety is often seen as reactive depression. Triavil treats both." And the *American Journal of Children* carried a large advertisement showing a tearful little girl. The message to doctors: "School, the dark, separation, dental visits, monsters . . . the everyday anxieties of children sometimes get out of hand. A child can usually deal with his anxieties. But sometimes the anxieties overpower the child. Then he needs your help. Your help may include Vistaril."

Testimony before the United States Senate alleged that the so-called anti-anxiety drugs are the most "overprescribed" of all psychoactive drugs.

How effective are the minor tranquilizers? A report titled "Medical Intelligence" in the *New England Journal of Medicine* answered: not very. Perhaps a little better than sugar pills. It stated: "A number of well-designed, controlled studies have failed to show consistent differences between drug and placebo therapy of anxiety."[7] It should not be assumed, however, that a sugar pill (placebo) has no effect. A Canadian psychologist at McGill University found that the placebo effect can be astonishingly powerful. This is particularly true when it is given to anxious people who have confidence in their doctors. In the *New England Journal of Medicine* report, the minor tranquilizers that have the most medical backing were said to be the benzodiazepines.

Another approach to converting anxious people into calm people is by use of the glamour machine of the 1970's, biofeedback. (I will describe its uses in more detail in Chapter 25.) Biofeedback machines are like mirrors that let you see instantly what is going on inside some part of your body. There are machines, for example, that record muscular tension. The most publicized let you see your brain waves in action.

If we are concentrating on a problem or are tense or anxious our brain waves tend to be very rapid jiggles, up to twenty-eight or more cycles per second. Those are called beta waves. If we are relaxed and calm the waves become slower (eight to thirteen cycles per second) and are also bigger. Those are called alpha waves. If we get really deep into reverie the waves get even slower and bigger (theta waves) until we fall asleep. In deep sleep we drop into delta waves (less than four cycles per second).

These brain waves carry electrical power. Barbara Brown of the Veterans Administration Hospital in Sepulveda, California, demonstrates this to skeptics with a toy train wired to a person's head. When she turns on the electric power of the brain waves, the train starts up.

By beeps, hums, or flashing colors, biofeedback machines give us instant information on the bodily activity being monitored. Take brain waves. The machine may be set to record beta waves as beeps and record alpha waves as hums. If we are tense and anxious, the beeps will predominate, but still there will be occasional hums. If we sit quietly with eyes closed for an hour and rejoice when we hear a hum we may well find that we have shifted gradually from a predominantly beta to a predominantly alpha state — and that we do indeed feel less anxious.

While the brain-wave biofeedbacks have gotten the most attention, some investigators favor muscular-activity feedback machines. They are a more direct and accurate indicator of tension, and can be carried readily on the body.

One way our bodies reflect the stress of anxiety is by teeth grinding, or bruxism. This usually occurs totally without our awareness, and causes dental damage. A Claremont College psychologist, John D. Rugh, rigged up for a group of dental patients portable machines that could be worn under the clothing to record muscular tension. All the patients were suffering from the ravages of teeth grinding. The machines were preset to emit a barely audible tone whenever there was muscular activity of the teeth-grinding sort around the jaw. All the patients were able to identify by the signal a number of life situations that provoked the teeth grinding. And most reported that they were able to reduce or end the teeth-grinding habit.[8]

TUNING OUT GLOOM OR HYPER-EXCITED STATES

Some people get down in the dumps because of a discouraging major event in their lives. Some are chronically doleful. We all have mood swings, but there are those who fall into a more or less chronic state of alternating between deep melancholy and an excited, euphoric mood. Their "mood-stat" is malfunctioning. Psychiatrists call them manic-depressives. Some investigators think those swings are related to the level of noradrenaline at certain brain sites.

The first major success in treating chronically depressed people was with electroconvulsive shock. Initially it was a rugged remedy. An electric current was passed between the patient's two temples. The convulsion that followed was sometimes so violent that bones were fractured. Today with anesthesia and muscle-relaxing drugs, the treatment is less dreadful. It usually relieves depression instantly and for several months. The patient is a bit disoriented for a week or so. Serious side effects come only when he is subjected to repeated courses. Then there may be a loss of mental capacity and the emergence of a rigid personality.

Meanwhile, a large number of antidepressant drugs have appeared and are widely prescribed. The efficacy of some has been about the same as placebos. But those of the tricyclic class, like imipramine, seem to be as effective as shock, and of course are easier to take.

One true wonder drug for stabilizing the tens of millions of people in the world who tend to have exaggerated mood swings has recently emerged, but with little thanks to the drug industry. Drugmakers may have preferred to ignore it because it is so plentiful in nature, and hence unattractive as a profitmaker. I refer to lithium, a natural alkali salt. For a few dollars you could buy enough sacks of industrial lithium to produce fifty thousand pills. It was only because of pressure from psychiatrists that some drug companies began, a few years ago, to put lithium compounds into drugstores.[9] Lithium pills cost about one tenth the price of most psychiatric drugs.

Lithium's principal effect is to calm manic-depressives while they are in the highly excited, euphoric, hyperactive stage. And it is taken only during this manic phase. But the calming tends in many cases to have the effect of easing both the highs and lows of manic-depressive attacks. In short it appears to be a mental-health stabilizer.

It has to be administered with extreme care because toxic effects from overdosage can be severe. But it is found in much of the world's drinking water, especially at spas. Well water in El Paso, Texas, contains a high level of lithium, and admission of city residents to neuropsychiatric hospitals is extremely low.[10]

A biochemist at the University of Texas was impressed by the relationship between lithium levels in the drinking water of dozens of Texas cities and mental-hospital admissions from those cities. The renowned psychiatrist Nathan S. Kline of Rockland State Hospital in New York asks why we should not add lithium as well as fluorides to drinking water. It would seem logical to do so, he thinks, if it can be established that our normal feelings would not be altered.

A different kind of dramatic mood swing, one that scientists have learned to create, involves our psychic energy. "Uppers" or "speed," such as the amphetamines, literally speed up greatly the firing of neurons in the arousal part of our nervous system. We feel charged up, often dangerously so. The "downers," such as the barbiturates, inhibit the firing, and so calm us. Recently the increased use of an ancient, deadly downer, heroin, has been attributed in part to the urgently felt need of amphetamine users to come down from their speed trips.

TURNING ON ECSTASY, TITILLATION, OR NAUSEA

The fact that the brain apparently has pleasure centers was discovered accidentally by the psychologist James Olds, another pioneer in brain probing. In experimenting with rats while he was at McGill University, he applied an electric current deep into a rat's brain as it explored a maze. He assumed that the electrode was in or near a pain center and would cause the rat, when stimulated, to shun a certain area of the maze. Instead, the rat seemed to develop a love affair with that area. Later, Olds invented ways that rats could electrically stimulate themselves, if they wished, by pressing a lever. Some rats seemed to delight so much in the sensation, whatever it was, that they pressed the lever eight thousand times in one hour.

Other investigators have found that rats will keep pressing the lever, with only occasional halts for food, drink, and naps, for up to twenty-one days. Here seemed to be a source of some sort of pleasure for which there was no satiation! It is conceivable that the animals were just compulsively scratching some kind of itch created by the stimulation. But human verbal reaction to the same kind of stimulation would indicate otherwise.

Humans, having more complex brains and a wider range of emotional reactions, have not matched the rats in obsessive pursuit of electrical delight. But we do have pleasure centers in the brain. And the pleasure involved often appears to have sexual overtones.

A male interviewer on Delgado's team talked with three different epilepsy patients on several occasions while each was having one of his temporal lobes electrically stimulated. Two of the patients were young women. One was an eleven-year-old boy. All expressed pleasure. One of the women, normally quite prim, started giggling, became flirtatious, and finally expressed a desire to marry the interviewer. The second woman, who was new to the interviewer, became extremely talkative, expressed fondness for him, and kissed his hands. The boy's behavior was even more intriguing. He was stimulated a few seconds every four minutes. After one stimulation he said, "Hey, I like those." After about forty-five minutes he expressed fondness for the male interviewer. In a later interview, while being stimulated, he seemed confused about his sexual identity and discussed his desire to get married. When the interviewer asked, "To whom?" the boy began to stammer out words and letters that ended with "y-o-u." Later he talked about his pubic hair and said, "I was thinkin' if I was a boy or a girl, which I'd like to be." Following another stimulation he said with pleasure, "You're doin' it . . . I'd like to be a girl." In a later interview when there was no stimulation a reference was made to his remarks and he became "markedly anxious and defensive."[11]

Olds's self-stimulator for rats inspired Robert Heath of Tulane to develop one for humans. It is a portable device that is strapped to the waist. By pushing buttons activating electrodes in the brain a person can modify his mood at will. With Heath's mental patients stimulation of the septal area seemed to bring the most pleasure. Patients said they felt great, and some clearly had sexual thoughts. One male pressed his septal button four hundred times in an hour. On the other hand, if the patients forgot and pushed the hippocampus button they said they felt sick all over.

One patient while pushing the septal button introduced a sexual subject and broke into a broad grin.[12] Another patient, out of apparent frustration, set a record for button pushing: he seemed to be trying to achieve an orgasm. Heath found that certain chemicals funneled into the septal area likewise produced signs of intense pleasure. Electrical stimulation of the septal region, he finds, is more likely to produce profound pleasure if a mental patient is suffering physical or emotional pain than if he is feeling well.

Some have speculated that with further investigation and refinement it might become possible for people rigged with push buttons to enjoy orgasms at a whim, and without the need of either a partner or masturbation. It is something one might choose to do during an interlude of commercials on TV. Russian scientists have reported evocating orgasm from a few of the people involved in a test of electrical stimulation of the ventrolateral thalamic region of the brain.[13]

A more commercially attractive titillator would be something that could be taken orally, like a pill. For thousands of years man has sought love potions. Ovid in *The Art of Love* cited a number of stimulators, including oysters. But until recently there has been no aphrodisiac meeting clinical proof. Claims have been made that LSD and "speed" pills stimulated or heightened sexual arousal. Both are dangerous. And the alleged effect can apparently be illusionary.

What may be the first true aphrodisiacs came from, of all places, t! laboratories of the National Institutes of Health, a U.S. government agency. Not surprisingly both were discovered by accident. Many congressmen would frown on federal spending for sex stimulators.

The first to be discovered, L-Dopa, was being used experimentally by

Frederick K. Goodwin to treat Parkinson's disease, a degenerative ailment.[14] L-Dopa appeared to increase the level of dopamine, an important neurotransmitter in the brain.

A reportedly far more powerful aphrodisiac bears the nickname PCPA. Its startling side effect was reported by a visiting team of Italian scientists. The members were studying the mechanism that controls sleep, body temperature, and other body functions at the National Heart and Lung Institute. Doses of PCPA given to animals caused great increases in sexual activity. One member of the team, Gian L. Gessa, reported that when rabbits injected with PCPA were put in the same cage with cats there were efforts to cross-breed without inhibition. He also cited findings of another research team that when PCPA was given to a woman suffering from cancer — this was before PCPA's sexual properties were known — her libido was heightened dramatically.

Gessa surmised that the sex drive may be a checks-and-balances phenomenon. The sex drive, he believes, is stimulated by hormones containing dopamine or its component L-Dopa, and is retarded by another brain hormone, serotonin. PCPA seemed to prevent the production of serotonin, and so in effect released the brakes on the sex drive.

In 1974, a spokesman for the National Heart and Lung Institute advised me that the team had since gone back to Italy. Whatever efforts had been made at the Institute to duplicate the team's findings had not been successful. The spokesman said that interest at the Institute in PCPA as an aphrodisiac has "languished." But since Gessa first reported his findings, Ernest Abel in his textbook on neuropsychopharmacology, *Drugs and Behavior,* reports other, later evidence that PCPA induces compulsive sexual activity in male animals of several species.

Gessa and his associate Alessandro Tagliamonte have continued their studies at the University of Cagliari in Italy. In 1975 they sent me new evidence from their own work and that of other investigators. In general the evidence suggested that in both male animals and human males, sexual behavior is inhibited by brain serotonin and stimulated by dopamine. In short, both L-Dopa and PCPA would logically seem to have an aphrodisiac effect. And a number of chemical inhibitors produce frigidity in women and impotence in men.

CHANGING THE SOCIAL ORDER

Even more than people, monkeys are status-conscious. They make no bones about drawing invidious lines in their colonies. Each monkey knows its rank and who in the colony it must defer to at the food tray, in sexual coupling, and in choosing a perch. The monkey at the top of the social order achieves its position by aggressiveness and maintains it by bullying.

The chief, usually a big male autocrat, expects more deference than an Oriental potentate. He stares directly at his subjects but they glance only furtively at him. They venture only by his invitation into the large part of the cage he has appropriated for himself.

How an individual monkey will respond to electrical stimulation of its brain is influenced by its status. Detlev Ploog of the Max-Planck Institute for Psychiatry in Munich found that in a clearly ordered colony of squirrel monkeys the response to identical stimulation varied from individual to individual depending upon its rank, sex, and role.

Delgado found much the same when he scrambled the social orders by reassigning monkeys to various cages. In one cage a female named Lina was the lowest-ranking of the four. As a general rule Delgado has found that stimulating an area of the thalamus tends to induce hostility. When she was so stimulated she showed agitated behavior, but whatever hostility she felt she kept in check. She attacked another monkey only once, whereas she was attacked twenty-four times. Later, she was put in a cage where Delgado knew she would be at least Number 2 in status. In the new setting, with the same stimulation, she attacked other monkeys seventy-nine times and was not threatened at all.

Experimenters at the United States Yerkes Regional Primate Center have found that they can make one male after another the dominant one in a group. They do it by electrically inhibiting the others. The female of the group switches her allegiance to whichever male becomes dominant.[15]

Perhaps Delgado's most fascinating experiment was when he maneuvered a real palace revolution by remote control. A male monkey called Ali had established himself as the truculent autocrat of a well-ordered colony of seven monkeys. He commandeered two thirds of the cage as his private territory.

From another room the Delgado group, with camera running, radioed electrical stimulation to Ali's caudate nucleus. The aim was to block out his aggressiveness. Rather quickly, Ali stopped his pacing and growling. He settled down, became benign and contemplative. And rather quickly the other monkeys sensed that something fantastic had happened. Soon they were crowding about Ali without fear, and he didn't seem to mind.

The liberation from tyranny lasted for about an hour. Within twelve minutes after the stimulation of Ali's brain was discontinued he managed to reassert his authority. The other monkeys fearfully retreated to their corner.

But this was not the end of tampering with Ali's authority. The human manipulators installed a lever near the food tray. When pressed, the lever would send inhibiting stimulation into Ali's caudate nucleus. As the days passed, a number of the monkeys pressed the lever accidentally or out of curiosity. Simultaneously Ali became less aggressive. Only a female, Elsa, figured out the connection between the lever and Ali's brief mild spells. What a find! Elsa became bolder, looked Ali straight in the eye. Delgado reports: "When Ali threatened her, it was repeatedly observed that Elsa responded by lever pressing."[16] On the days when the lever was available to Elsa there was a drop in the total number of Ali's aggressive acts toward his subjects. Something closer to peaceful coexistence came to prevail.

Would we have a more peaceable world if we could all be Elsas? Let's strain our imaginations, think the unthinkable. If each of us had an electrode implanted in the aggression-inhibiting area of his brain, and if each of us had a push-button, microwave pocket transmitter which we could focus at an obstreperous person in our family or work group, we would have an intriguing situation. But doubts arise. Who would give the orders that would make the societies run? And an aspiring Big Brother would certainly cheat and have his own electrode removed, and those of his police as well. And his police would certainly have extrapowerful transmitters that could keep everybody within half a mile in a pacific — and hence submissive — mood.

PRODUCING ROBOTLIKE BEHAVIOR

In some of his preliminary testing with several brave bulls Delgado triggered a variety of stereotyped physical movements. He caused a bull to move in a circle, to turn its head, to lift a certain leg, and to utter "moo." The latter so intrigued him that he repeated the stimulation one hundred times. One hundred consecutive "moos" were uttered.

With humans he and his associates have stimulated several areas involved in motor activity. They were able to evoke cries of sustained vowel sounds. He caused one woman patient in his group, when she was alone in her own room, to turn her head and move her body as if she were looking for something. This was repeated. When she was asked what she was doing, the woman always had a plausible explanation. Apparently she had no idea she was responding to the electrical stimulation of her brain. Her answers were: "I am looking for my slippers," or "I heard a noise," or "I am restless," or "I was looking under the bed."[17]

In monkeys Delgado has triggered complex movement patterns in short order. This would seem especially astonishing when compared with the exhausting weeks of work that operant conditioners spend just to teach a pigeon to do a figure eight. The same difficulty arises if you try to train an animal to walk by attaching wires to the nerve bundles that activate its muscles. More than a hundred wires would have to be used. But apparently when you wire up to a brain structure the brain has the sophistication to handle the whole series of actions as a pattern. Delgado achieved his most dramatic results by stimulating the red nucleus in the brain of a monkey named Ludy. After five seconds of stimulation, Ludy interrupted whatever she was doing, changed her facial expression, turned her head to the right, stood up on her two feet, circled to the right, walked on her two feet in perfect balance to a pole, climbed the pole, returned to the floor, growled, threatened or actually bit a subordinate monkey, stopped being aggressive, approached the rest of the group in a friendly manner — and resumed spontaneous behavior. This sequence of acts took about fourteen seconds. Under stimulation she always performed this same purposeless series of motor movements in the same order, with some improvisation in how she did them. The sequence of actions persisted after twenty thousand stimulations.[18]

Some years ago the science writers Ruth and Edward Brecher reported an attempt to automate a donkey by the use of electrical brain stimulation. The project was undertaken by a corporation seeking a research and development contract with the U.S. Department of Defense. Why the military was thought to be interested can only be speculated. The corporation classified the project as secret and even refused to confirm that it actually existed. But the Brechers were sure enough of the facts for *Harper's* magazine to publish their account of the experiment.

The project was to send a donkey on a ten-mile trip over rough terrain solely by brain stimulation. An electrode was planted in a pleasure center of the donkey's head. The electrode kept titillating the donkey as the donkey walked straight toward the goal. When sunlight struck a prism attached to the donkey's collar a photoelectric eye switched on the reward-giving electrode. The current switched off whenever the donkey went off course or stopped walking. The Brechers related: "Thus accoutered, the joyful donkey trotted straight ahead up hill, down dale, even crossed a mountain, neither straying nor lagging, to its predestined goal — a substation some five miles away. There the prism was reversed — whereupon the donkey retraced its arduous course over the mountain and back to its starting place." They added that when movies of "Project Donkey" were screened at the Pentagon as part of the corporation's contract application a scientist murmured: "There, but for the grace of God, go I." A brain physiologist I consulted mentioned the secret experiment but said he had the impression that there had been some disappointing aspects to it.

Here, more briefly, are some other ways scientists are learning to modify mood and behavior:

Wiping out the maternal instinct. One instinct widely shared by Man and mammals is the maternal one. The mother of a baby rhesus monkey spends months hugging and tidying her baby and cooing to it. She is in despair when the baby disappears from her view.

Delgado reports that when a ten-second electrical stimulation was applied to the mesencephalon of a mother monkey named Rose she completely lost interest in her baby for about ten minutes. "She ignored his tender calls and rejected his attempts to approach her." The baby seemed disoriented and sought warmth with another nursing mother in the colony.

If family bonds can indeed be disrupted electrically, conceivably other bonds can too. If so, and with further advances in brain technology, totalitarian leaders — again just conceivably — might be greatly interested in the idea of instituting some kind of broad application. A common strategy of totalitarian regimes for maintaining control is to take over child socialization from the family as early as possible, and also to try to disrupt preexisting small-group affiliations. The isolated individual is more malleable.

Bending the mind. For thousands of years humans have known that certain plants, peyote is one of them, contain substances that can create hallucinations which often are exciting. They commonly have been used in religious rites and orgies. They disrupt the normal electrochemistry of the brain. In the past quarter century scientists have uncovered a host of mind-altering drugs that can be synthesized. Most notorious is the acid called LSD, which is soluble in water. It is so explosively mind-altering that, reportedly, a pound of it placed in the reservoir of a great city could, temporarily, cause psychotic-like behavior in millions of people.[19]

Other drugs act simply to distort one's sense of time. Still others confuse or disorient. In the United States, agencies of the federal government have subjected at least four thousand people to mind-altering experimentation. A national commission study in 1975 found that most of the experiments conducted for the Central Intelligence Agency were done without the subjects' awareness and in at least one case allegedly led to suicide. Some years ago the Army Chemical Corps began stockpiling a drug called Agent BZ for possible combat use. It disorients. Spokesmen called it a "benevolent incapacitator."

Altering the mood to eat or sleep. By electrical stimulation Delgado has made gluttons out of cats. He speculates that perhaps people who chronically overeat may have abnormal nerve activity in the same area he was stimulating. He has also caused monkeys to ignore bananas, which they are usually very fond of.

As for sleep, Nathan S. Kline, an authority on psychotropic drugs, points out that the need for sleep can be reduced by monoamine oxidase inhibitors. He notes that some of the Mogul emperors, by rigorous training, developed horse soldiers who conquered large parts of the world on a travel regime that allowed three and a half hours of sleep a day. Current physiological evidence, he states, indicates that three and a half hours may be plenty. Sleeping during the dark hours may have had a survival value because of Man's poor night vision. He suggests that scientists may soon eliminate the need for humans to sleep at all: "It now looks as though we may be able to simulate or induce the bioelectric-biochemical activity required" to keep humans functioning without sleep. He adds: "Constructive use of these additional billions of man-hours every day is indeed a challenge."[20]

Indeed it is. But what about the fact that we are getting into an era of increased leisure time — and the fact that many people love the process of sleeping?

In the future we will undoubtedly hear much more about modifying brain function by chemicals introduced into the blood stream than about electrical implants. Implanting is so much more expensive. Also, except in institutions where coercion or promises of release are possible, implants have to be truly voluntary, and most people find the idea of electric charges going off in their heads unattractive. On the other hand, most people have very positive attitudes toward producing change by taking pills. And they have already accepted medication via fluoridation of their drinking water. Other chemicals may in the future be added to their water without their awareness. Buckets of pacifying chemicals might be quietly poured into reservoirs by the authorities in times of ugly social unrest. And conceivably buckets of suggestibility chemicals might be added whenever the authorities were about to launch a massive propaganda campaign. Or the specific water drinkers could be pinpointed. The chemicals could be used to forestall rebellion on the campus, strikes in industry, or riots in ghettos.

So what do we conclude about this new technology for modifying mood and behavior by electrical or chemical means? It certainly makes broad assumptions about the physical, mental, and emotional plasticity of Man.

Experimentation that advances knowledge of the cause and possible cure of human ailments certainly should be welcomed. Today's scientists have done wonders in penetrating the mysteries of various parts of the human brain and in finding chemical clues to such ailments as insanity, epilepsy, and Parkinson's disease.

If stern standards are met, many medical applications of brain research can be applauded. The use of implanted devices that sense the approach of an epileptic seizure, then head it off via an electric stimulation by a computer, is another laudable use. Still another is the use of self-stimulators for people who are terminally ill and in intractable pain.

Drugs that prove effective in helping a dying person to enjoy a sense of euphoria or even occasional happy hallucinations could, I think, be justified.

And one of the great medical advances of this era is the use of strong tranquilizers to help the mentally ill live more normal lives.

We come, however, to gray areas. The daily use of tranquilizers to help keep an erratic pulse normal may be more justifiable than their use by millions of people to ease everyday tensions. Some scientists have expressed concern that such habitual pill taking slows the reaction time of people while they are working or driving cars. And is initiative affected? Similarly, would use of anti-aggression pills have the long-term effect of lowering ambition and creativity?

A broader question is whether it is socially healthy for millions of people to expect to achieve continual bliss by way of psychotropic drugs. This would include aphrodisiacs that undoubtedly will become nationally advertised unless effective curbs are instituted.

As for electrical brain stimulators for instant regulation of mood and behavior, the control of the stimulation handled by some other person, seen or unseen, is hard to justify, except in experimentation and with informed consent. It might be justifiable upon request in medical situations.

The use of self-stimulators, for those who could afford them, seems to pose no ethical or moral problem. It possibly would not be socially sound for large numbers of people to have orgasms by push button. Widespread achievement of push-button orgasm in addition to masturbation and use of vibrators might promote the already rampant self-preoccupied *me* trend and diminish social interaction.

There is also the unresolved medical problem of whether any physical intrusion of the brain creates scar tissue that may subtly alter personality or body efficiency. The respected science editor Albert Rosenfeld offers this interesting speculation:

"One can easily imagine people in the future wearing self-stimulating electrodes (it might even become the 'in' thing to do), which might render the wearer sexually potent at any time; that might put him to sleep or keep him awake, according to his need; that might curb his appetite if he wanted to lose weight; that might relieve him of pain; that might give him courage when he is fearful, or render him tranquil when he is enraged."

It is a seductive image. Would such happy button pushers gradually start having identify problems? Or become a pain in the social body? Only time would tell.

The same questions arise over the prospect that the normal persons of the future may carry greatly enlarged pillboxes. There might be pills to make them more aggressive, optimistic, creative, euphoric, dynamic, placid, or sexed up. They might well become creatures of artificial, externally created sensations. Who is the real self? Is there deprivation when a person misses many of the natural moods that would go with his disappointments, triumphs, and concerns? And how are we fellow residents in his community able to know what he is like, really, stripped of pills? A further question that concerns some scientists is whether continual, artificially induced mood swings may not be permanently damaging.

Looming above all these concerns, however, is the possibility that control of our moods and behavior may fall into other people's hands, through such measures as placing chemicals in the water or implants in the brain. Delgado established in his experiments that electrical stimulation always overrode human will power. The subject's fingers would start fluttering under stimulation. The person would be challenged to make a fist — and look dumbly as his fingers continued to flutter.

In a thoroughgoing dictatorship certain selected people could be required to submit to having two electrodes planted in their brains. One electrode would go to a pleasure center; the other to a pain center. The promise of ecstatic feelings might make the requirement not seem too distasteful. Once installed, the regime would have the instant capacity to administer rewards and punishments to get blind compliance. The controller would simply have to ask: "You wouldn't want another of those terrible headaches, would you?"

Some have speculated that it might be economically attractive for a dictator to program large numbers of people to make simple motor responses like those of Delgado's robotlike monkey. An electrical engineer, in speaking some years ago at a national electronics conference at the Illinois Institute of Technology, tossed out this idea:

A baby could have a socket holding electrodes planted on or in his brain within a few months of birth and "the once-human being thus controlled would be the cheapest of machines to create and operate. The cost of building even a simple robot, like the Westinghouse mechanical man, is probably ten times that of bearing and raising a child to the age of sixteen."[21]

5

New Personalities
for Old

*Some techniques for modifying behavior are
relatively reversible . . . ; others are in some sense
irreversible.*
— Gardner C. Quarton, psychiatrist

In the last chapter we were talking primarily about temporary or short-term changes in behavior. In this one we will discuss the ways in which a number of scientists are attempting to mold or remold basic aspects of the human personality.

Some are focusing on very early childhood, when the potential for moldability is highest. At birth an infant has a very large brain for the size of its body; but the baby is still close to being mindless. Every cell in its body contains the genetic code that will largely dictate its physical characteristics. The code will later influence somewhat the new individual's mental functioning.

The ancient part of the brain concerned with self-preservation is functioning at birth. But the bulk of the brain of a newborn human baby can be thought of as a blank ball of wax awaiting impressions throughout its mass. Inside the ball of the brain the billions of nerve cells are tied into a network of chemical-electrical switches. In these cells the impressions received from the five senses — which capture one's life experiences — will be processed and stored. At birth the nerve cells are in place but the growth of the billions of fingers that reach out to make interconnections with other cells have barely begun. Thus, capacity for thought at birth is apparently slight.

The human baby in its first several weeks of life is mainly preoccupied with survival. The infant is incapable of a true social smile until about the sixth week. Eye coordination comes slowly so that for the first few months the baby may smile at any nodding object with the configuration of a human face. That configuration can be a head covered with a grotesque mask, or it can be a dummy's head.

There appear to be critical periods in human development when the brain of an infant or child becomes extraordinarily receptive to developing certain aspects of behavior and personality. Benjamin Bloom, an educational psychologist at the University of Chicago, found that each human characteristic has its own growth curve. The easiest time to make important changes in a child's intelligence, for example, is before the age of four. By that time intelligence is stabilized to the point at which IQ measurement takes on noteworthy significance.[1]

PROFESSIONAL IMPRINTERS?

A number of respected scientists believe that the brains of young humans accept some *imprinting* of behavior patterns in those critical periods. By imprinting I mean a type of learning that can occur only within a limited period of time early in life and is relatively unmodifiable thereafter. That animals have critical periods for imprinting is well established.

The Austrian ethologist Konrad Lorenz pioneered the classic experiments on imprinting in animals. He arranged for himself to be the first moving object that mallard ducklings encountered after hatching. Lorenz waddled in a squatting position so that he had the height-width ratio of a mother duck and he quacked like a mother duck. The ducklings fixed on him as their "mother," followed him in file wherever he went, and came flying to him when he quacked.

The psychologist Harry Harlow, while at the University of Wisconsin, imprinted baby monkeys to accept as their "mother" a foam-rubber dummy monkey with a built-in feeding bottle. Kittens have been imprinted to accept as fellow kittens rats placed in the same cage with them. The critical importance of timing in the development of certain skills was dramatized by two Harvard scientists who studied the way kittens' eyes develop the ability to recognize shapes and patterns. This ability — which is a physiological development, not imprinting — apparently develops in the fourth week. If the kittens are blindfolded during this week they become, for all practical purposes, blind for life.[2]

For humans, the child psychologist Philip H. Gray placed the most critical period for imprinting behavior patterns at "from about six weeks to about six months," or roughly at the onset of learning ability.[3] Others place it later. Many psychologists have difficulty reconciling the idea of imprinting as a very rapid and irreversible effect with the traditional, more fluid concepts of human learning. Those who recognize human imprinting consider it a more complex process than Lorenz's experience with ducks. The noted child specialist E. J. M. Bowlby, however, insisted that the way the human infant develops attachment behavior legitimately comes "under the heading of imprinting."

Awareness of one's sexual identity is another area of learning that seems to come under imprinting. John Money of the Office of Psychohormonal Research at Johns Hopkins is world-renowned for his work with hermaphrodites. Accidents happen, sometimes because of malfunctioning hormones. Baby boys or girls, because of their visible genitalia, may be assigned a gender that is not consistent with their chromosomes, gonads, hormones, or internal structure. When the ambiguity is discovered surgery is often performed to make the child's gender consistent with its internal organs.

Money has found that when the children are under the age of two and a quarter years at the time the reassignment is made, they negotiate the change "without even mild signs of psychologic nonhealthiness." However, if the change is made after two and a quarter years the chances are slim that the child can be rated psychologically healthy.[4] Money has often used the term *imprinting* in his writing. In his recent book, *Man and Woman, Boy and Girl,* he used the term *imprimatur* instead. I am told he did this so as not to antagonize the people who believe that the phenomenon of imprinting, in the strictest sense, occurs only in animals. Johns Hopkins is the world's leading center for transsexual operations on adults.

John Money has alluded to the acquisition of language as another area in which imprinting of some sort occurs. True bilingualism — or capturing the rhythm of languages — becomes more and more difficult with age.

José Delgado also talks of what he considers to be imprinting, and suggested to me another reason why the time element is critical: "Imprinting decreases at certain ages. Pathways of the brain close at certain ages for every function. You can't become a good musician if you start too late. I did not start to learn English until I was twenty, which is why I can't get rid of my Spanish accent. My children have lived in both Spain and America and have no accent."

Whether you call it imprinting or not, we can be fairly certain that humans are particularly impressionable in their development at certain times. And the most impressionable period for many characteristics, including "personality," is early childhood.

The most decisive period in a child's intellectual, social, and emotional development is between eight and eighteen months, according to Burton White, a Harvard specialist in child development.[5] This is when the child is moving out of the crib, exploring, learning language. How a child is handled during these months is more critical to developing general ability than at any other time.

A leading child psychologist, Eleanore B. Luckey, has accepted the concept of imprinting as a development that occurs when the child is extraordinarily impressionable. She has made the interesting proposal that the periods of maximum impressionability be pinpointed for various characteristics, and that a new profession — particularly suited to women — be created: professional imprinter.

The imprinters would go into homes and serve as consultants to natural mothers. If, for example, it is established that personality fixing proceeds most rapidly between the eighth and tenth month, an imprinter specializing in personality might be on hand several hours a week during that period to help optimize the traits particularly desired by the family. A good deal still needs to be learned before such a technology could be really effective.

A mild form of professional imprinting has now begun at the New York Medical College. In a three-year controlled experiment, the college is conducting a weekly class in "successful mothering" for about twenty mother-baby pairs. At the beginning of the experiment, the infants were about four weeks old. At the end of three years they will be compared in emotional and intellectual development with a control group. Nina Lief, the project director and a psychiatrist, contended that "dropouts begin in the cradle." She believes that the more a mother talks to her baby the sooner the baby will talk. But she cautions, "Don't expect real sounds until he is six months [old]."

Some scientists question whether the modern family is indeed competent to have full charge of child rearing. Robert S. Morison, a distinguished neurophysiologist, believes in imprinting. He contends that it is idle to talk of a complex society of equal opportunity as long as that society "abandons its newcomers solely to their families." Many families, he argues, have "haphazard educational procedures" for the children's most impressionable years.

Delgado, too, believes that expert help is needed: "We should try to establish at the earliest possible moment of the baby's life a program of psychogenesis." By that he means "the use of available physiological, psychological, and psychiatric knowledge for the formation of the child's personality." Delgado agrees there is a strong possibility that professional imprinters will emerge. He adds, though: "But to imprint for what? What kind of humans do we want to construct?"

Many characteristics have been suggested as desirable candidates for imprinting — love of peace, love of nature, love of one's fellowman, love of knowledge. But so far, the experts have contented themselves with trying to promote emotional well-being and intellectual competence. Perhaps they will agree eventually on which of the more specific characteristics to work on.

Some scientists, incidentally, believe that personality shaping can be started chemically while the baby is still in the womb. In 1977 *Nature,* the noted British scientific journal, carried a report on suggestive findings by the endrocrinologist June Reinisch at Rutgers University. She was studying the effects of synthetic hormones used to try to prevent miscarriage. (The use of some of the hormones has become controversial.) Those mothers treated with progesterones had produced children who on personality tests appeared to be strong on independence and self-confidence. The children whose mothers had been treated with estrogens, in contrast, appeared in tests to be less self-reliant, more inclined to identify with a group.

SHAKING UP PERSONALITIES

The already-set personalities of young people can be remolded or at least profoundly redirected. A number of techniques have recently been explored. One, a kind of youthful brainwashing, emerged as a side effect of a federally funded program in Florida. The aim was to rehabilitate adolescents believed to be drug users. The program, called the Seed, apparently was quite successful in getting the young people off drugs. But the controversy over personality changes accompanying this rehabilitation caused the Department of Health, Education and Welfare to back off.

The youths or "Seedlings" pressed into this program were isolated from family, friends, and the outside world. They were stripped of all identification and thrust at a low position into a precise social structure. Then they were subjected to carefully managed, intense, peer-group pressure from fellow Seedlings already in the program. A Seedling could advance upward in the structure only by rigid right-thinking. There was a good deal of talk about goodness and love but there was also a deliberate effort to smash the new Seedling's psychological defenses and make him feel totally dependent on the group. He was instilled with the urgency of being constantly aware of the group's wishes. One Seedling reported that Seedlings slept in the same room with him; they accompanied him whenever he traveled by car and even when he went to the bathroom.

During the U.S. Senate hearings on the program the statement of a guidance counselor at the North Miami Beach Senior High School was introduced. She had encountered many returned Seedlings. The statement read in part:

"When they return, they are 'straight,' namely quiet, well dressed, [with] short hair, and not under the influence of drugs compared to their previous appearance of [being] stoned most of the time. However, they seem to be living in a robotlike atmosphere, they won't speak to anyone outside of their own group. They sit in a class together and the classes become divided [with] Seedlings opposing non-Seedlings. . . . Seedlings seem to have an informing system on each other and on others that is similar to Nazi Germany. They run in to use the telephone daily, to report against each other to the Seed."[6]

PERSONALITY CHANGES BY ELECTRICITY

Apparently, personality transformations can be triggered by the prolonged use of electrical stimulation, if accompanied by counseling. Electrodes can remain in place for months in the human brain without discomfort. José Delgado reports that lasting changes in behavior, including decreased friendliness, could be induced by reticular stimulation in gibbons.

Robert Heath of Tulane University and his collaborator Charles Moan found that a conspicuous element of one patient's personality underwent lasting change because of electrical stimulation. The discovery was a result of an interesting treatment they arranged. This included stimulation of the septal pleasure center in the brain. The patient had long been suicidally depressed and had been a drug addict. He was a homosexual and tended to be chronically angry. He was first shown a stag film of nude males and females frolicking in intercourse. He reacted with anger and disgust. Later the pleasure center of his brain was subjected to a series of stimulations that put him into a state of euphoria. The euphoria caused him for several days to exhibit an improved disposition. While he was in this new mood the same stag film was rerun for him. He became sexually aroused and masturbated to orgasm. During the following week his interest in women continued. This was viewed as encouraging by his counselors. And they answered his questions about sexual techniques. (He had never in his life had a heterosexual experience.) Stimulation of his septal pleasure center continued periodically and he eventually expressed a desire for a heterosexual relationship. At this point it was decided that the introduction of a sympathetic, cooperative young prostitute might well constitute sound therapy. She seduced him as they lay side by side in a private room and he reported experiencing intercourse climaxed by orgasm. During the following year counseling continued on an outpatient basis. He developed a close sexual attachment to another girl, and reported that he had become heterosexually oriented.[7]

If such a technique were to become easily available some legislature might pass a law requiring everyone convicted on a homosexual charge to undergo the treatment. That, of course, would be contrary to the wishes and beliefs of the majority of the homosexual community.

NEW PERSONALITIES BY SURGERY OF THE MIND

The fact that personality can be transformed by surgery has been known ever since the potentates of the ancient Middle East put eunuchs in charge of their harems. In recent years brain surgeons have become adept at effecting a variety of personality changes by either cutting or burning inside the brain.

Since the brain is the organ of the body fundamental to individual identity, modify it and you modify identity. The brain can also be modified in its function on a long-term basis by the use of chemicals. For example, the equivalent of castration can apparently now be achieved chemically by treating a person with an anti-androgen drug called cyproterine acetate, which was developed in West Germany. Some animal studies have shown that it can cause permanent atrophy of the secondary sex glands. The castration effect can also be created surgically by destroying brain areas connected with the sexual drive.

Surgery of the psyche is designed to modify personality, ways of thinking, and ways of behaving and feeling. It is also known as mental surgery, lobotomy, sedative neurosurgery, and most commonly, psychosurgery.

In its earliest and still-used form it was the operation called lobotomy. This involves a simple severing of fibers between the frontal lobes and the deeper portions of the brain. Experiments with a chimpanzee had shown that destruction of certain frontal regions of the brain eliminated temper tantrums. A Portuguese neurologist, Egas Moniz, happened to hear a reference to the chimpanzee experiment at an international meeting and went home and performed lobotomies on twenty long-term psychotics. He claimed improvement in most of the cases and won himself a Nobel Prize.

Soon the craze for lobotomies spread. The craze was helped by a challenge that World War II had created. After the war tens of thousands of severely disturbed veterans were swelling veterans' hospitals. The U.S. Veterans Administration backed the operation in the hope of getting the patients home.

Many of the operations performed during this period did not involve opening the scalp. Some surgeons simply ran an ice pick or a device resembling an ice pick up past the eyeball and wiggled it around for a while. This was to sever fibers connected to the frontal lobes. Certain people doing the "operation" were not even surgeons. One doctor who did several thousand lobotomies was Walter Freeman of George Washington University. He said it was a simple matter to perform a dozen supraorbital lobotomies in a few hours at the operating table.

Here, wonder of wonders, was a permanent pacifier. It seemed to transform the agitated, the violent, the distressed, the compulsive. Great numbers of patients did seem transformed enough to be sent home from mental hospitals.

But then the reports of frequent, undesirable side effects started coming in. Some lobotomized subjects became so placid they were barely able to bestir themselves. Many were dubbed "vegetables." Others dramatically lost their inhibitions against doing what they had previously considered to be immoral. Loss of creativity and loss of intellectual capacity were reported. In general there frequently seemed to be a blunting of what one lobotomy specialist called "the higher sensibilities."

As recently as August 1975, the *Archives of General Psychiatry* carried a long-term assessment of forty-three lobotomies on mentally troubled people. The principal author was the psychiatrist whose evaluations initiated the operations. Most of the patients were said to show no longer the specific symptoms that had inspired the operation. These symptoms were obsessive, compulsive, or hypochondriacal behavior. However, most still had problems. Seventeen had incurred substantial gains in weight.

By the mid-1950's there was a growing wariness of lobotomy, and two additional developments contributed to a sharp drop in its use for several years: the introduction of strong tranquilizers and electroshock. Both seemed to have the effect of changing distraught personalities to more bland ones.

Starting in the mid-1960's, however, surgery of the psyche was on the rise again. Several developments account for the increase:

• The perfection of the stereotaxic (head-clamp) method for precise entry of surgical instruments deep into the brain.
• The growth in knowledge about the role of the deep-down limbic system in controlling emotions and about the dramatic changes that can be triggered there.
• New, more adroit, less tissue-destructive techniques for performing modified lobotomies.

Today brain structure (and personality) is being changed by surgical cutting, by burning brain cells with the tip of implanted instruments, by injecting olive oil, by planting radioactive seeds, by ultrasonic beams, by freezing local areas, and by proton beams that explode their charge at a specific distance. Laser beams and cell destruction by chemicals to achieve brain modification also are being investigated.

Reports of long-term reduction of chronic anxiety in some cases have come from psychosurgeons who use differing techniques and who operate on different areas of the brain. The British have done a considerable amount of work on altering anxiety and depressive states. One team that performed more than two hundred operations found that nearly half the patients underwent a change of personality.

In one publicized case in England a young salesman with an apparent compulsion to gamble was arrested for larceny. A medical report was prepared on his condition as observed by psychiatrists at a Lancashire mental hospital. The report recommended a brain operation to cure him of his compulsion. The magistrate ordered the operation. There was a public uproar. England has millions of steady gamblers. The magistrate withdrew his order.

In animals, at least, operations on the amygdala sometimes produce spectacular hypersexuality. Male cats for example try to mount any other four-footed animal they can climb on. Significant increases in aggressiveness have in some cases occurred after destruction of a portion of the hypothalamus. The psychiatrist J. M. C. Holden studied the results of several hundred lobotomies performed in the Middle West. He reports that a frequent effect of extensive cutting is an "irreversible change in mood, emotion, temperament and all higher mental functions."

It is the successes in transforming aggressive, violent, or agitated people by surgical intervention deep in the brain that has provoked the most speculation.

Surgeons in Mexico, Japan, France, Denmark, and the United States — to mention just a few countries — have been treating violent behavior by destroying a part of the amygdala. That limbic organ is the most conspicuous seat of our emotions. Others in Japan have engaged in "sedative surgery" by destroying tiny bits of the hypothalamus.[8]

Louis Jolyon West, a noted brain surgeon at the University of California, Los Angeles, reported an interesting success story. A waitress who had repeatedly tried to slash herself had had to be confined to a straitjacket for several years. After an operation on her amygdala, her rages stopped and she was able to find a steady job.[9]

The Boston brain surgeon Vernon H. Mark and his colleagues performed more than a dozen operations on the amygdalas of patients with a history of violence. They believed the violence to be associated with epilepsy. One case in which Mark was the principal figure was that of a twenty-one-year-old girl named Julia. She was an epileptic and she also had made many seemingly senseless assaults on people, some of them total strangers. The surgeon, in consultation with a psychiatrist, came to believe that some of her assaults and rages were substitutes for the usual convulsions of epilepsy.

Following the removal of a part of her amygdala she had fewer epileptic seizures and only occasional exhibitions of abnormal anger, though she was still judged a psychotic. She was able to live at home, sing in the church choir, and pass her high school equivalency examinations. Others, however, reported that after the operation she became more despondent, stopped playing her guitar, and showed less interest in intellectual discussions.[10]

Studies of the results of amygdalectomies upon monkeys are disquieting, in the opinion of Stephan L. Chorover, a physiological psychologist at the Massachusetts Institute of Technology. After the operations the monkeys became less hostile toward their human handlers but acted in many inappropriate ways when returned to their troop. Their inability to cope made them social isolates and they either starved to death or were killed by predators.

Is there a lesson here for humans as far as tampering with the amygdala is concerned? Chorover thinks so, and cites the case of a brilliant engineer who had a tendency to go into rages. He was referred to a Boston surgeon who suspected from studies of his brain-wave activity that his behavior might be a form of epilepsy. Chorover contends that the existence of a causal connection between epilepsy and violence remains an open question. At any rate the engineer finally agreed, after weeks of patient explanation by the surgeon, to have cuts made in his amygdala. Four years after the operation the surgeon stated that the engineer had not had a single episode of rage since the operation but that he did continue to have "an occasional epileptic seizure with periods of confusion and disordered thinking."

Chorover reports, however, that this brief summary blurred over the true situation. The engineer's family invited a Washington psychiatrist to make an independent review of the case. The psychiatrist reported that after the operation the engineer lost his job and his wife, and was so confused — he was unable to work and was incapable of taking care of himself — that periodically he has had to be hospitalized.[11]

There has been a tendency on the part of some surgeons to justify operating on the ground that the violent behavior is somehow related to one or another brain disease. That gets them out of the controversial behavior-control business. One enthusiast of psychosurgery, Orlando J. Andy of the University of Mississippi, went so far as to propose that all abnormal behavior results from structurally abnormal brain tissue.

Proving this relationship is pretty hard to do, however. Most of the limbic system is difficult to x-ray. Also, most of the electrical activity in that area is hard to pick up on the conventional brain-wave instruments placed on the scalp. Only implants deep into the brain at the suspected site can pick up dependable brain-wave clues.

If an assertive person is made into a placid person by surgery it is an irreversible change. He isn't the same person anymore. How much manipulation of personality by intrusion into the brain is justifiable? Should it be restricted to mental-hospital patients with certifiable brain damage who want a change, or if they are mentally incompetent, whose closest kin request the operation?

Do we really know enough about brain function and the often-delayed effects of altering the brain to justify irreversible surgery except in extreme cases? By extreme, I mean people with provable brain damage and people suffering constant unrelievable pain. I don't think we do know enough.

We should confine psychosurgery to these extreme cases, and then only after a hospital committee has reviewed all the facts.

A line must be drawn. If we get into the business of having surgeons operating on the brains of people judged by them or by governmental authorities to be too aggressive or belligerent for the good of society, then psychosurgery can become a convenient tool for social control. We will take a look at this potentiality early in the next chapter.

On Making Man More Tractable

6

If America ever falls to totalitarianism, the dictator will be a behavioral scientist and the chief of police will be armed with lobotomy and psychosurgery.
— Peter R. Breggin, psychiatrist, in Congressional testimony

Perhaps the riots of recent years — on the campus, in the ghetto, and against the war in Vietnam — gave the movement a push. At any rate we have been seeing a widespread, anxious search for neat ways to control unruly people. The appealing new approach is to handle such people quietly, scientifically.

The Russians have been in the lead in using this approach, although often as a sham. They have had doctors examine outspoken dissidents and declare them mentally ill. Then the dissidents were sent off to asylums. The Russians also are reported to have devised a simple technique for permanently enfeebling the brains of dissidents they have detained. A Canadian psychiatrist at the University of British Columbia Medical School testified on this at a hearing before the Subcommittee on Internal Security of the U.S. Senate. He stated that the Soviet police had discovered that the use of massive doses of reserpine caused some atrophy of the brain. His charge, he said, was based on talks with Soviet psychiatrists and former Soviet residents. Reserpine ordinarily is used to treat hypertension.

The yearning for a placid, conforming populace in the United States was best articulated by former Vice-President Spiro T. Agnew: "We're always going to have a certain number of people in our community who have no desire to achieve or who have no desire to even fit in an amicable way with the rest of society. And these people should be separated from the community, not in a callous way, but they should be separated as far as any idea that their opinions shall have any effect on the course we follow."

Many managers of institutions that inevitably deal with unruly people seem to judge their success by the degree of orderliness and placidity achieved. I refer particularly to managers of schools, prisons, corporations, homes for the aged, and mental hospitals.

Seymour Halleck, a psychiatrist at the University of North Carolina, has observed that many of his colleagues evaluate the success of treating emotionally distressed people by only one criterion: "whether it makes the patient more placid."

And the psychiatrist H. L. Newbold, cited earlier, seems to feel that taming Man by programming him for tractability should be a major objective of psychiatrists. He points out that administering psychochemicals makes "the patient . . . more tractable and thus reprogramming by the therapist is made easier."[1] He calls "tractability" of "pressing interest." Man gets into a lot of trouble, he says, because he is biologically programmed to be expansive (aggressive). Overexpansive children might well be kept on tranquilizers while they are being programmed for life. This enables them to fit more easily into the social norms.

A number of people and organizations have been urging that some sort of system for spotting the violence-prone be adopted. Certain law-enforcement and genetics experts — without sufficient warrant — have proposed that all newborn male babies be screened for an extra Y chromosome. About one in five hundred males has the XYY chromosome. The fact that in some surveys a somewhat higher proportion of XYY's have showed up in prison populations has inspired the scanning proposals. But how significant is a seeming variance when you consider that a great many people are in prison for cupidity, not violence.

A man who was President Nixon's internist at one point, Arnold A. Hutschnecker, proposed that we rid society of violent people by the mass screening of six-to-eight-year-olds. Psychological tests would be used. The idea would be to detect those with real or potential tendencies toward violence. Also, he would assign children identified as "severely disturbed" to camps for special training. According to a report by Charles Witter in *Trans-Action,* the social science journal, Hutschnecker's proposal was written on White House stationery and was sent to the Secretary of the Department of Health, Education and Welfare for suggestions. And according to the same report, it was studied for three months at the National Institute of Mental Health. The House Subcommittee on Privacy got wind of it and discussed holding hearings. Then HEW responded to the White House unfavorably.

Kenneth Keniston, an authority on alienation, has sardonically envisioned that within a decade Congress would pass a Remote Therapy Center Act. This would set up 247 centers, mostly in the Rocky Mountain area. Each would hold one thousand patients. Each would treat inner-city dwellers identified by trained spotters as suffering from "aggressive alienation syndrome."

The brain surgeon Vernon H. Mark and Frank R. Ervin, a psychiatrist, in their controversial book *Violence and the Brain* call for an early-warning test on violence. They propose to set up a basic standard of behavior on the violence dimension that individuals with a normal brain can meet. Those who can't, they suggest, could somehow be diagnosed by testing further for malfunction of the limbic system of their brains.

PACIFYING BY BRAIN SURGERY

Some of the proposals go beyond the mere screening of unruly people. If other techniques for treating them are unproductive, psychosurgery would be the treatment of choice. The psychiatrist Peter R. Breggin, a critic, says of psychosurgery: "For the would-be despot — be he politician, prison warden, mental hospital superintendent, law enforcement official, bigot or simply family tyrant — psychosurgery is a dream come true."

In 1967, when rioting in the ghettos was attracting national attention, Ervin and Mark, along with their Harvard colleague William H. Sweet, a brain surgeon, wrote a letter to the *Journal of the American Medical Association.* They pointed out that while social conditions unquestionably were a factor in provoking the riots, it was time to look at the individuals who were conspicuously violent. They should be clinically examined. The goal of such studies, they said, would be to pinpoint, diagnose, and *treat* those people with low violence thresholds "before they contributed to further tragedies." Mark, for one, has backtracked somewhat by pointing out that the police, too, are often to blame for violence during riots. But their letter was quoted in *Ebony* magazine in an article titled "New Threat to Blacks: Brain Surgery to Control Behavior."

At the Second International Congress of Psychosurgery in Copenhagen, M. Hunter Brown, a neurosurgeon from the Los Angeles area, proposed pilot programs for the precise rehabilitation of the "violent" prisoner-patient. He hoped his state of California would provide programs that would include diagnosis, trial use of the newer drugs, "and finally neurosurgical intervention to specific targets as indicated. Until then, humanity must mark time."[2]

A neurophysiologist, Robert Livingston of the University of California, San Diego, has called present penal systems self-defeating and has proposed "proper intervention" in the brains of prisoners who have temporal lobe defects.[3]

Around the world thousands of operations are being performed each year on the brains of unruly or deviant persons. A few of the surgeons profess to operate only — or mainly — on persons with clear evidence of organic abnormalities in the brain, such as epilepsy. It should be noted, incidentally, that it is the rare epileptic who is violent toward others during his seizures. If he is unruly at other times it may be coincidental. A criminal who has been pistol-whipped a few times may well develop a brain pathology. But in general the evidence of any correlation between provable brain pathology and tendencies to violence is skimpy.

Other surgeons say they are just trying to help people with tendencies to violence stay out of trouble. And some make no bones about the fact that whatever the condition of the brain they are in favor of operating on it, primarily on the amygdala. Their aim is to help make individuals less troublesome for the people who want to, or must, manage them.

The trouble can be as simple as that of a destructive, hard-to-manage child whose mother seeks help. Walter Freeman, sometimes called the dean of psychosurgeons, cited the case of a toy-smashing six-year-old who was a real problem to her mother. After two extensive lobotomies, he reported, the child became "quite withdrawn, but less troublesome."

Peter Breggin notes that the majority of the many psychosurgery operations performed by Orlando J. Andy, director of neurosurgery at the University of Mississippi, have been on children. They were classified as hyperactive, aggressive, or emotionally unstable. Breggin cited a letter from a psychologist working with Andy who stated that the operations on the children were to "reduce the hyperactivity to levels manageable by parents."[4]

More common, however, is the brain operation performed on people in institutions that make them easier for the custodians to manage. One of the largest groups in the world on which surgery of the amygdala has been performed to control aggression and hyperactivity is that operated on by V. Balasubramaniam at the General Hospital in Madras, India. He has described the surgery as "sedative neurosurgery," designed to make the patient "quiet and manageable."[5] Balasubramaniam cites violent patients who could be safely left in the general wards after the operation. One formerly assaultive patient became helpful around the ward and even looked after other patients. Other surgeons have cited postoperative patients who were more willing to handle such chores as floor mopping.

Robert Neville has headed a task force formed by the Institute of Society, Ethics and the Life Sciences to study the full social impact of psychosurgery. He has concluded that psychosurgery has a dangerous attraction for institutions as "the cheapest and easiest treatment to adopt for controlling patients."[6] This attraction, he believes, raises the hazard that psychosurgery can be improperly used to subdue aggressive dissidents.

Breggin puts the appeal more flatly. The situation poses a potentially disastrous situation for children in state institutions or adults in state prisons. Both, he says, "are entirely under the control of authorities whose major intention is to manage them in the most economical and most efficient manner."

He cited in the *Congressional Record* medical reports by Walter Freeman in which Freeman asserted that there is no compelling need to operate on a hospital patient if the ward notes read: "Gives no trouble in the ward." In 1971, Freeman commented further that psychosurgery is "the ideal operation for use in crowded state mental hospitals with a shortage of everything except patients."[7]

The appeal to law-and-order people of a quick, surgical taming of unruly people is seen in the fact the U.S. Department of Justice granted nearly $200,000 to the Neuro-Research Foundation in Boston. The foundation was set up by Mark, Sweet, and Ervin. The money was to be spent in trying to identify violent prisoners who, in routine screening, might also appear to have brain disorders (and thus presumably be candidates for psychosurgery). Recent criticism of psychosurgery on confined people, however, caused the Justice Department to beg off when Mark, Sweet, and Ervin asked that the fund be increased to $1,300,000.

A number of state law-enforcement agencies, particularly in California and Illinois, have shown themselves to be fascinated by the surgical approach to coping with unruly prisoners. In 1971 the director of corrections for California came up with a broad-scale proposal for "the neurosurgical treatment of violent inmates" who seemed on test to have brain disorders that might account for the "violence."

Stephan L. Chorover of MIT came upon a portion of an official affidavit concerning one inmate who presumably would be considered a candidate. He was being transferred to the prison at Vacaville, California, for psychiatric evaluation. The man was described as "aggressively outspoken, always seeking recruits for his views that the institution and its staff were oppressing inmates." This man was teaching other inmates karate. He had been involved in a work stoppage. He sought from the outside books attacking society. He set a fire in May to demonstrate his political views. In short, he appeared to be a strong-minded dissident.[8]

Three years earlier, three inmates of the Vacaville prison hospital who had a history of violence agreed to have part of their amygdalas destroyed. The results were not particularly impressive. The one who showed the most initial improvement was paroled, and was soon back in jail on a robbery charge. He complained that he had begun losing control of his emotions.[9]

It is undoubtedly true that surgery in parts of the deep limbic system, such as the amygdala, *can* dampen emotionality. But the techniques are still primitive. Furthermore, as indicated, there is an intricate interplay between the various areas of the brain. Ayub Kahn Ommaya, as research director of the U.S. National Institute of Neurological Diseases and Stroke, testified that "every part of the brain requires the other parts to function."

We have relatively little knowledge of how digging around in one area discombobulates the whole brain system.

Ommaya also expressed doubt that the presence of abnormal brain waves necessarily indicates structural damage in the brain. Early enthusiasts of psychosurgery cited substantial rates of brain-wave abnormality in prison populations as evidence of brain damage. They did this without checking the brain waves of a control group of nonincarcerated people. Many individuals who reveal brain-wave abnormalities are not noticeably prone to any abnormal behavior.

We have then, with psychosurgery, a new technology that was developed largely for humanitarian reasons. But it has become widely used in the world to make people more tractable, whether or not they behave in ways dangerous to others or themselves. And it has the potential for the manipulative control of Man, including the elimination of dissenters. All this raises serious questions about when — if ever — it should be permitted.

Peter Breggin has called for a total outlawing of all forms of psychosurgery as unjustifiable mutilation. A number of bills have been proposed in the United States Congress either to outlaw or restrain psychosurgery. One would impose a fine of $10,000 on any psychosurgery performed at a federally financed institution.

My feeling is that outlawing psychosurgery, at this stage, would be a bit drastic. A flat outlawing would void the humanitarian goal of offering hope of relief for some hopelessly sick people willing to have anything done to ease their misery. As Henry K. Beecher, a Harvard anesthesiologist, put it: "Talk about civil liberties — it would seem that hope for relief is a rather important one."

We are dealing, however, with a very imprecise and misused technology. We do not understand the functioning of the brain well enough. Far too little experimentation has been done on primates. (One of the best primate studies now under way is at the Yerkes Regional Primate Center at Atlanta, Georgia.)

I believe there should be a moratorium for five years on all psychosurgery on humans that does not meet stern criteria of diagnosis and supervision. Irreversible psychosurgery at this experimental stage should not be permitted on confined persons even if they give consent — without exceptionally thorough independent review — since the coercive atmosphere makes truly informed, willing consent difficult to obtain. For example, refusal to volunteer might affect prisoners' chances of parole.

This position — that consent alone is not enough — was upheld in 1973 by a three-judge panel in the Wayne County Circuit Court in Michigan. The case involved a prisoner with a long history of violence who had been selected for experimentation and had agreed to it. The judges became convinced during the lengthy testimony that the predictable effectiveness of psychosurgery for controlling violent conduct is unknown; that scientists do not know enough about brain structure to understand why the surgery should work, much less whether it might work in specific cases; and that this kind of surgery has the inevitable and irreversible effect of blunting the emotions.[10]

During the moratorium no hospital in the nation should permit psychosurgery on *anyone* unless the case is first reviewed by a committee consisting of people not associated with the hospital. An advisory commission to the Secretary of Health, Education and Welfare has recommended (in 1976) that psychosurgery still be treated as an experimental procedure. Also, no operation should be permitted that is not clearly therapeutic — rather than manipulative — in intent. Finally, a national registry of people — identified only by code number — who undergo, or have in the past decade undergone, various kinds of psychosurgery should be started, and the records should include an evaluation of the results. This would provide researchers with many of the answers we need to know before psychosurgery, under tight regulation, can be accepted as a standard, permissible therapeutic procedure. Brain surgeons are usually too busy to make more than quick superficial checkups on patients into whose brains they have intruded.

For the present, people who are worried about their tendencies to be violent would, I feel, be better advised to take the pill route. This is usually not irreversible. There are the moderate tranquilizers. There are also specific anti-aggression drugs, mentioned previously, available or in development. The absence of a certain enzyme in males creates an excess of uric acid, which is known to cause hyperaggressiveness.[11]

Furthermore, there is lithium. Michael Sheard, a psychiatrist at Yale, tried it on twelve inmates of the Connecticut State Prison at Somers who had long histories of violent behavior. For one month they took one lithium carbonate capsule every day. The next month they, unknowingly, took placebo pills identical in appearance and taste. The alternation went on for a number of months. Sheard reported a notable reduction in violent behavior during the lithium periods and with no detectable serious side effects.[12]

MAKING PRISONERS SHAPE UP, SCIENTIFICALLY

Behavior shapers other than surgeons have been actively promoting tractability in prisons.

Reenter the operant conditioners and brainwashers. Also enter the group therapists and drug manufacturers.

The traditional way to cope with nonconforming prisoners was to put them in the Hole on bread and water. In the 1950's that gave way to a reformist ideal that no prisoner was beyond redemption and that all, even the worst, could be helped by psychological counseling. Individual talking-out of problems, however, proved to be time-consuming. Meanwhile, crime was increasing.

The law-and-order people began calling for no-nonsense approaches to handling prisoners. They talked more of "management" of prisoners, less of rehabilitating them. They became intrigued by a number of "managing" techniques coming out of the behavioral and medical sciences.

Many prisons found that by the use of such powerful tranquilizers as Prolixin they had their troublemakers walking around the yards like zombies. A prisoner at the Arizona State Prison wrote in the *New York Times* that he had seen electroshock treatments — usually reserved for treating depression — used to reduce the vigor and vitality of jailhouse lawyers and inmates considered to be "political radicals" by prison officials. The prisoners called it "Edison's medicine."[13]

There was blunt talk of brainwashing prisoners to soften them up for rehabilitation through an array of behavior modification strategies. A number of sources, including Jessica Mitford in her *Kind and Usual Punishment,* tells of a lecture on brainwashing given to a conference of wardens by an MIT psychologist. He had studied brainwashing techniques used on United States prisoners by Asian Communists. In his lecture he suggested that certain of the techniques might be used effectively in our prisons. According to Mitford, the wardens were urged especially to consider the systematic isolation of a target prisoner from his friends. The wardens were briefed on the promotion of mutual distrust, achieved (to cite one method) by tricking the prisoner into writing statements that are then shown to the others. This was to convince him that he could trust no one. The wardens were also told about a concluding technique of putting the softened prisoner into a new, ambiguous situation in which he was confused by unclear standards. Then pressure was put on him to change in the directions desired.

Mitford tells of evidence that this brainwashing approach had been put into effect in at least one federal penitentiary — the one at Marion, Illinois. Members of the Federal Prisoners' Coalition smuggled out a detailed report of an alleged experimental program startlingly similar to that proposed to the wardens. It was applied primarily to troublemakers. And it was a part of a personality-restructuring program worked out by Martin Groder, the prison psychiatrist. He had become the Skinner of prisoner modification.

A prisoner was harassed, by the methods described above, until he agreed to submit to Groder's thought-reform program. Then he was subjected to intense pressure from a team of reformed prisoners. They probed his fears. They systematically attacked his whole framework of beliefs. The idea was to bring him under thought reform so that he could be "reborn" as a highly probable "winner in the game of life." Once reborn, he moved into plush quarters and became himself a member of an attack team against other prisoners still waiting to be reborn.[14]

Meanwhile, the operant conditioners, their briefcases bulging with aversive-rewarding strategies for shaping behavior, began showing up in U.S. prisons. Many were wholly or partly supported by grants from the U.S. Bureau of Prisons or from the Law Enforcement Assistance Administration of the U.S. Department of Justice.

The LEAA was requested to provide a list of behavioral research projects it was supporting. In response, it provided a computer printout of four hundred projects that might involve behavior modification.

At a federal prison for women in Alderson, West Virginia, female inmates were operantly conditioned by being locked in a cell with only a bathrobe and chamber pot. This aversive therapy was to motivate them to earn their way to clothing, furniture, and normal prison life by exhibiting desired behavior.[15]

A more celebrated application of behavior modification occurred at the Medical Center for Federal Prisoners at Springfield, Missouri. The program was designed to "rehabilitate" problem prisoners sent there involuntarily for treatment. Reportedly some were not a "problem" because of violence but rather because of their cunning in organizing escapades to harass prison staff members.

The conditioning strategy was cold-turkey therapy with a promise of relief and rewards. On arrival the participants were thrown into solitary confinement in dark cramped cells — six by ten feet — and deprived of privileges. Real aversive therapy. Each prisoner had to build up twenty consecutive "good days" before he could start enjoying the first of three levels of rewards. One prisoner disliked the game so much that he spent forty-two days in total darkness, except for exercise and showers.

A prisoner compiled a "good day" by stopping the use of abusive language, by keeping his cell clean, and by observing all the rules. He was closely monitored by prison aides with checklists in hand. Within a year, it was hoped, each participant would become an obedient citizen enjoying all available privileges. Then he would be returned to his "home" prison. He was to be inspired to progress by the title of the program: START (Special Treatment And Rehabilitative Training).

Most behavior shapers believe that positive reinforcers produce better long-term effects than punishment. But in their dealings with prisoners they found the subjects generally suspicious. Also they found themselves working with obedience-minded wardens quick to order aversive treatment.

Several newly developed drugs, it turned out, can have a more aversive effect than weeks spent in solitary confinement. The drug apomorphine can cause uncontrolled vomiting for up to an hour. Even more motivating is an injection of Anectine (succinylcholine). It has been used extensively by researchers as "aversive therapy" in the prisons at Vacaville and Atascadero, California. The drug is derived from the South American arrow poison curare.

Anectine takes the misbehaving inmate to Death's Door. It paralyzes his muscles for breathing. He is suffocating but fully conscious. It is during this suffocating phase that the therapist reminds him of his inappropriate behavior and that this kind of aversive therapy may be helpful in enabling him to behave more appropriately in the future. Once the terror-stricken prisoner seems to get the point, the drug's effect is counteracted.

In 1973 and 1974 some of the techniques and tools used for reshaping the behavior of prisoners came under attack. A court declared that the use of the vomiting drug in an Iowa prison is cruel and unusual punishment unless the inmate gives written consent. The American Civil Liberties Union won a court order ending some aspects of the Marion brainwashing program. The Federal Bureau of Prisons, faced with a lawsuit from inmates and a Congressional hearing, stopped its START program at Springfield. LEAA faced charges that federal police were getting into the mind-bending business and nervously announced that it would at least stop funds to support medically oriented programs for systematic behavior modification. It was speculated that any half-bright state administrator of prisons could still salvage LEAA grants by renaming his programs. Six months after the announcement not a single behavioral or biomedical research program had in fact been canceled.

The director of the Bureau of Prisons made it immediately clear that although START was being dropped, the bureau was still very much interested in the use of behavior modification. He saw it as an "integral part" of many prison programs. "We're going to start programs in all our penitentiaries' segregation units," he disclosed. "Only they won't have titles that carry such emotion."

Aversive therapy continued to flourish in state prisons. But in general the name of the game now was Change the Name — and ease up on the more blatantly coercive projects.

Monitor, a bulletin of the American Psychological Association, reported a scramble among psychologists to dump the phrase "behavior mod." It was getting all mixed up in the public's mind with psychosurgery, drugs, electroshock, and coercion.

One behavioral spokesman, Leonard Krasner, proposed "environmental design" or "behavior influence" as a possible substitute for behavior mod. Other worthy synonyms mentioned in the report were "behavior therapy," "behavior shaping," "social learning," and "applied behavioral analysis."

Krasner criticized B. F. Skinner for persisting in using the word "control." Skinner was reported to be opposed to dropping it. In his view undertakers did little to alter their image by changing their job title to funeral director. The APA *Monitor* report concluded: "Perhaps the best that psychologists can do for the time being is to speak softly, and for heaven's sake, don't carry an electric cattle prod."

However, by 1976 the APA was, as mentioned in Chapter 1, using the phrase "behavior control" as the theme for the Master Lecture Series at its annual convention.

By whatever name, the use of scientifically developed behavior-modification techniques proceeded to go forward in at least a dozen federal penitentiaries. Officially, electroshock, aversive drugs, and the like were taboo. Instead, a host of strategies were tried that sprang from the fertile mind of Martin Groder, the man who tried brainwashing at the Marion penitentiary.

An editor of *Human Behavior* reported that Groder called his overall approach Asklepieion. He explained it as an "intensive treatment community." It combines "transactional analysis, gestalt therapy, primal-scream therapy, psychodrama and confrontations and marathon groups derived from Synanon games."[16] (None of these techniques involve behavior modification as practiced by operant conditioners.)

The main center where these Groder strategies were supposed to be tried was a somewhat mysterious federal installation near the backwoods village of Butner, North Carolina. The installation would cost more than $13,800,000. Groder was named the warden.

Its experimental area for testing behavioral modification techniques was to house two hundred inmates in relatively comfortable surroundings. There were to be no cells in the plan, just "behavior modification units." Yet a number of prisoner associations around the country began protesting as if in dread.

Before the center finally opened in 1976, after delays, its name was changed twice. At first it was to be called the United States Behavioral Research Center; then the Federal Center for Correctional Research. By the time it opened, it was simply named the Butner Federal Correctional Institution.

By opening time, Groder with his provocative ideas had dropped out as warden. At last reports, the behavioral-research part of Butner, which houses two hundred hard-core repeat offenders, was essentially voluntary. Brutality was out. A Bureau of Prisons psychologist stated that the inmates are being offered a choice of programs, "sort of like a college catalogue." The asserted aim now is to test a central behaviorist contention that environment shapes behavior.

All this suggests that it may be possible for behavior modifiers, under pressure from prisoners and libertarians, to design useful programs. Many prisoners want a chance to feel better about themselves. And at least some want help in improving their chance of going straight outside. If they are treated as responsible persons the chances are at least greatly improved that they will respond responsibly. A first requirement, however, is that the goal be to help the prisoner, not manage him more efficiently.

A case in point is the behavior of the inhabitants of a Denver facility known as the CAT house. That is short for the Colorado Closed Adolescent Treatment Center. There the youngsters have been running their own behavior-modification program, with guidance from adults. Reports are that for the most part it has worked out well.[17]

HOW TO MAKE THE AGED AND MENTALLY AILING MORE MANAGEABLE

Institutions dealing with the aged, the mentally retarded, and the mentally disturbed also have shown interest in scientific ways to promote tractability in their subjects.

The British neurosurgeon Geoffrey Knight has reported that he first was inspired to perform lobotomies after reading about the increased admission rates of old people to state hospitals. He has performed more than a thousand of these operations.[18]

However, for the institutionalized aged, the retarded, and the disturbed, the primary pacifier of choice is the tranquilizer. At Congressional hearings, an official of the National Council of Senior Citizens protested the abuse of tranquilizer drugs in nursing homes. He charged that there was a growing tendency to give patients ever larger doses of tranquilizers "for the sole purpose of keeping them quiet and easy to manage."[19]

There are excellent reasons to prescribe tranquilizers for some people in nursing homes. But tranquilizers have a definite economic attraction to those nursing home managers concerned with profits. And the effects of the pills can deprive the older patient of interesting life experiences. This is especially true when patients are kept so doped up that they lie around in bed most of the day. The director of the National Council of Senior Citizens pointed out: "If you are running a [nursing] home for profit, and you have got a tranquilized patient body, then the chances are you can save money on staff and food, by keeping these poor people like zombies."

More recently, in August 1975, Senate hearings were held on the operation of institutions for the mentally retarded. The chairman, Senator Birch Bayh, said he was "dismayed" by the testimony on drugging patients. One witness, James Clements, the director of the Georgia Retardation Center and a pediatrician, charged that drugging retarded patients in order "to subdue behavior" was the rule, not the exception. He added that the behavior being subdued resulted from the environment of a typical institution for retardees. Untrained ward attendants, often with a ninth-grade education, hand out drugs merely "to stop behavior they don't like." At some institutions more than three quarters of the retardees are kept on drugs. Senator Bayh said it sounded to him like chemical handcuffing.

There have also been charges that tranquilizers are abused in hospitals for the mentally ill. Again, it is alleged that the primary aim is to subdue problem patients that annoy staff members.[20] Such use of drugs can actually prevent the patient from undergoing life experiences necessary for him to learn again to cope with society.

TAMING UNRULY PUBLIC SCHOOL CHILDREN CHEMICALLY

The uproar began in Omaha. Thousands of children in the public schools were reportedly being drugged during school hours to reduce their unruliness. It was for the children's own good, so the argument went, since they had some sort of brain malfunction. The pills helped quiet their brains so they could be better students in the early grades. And they would stop disrupting the work of other students.

At a world conference of psychiatrists the problem of children with a certain pattern of restlessness had been given a medical label: "the hyperkinetic syndrome." Other psychologically oriented doctors called it "minimal brain dysfunction" or "minimal brain damage." But others simply said that the children were "hyperactive" and let it go at that.

The main symptom was that the children were fidgety. Also they were easily distracted. They had short attention spans. They were impulsive.

Who diagnosed these young sufferers in Omaha? Mainly their teachers. How did the teachers learn to make these diagnoses? Many had heard talks by representatives of drug companies, or were briefed by other staff members who had heard the talks. The teachers told parents that doctors now had pills that could help their children improve their concentration and perform better in class. If the parents were concerned they should seek medical help.

Thus it was that several thousand Omaha grade school students were carrying pills in their pockets or lunch buckets. Almost all were boys. Some left the pills with the teacher. And sometimes the teachers reminded them if they forgot to ask for their pills.

What were these pills? That was the surprise. Mostly they were similar in content to "speed" pills, which people can get arrested for carrying without a prescription. They were either amphetamines or Ritalin, which has a similar effect. In short, the pills were stimulants. Some of the pills were listed by the U.S. Food and Drug Administration as "not recommended for use in children under 12."

Although these stimulant pills have the effect of charging up adults, often dangerously, they seem to have an opposite effect on hyperactive prepubescent boys. And in the last few years they have been tried as a way to calm hyperactive or aggressive teenagers.

A recent edition of the *Physicians' Desk Reference* says this, among other things, about the possible side effects of Ritalin: "In children, loss of appetite, abdominal pain, weight loss, . . . insomnia, and tachycardia may occur more frequently. Toxic psychosis has been reported." Some doctors also believe it lowers the threshold of epileptic seizures. But many believe it is relatively nontoxic. In Sweden the use of Ritalin is rigorously regulated.

The Omaha situation was brought to national attention by Robert Maynard of the *Washington Post,* who heard that an interesting controversy was raging in Omaha. Maynard journeyed there to investigate. He got the impression that a lot of youngsters had been classified as hyperkinetic too loosely. He also noted charges from ghetto parents that Omaha was trying to drug black children into quiet submission. Some Omaha doctors claimed that the pills did more than simply calm hyperkinetic students. The pills also made them happy. One doctor told of a mother of a pill-taking boy who came home from a meeting and found on the table the homework her child had finished. The child had written a note saying: "Thank you, mother. I feel so much happier."

The public discussion created by the report from Omaha flushed up the discovery that drugged schoolchildren were a fact of life in hundreds of American school systems. The most commonly mentioned estimate then was that at least 250,000 grammar school pupils were on amphetamines or Ritalin. And the number was growing rapidly. Now the estimates range from 500,000 to 2,000,000 (the latter figure appeared in a 1976 report in *Science Digest*).

There were also reports that in many school systems the proportion of pupils on stimulant drugs ran between 10 and 15 percent.

The most comprehensive study of school practices regarding pill taking in the schools, at this writing, was made in the Grand Rapids public school system. The investigators were a sociologist, Stanley S. Robin, and an educational researcher, James J. Bosco, both of Western Michigan University.[21] They randomly sampled 150 teachers. Sixty-five percent of the teachers knew of pupils on Ritalin. Forty percent of the teachers reported that they recommended consultation with a physician when a child appeared to be "hyperactive." A few teachers had suggested Ritalin by name.

Later, Bosco spoke more broadly about the national scene at a meeting of the National School Boards Association in Houston in 1974. He said:

"Stimulants and other drugs are currently a fact of life in the management of hyperkinetic children in classrooms." (Note the familiar word "management.")

Officially, school personnel are only being helpful when they suggest to parents that their children might go to a doctor to see if they might benefit from stimulant drugs. But in some cases there appears to have been a clear element of coercion: threats to hold back a child or put him in some class with a disabled label.

The *Providence Journal* cited a mother who felt she had been forced by school officials into drugging her child. A mother in Little Rock, Arkansas, claimed that she felt so harassed by requests that her child be medicated that she moved to another state. There, there were no further problems. She alleged that she was warned that school officials at her school in Little Rock were considering using her child in a trial court case. The case would have tested whether children could be put on medication without the parents' consent.[22]

I know of no such case reaching the lawsuit stage yet. But in Taft, California, an isolated oil-and-agriculture town of 18,500, parents have taken the offensive in a test class-action legal case. In *Benskin et al.* v. *Taft City School District et al.* the parents of eighteen children have alleged that school authorities coerced them into putting their children on the psychoactive drug Ritalin.[23] At this writing the case is awaiting trial. The parents charged that teachers had repeatedly told them their children needed Ritalin, or higher dosages of it. And they charged that their children had suffered serious adverse reactions, such as cramps or crying spells or listlessness, of which they had not been warned.

The primary legal charge is that the school did not give the parents the right to informed consent.

The parents of one child were called into conference because their child was withdrawn and stared out the window. In another case the offense was that the child did not sit still.

It appears that the school physician, who often sat in on parent-teacher conferences, was a Ritalin enthusiast. She believed that the drug was an enhancer of learning ability in children who were overactive. A number of other top school officials were also said to be enthusiastic about drugging children. Some years ago the school system became eligible for federal funds with which to do a study project. The money was used to try to find out why children didn't learn. Teachers were asked to fill out forms on children who appeared to be headed for learning problems. And they were given a lot of ideas about what symptoms to look for. A teacher who sets out looking for problem children can usually find them. Of about five hundred children screened, approximately seventy-five were classified as having learning disabilities. A number of the children so labeled are named in the lawsuit as being among those allegedly coerced into taking Ritalin.[24]

In the lawsuit the parents allege that taking the Ritalin was in essence made a condition of attending public school.

It is pertinent to recall at this point that in the United States parents are required by law to send their children to school. The schools are an institution of government. That being so, are schools in general exerting, however subtly, any kind of governmental pressure to get children on behavior-modifying drugs?

In Omaha there may well have been witting or unwitting pressure, as in Taft. Teachers in Omaha were invited to identify children they thought could benefit from drugs. And there were reports of a systematic screening of children for referral to private physicians by the school system's health and psychological services.[25]

When a teacher talks with a parent it often isn't just a friendly discussion. Perhaps the child has sassed the teacher. Perhaps the parents are themselves deficient in education. Certainly the teacher is in a position to give the child high or low grades. Edward T. Ladd, professor of education at Emory University, stated in discussing pills in classrooms: "When teachers make suggestions, many parents take them as recommendations, while recommendations, in turn, are often taken as witting or unwitting threats."

Another related question is how much are educators who encourage the use of pills motivated by a genuine desire to help the pupil? And how much are they motivated by a desire to create more orderly classrooms? Motives of course are hard to untangle.

Bosco cites a 1974 report from the *American School Board Journal* stating that pediatricians feel they are under pressure from educators and teachers. The education people are concerned about children who "act up in school or are difficult to manage." And he adds that he has picked up the same vibrations.

"In my conversations with physicians," he says, "the charge of inappropriate teacher demands that the child be placed on stimulants is not unusual." Many family doctors or pediatricians tend to accept the teacher's judgment and after a cursory examination prescribe stimulant drugs on a trial basis. Ninety-six percent of the teachers in the Robin-Bosco study were confident they could identify hyperkinetics in their classrooms.

Bosco and Robin advised me that actually teachers were largely uninformed about Ritalin's specific properties. They added that university instruction and textbook presentations about hyperkinesis and its treatment are thin, haphazard, or absent.

Perhaps for good reason. The medical experts themselves are pretty hazy about how to diagnose a *medical* condition of "hyperkinetic syndrome" or "minimal brain dysfunction" or "hyperactivity" beyond alleged symptoms like restlessness and short attention span. No two children seem to have the same set of symptoms.[26]

The *Journal of Pediatrics* reported a study in which three skilled diagnosticians studied a group of children referred to them for treatment as hyperactive.[27] The diagnosticians concluded that nearly 60 percent of the referred students might not be hyperactive. And the diagnosticians disagreed among themselves in nearly a third of the cases.

Medical Opinion carried an opinion by Gerald Solomons, director of the Child Development Center at the University of Iowa: "There is no accepted definition of minimal brain dysfunction. Where you have a diagnosis as nebulous as this is, you run a risk of labeling every child who is not conforming or keeping order in class."[28] He cited a study in which second-grade teachers, on a checklist, rated approximately one half of their little boys as "restless," "disruptive," "have short attention spans," "easily distracted." All are alleged symptoms of some brain disorder! Although Solomons believes that there is a condition called minimal brain dysfunction, his center prescribes drugs for only one in ten of the hundreds of children referred to the clinic each year.

It is significant that about six times as many little boys as little girls get classified as having a hyperkinetic syndrome. In general, from about age one on, normal boys are more active and unruly than girls. Elementary schools are dominated by female teachers. Many of them place a very high value on students who are neat, quiet, and eager to recite. At that age most of the volunteer reciters are girls.

Sydney Walker III, director of the Southern California Neuropsychiatric Institute, contends that hyperactivity is not a disease but a collection of symptoms. He adds that exhaustive diagnostic tests are required before the hyperactive label can be used on children accurately in any medical sense.[29]

Scientists disagree on whether something like "minimal brain dysfunction" can be spotted — and related to classroom behavior — by brainwave analysis or studies of brain chemistry. The brain investigator Paul Wender thinks he can demonstrate an organic problem on only one out of ten boys to whom Ritalin or amphetamines are being given. One organic ailment producing hyperactive symptoms is not in the brain at all, but is caused by pinworm.

Others blame hyperactivity on social conditions. An inner-city youngster who comes to school without breakfast may be restless from hunger. Disorganized parents can cause hyperactivity in their children. Barren, overcrowded classrooms, where students are allowed less movement and freedom of speech than prison inmates, can contribute to fidgetiness. The psychologist Stephan Chorover suggests that hyperactivity "is really more a teaching disability than a learning disability." In some areas of high air pollution, lead poisoning can cause symptoms of chronic restlessness. Abnormal sugar levels in the blood — both high and low — are also suspect by some investigators.

Perhaps the most startling possible cause of what is called hyperkinetic behavior is artificial food colors and flavors. Ben F. Feingold, an allergist, studied hyperkinetic children at the Kaiser-Permanente Medical Center in central California. He began wondering if these children might already have been drugged. His search narrowed to the thousands of artificial colors and flavors now added to food products, including soft drinks. Since these chemicals are eaten as food rather than prescribed as drugs, they have never had to meet the strict testing standards of drugs.

When Feingold began putting children labeled hyperkinetic on a diet free of these artificial additives, the troublesome symptoms of many of them disappeared within a week![30] He hypothesizes that some youngsters are more affected by these additives than others. The National Institute of Education found considerable merit in Feingold's contention about food additives.

A number of reports in educational and scientific journals have alluded to the improved classroom performance of children on stimulant drugs. Now, even this is challenged. At the 1974 annual meeting of the American Medical Association, an Ohio State University professor questioned the conventional wisdom. Herbert E. Rie, professor of pediatrics and psychology, had watched eighty children for three years. They had been diagnosed as hyperactive and were on a regimen of stimulant drugs.

He reported that while the drugs seemed to calm behavior and to increase the attention span they did not improve school performance as measured on the Iowa Test of Basic skills. He commented: "The kids look like they are doing better — they're out of people's hair — but they are not performing a bit better."[31]

This might be due to the fact that kids on stimulants often seem to be flat, humorless, almost emotionless. "We believe," he said, that children have to be "excited and involved to learn." He speculated that the apparent dulling effect produced by the drugs balanced out the improved ability to concentrate.

A 1975 report from psychiatrists at the University of Auckland, New Zealand, found that in only three out of twelve tests did stimulant drugs do better than placebos in enhancing the mental functioning of hyperactive children. When the dosages were strong, mental functioning was worse than with placebos.

It might also be argued that the use of stimulant drugs defeats a major goal of good education: to help children become more adept at self-control. If drugs are controlling their dispositions for them they may grow into adolescence deficient in this crucial human skill. A child psychiatrist at the University of Iowa cited an example of the complaints he hears.

The father of an adolescent who had been fed stimulants for several years said in effect: "Help me, I don't know what to do with him. He is taller than I am and has the self-discipline of a six-year-old."

Steady, long-term use of stimulant drugs appears to have a physical side effect relating to weight. A study by the Baltimore County Department of Health found that children on stimulant drugs gained only 60 percent of their expected weight gain each month.[32]

Still, at the 1976 Master Series Lectures of the American Psychological Association a professor from the University of Chicago School of Medicine gave a generally pro-drug talk. The speaker was Daniel X. Freedman. His topic was "Pharmacology and Behavior Control." He argued that drugs which decrease hyperactivity are beneficial to hyperactive children. Untreated hyperactive children, he suggested, may develop character disorders or become alcoholics as adults. And he contended that side effects did not seem to be as drastic as some had claimed.

I have seen no evidence that amphetamines have any addictive effect on the children receiving them. However, hundreds of thousands of children may well be getting the impression that the way to solve the problems of life is to take pills.

Several states have passed laws setting forth what teachers can and cannot do about administering drugs during school hours. All states should have such laws. Some school systems have drawn up firm policy statements on the ethical behavior of teachers that place limits on their participation in treating seemingly hyperactive children. All schools where stimulant drugs are in use during school hours should have such printed policies. In addition, Bosco and Robin feel that there is an urgent need for schools involved in such programs to keep better records. They are embarked on a large-scale study into the way the teacher-parent-doctor relationship seems to work in considering whether to medicate, how treatment is monitored — if at all — and how it is terminated.

Undoubtedly the stimulant drugs are only the first that will find their way into common use or will be promoted by educators. If you doubt this, consider James Basco's forecast, made to that annual meeting of the National School Boards Association:

"In my crystal ball, I see that the teacher of the 21st century will not be trained in schools of education as we know them today but rather will be trained in a school which is an amalgam of contemporary schools of education, medicine and pharmacy. A considerable portion of the teacher's training will be devoted to understanding physiology and psychopharmacology, which will equip the teacher to administer drugs which affect learning and learning-related behaviors. Drugs will be used for a variety of reasons, such as for pupils who are having difficulty with perceptual discriminations, for pupils with attention malfunctions, or for some who have difficulty remembering mathematics facts."

Will parents be consulted or asked to approve? They certainly should be. (For additional reasons why we may see more use of drugs on school-children, see the next chapter.)

Bosco's opinion is not a far-out one. The *Journal* of the National Education Association carried an article speculating that in the future there would be jobs on school staffs for pharmacists and biochemical therapists.

It may well develop that neuropsychiatrists will soon be able to relate beyond doubt a number of *demonstrable* organic conditions to specific behavior of children in school. For such children the carefully supervised administration of some drug may well deserve recommendation to the parent. But until then teachers and doctors should be careful about tossing around such stigmatizing phrases as "minimal brain damage," "hyper-kinetic syndrome," or "minimal brain dysfunction."

For most of the highly charged little boys in the nation's classrooms methods others than notorious drugs should be tried in order to improve their classroom performance and deportment. Parents should be given information about the possible role of artificial food additives in producing hyperactivity in their child. And they should be given a leaflet of information on how to reduce the additives in the child's diet. Let the soda-pop and food processors wail.

To continue with approaches other than drugs:

- Loosen up the classroom environment. For example, try open or semi-open classrooms, which give the active child more chance to walk and talk.
- Give the child warm personal attention and counsel instead of crossing him off as a disrupter.
- Provide soundproof Relaxation Rooms, where highly active children can go for ten minutes before the beginning of the morning and afternoon sessions. Encourage them to whoop or pound off their excess energies. They might also go into several minutes of deep breathing. And the session might be ended with a few minutes of eyes-closed meditation.
- Offer them a free glass of milk in the morning. It may well be appreciated.

We have noted in this chapter a variety of scientific techniques that are being developed in a variety of situations to make people less unruly. They include cutting into the brain, a number of drugs, and operant conditioning, both harsh and benign. All have in common a manipulative approach to people's behavior. The professed goal in each instance is to help the individual. But I think we have seen convincing evidence that easier manageability is a primary consideration in the back of the minds of most of the advocates, whether they are wardens, nurses, attendants, or teachers. It is time to make searching studies of the motives of all people and institutions fascinated with the new techniques that have as a major effect a heightening of tractability.

Until such studies have been made we should bear in mind the caution of the neuropsychologist Elliot S. Valenstein: "We would be in serious trouble if a number of influential people became convinced that violence is mainly a product of a diseased brain rather than a diseased society."

Building Brighter— 7
or Duller—
People

It is my judgment that within five to ten years there will be available a regimen combining psychological and chemical measures that will significantly increase the intelligence of man. This troubles me, and should trouble you.
— David Krech, a leading authority on learning

Krech made that statement in 1968. His forecast is coming true. Some scientists are predicting that IQ's can be raised on the average by twenty points. And Krech still is troubled.

Modifying the intelligence of individual humans is now clearly possible. This can prove to be a momentous turn in human events. Modifying the capacity for specific skills also seems possible.

The ironic aspect of this impending capability is that brain scientists still have only the dimmest notions about how the brain produces intelligent actions. A student of the brain, Shinshu Nakajima, put it this way: "Intelligence was once a miracle, then it was a mystery. When scientists believed to have grasped it, it became a mirage."

Some suggest that Man will never know in any detail how and where his brain stores away particular bits of information — and uses that information many years later to solve a problem.

A man-made computer can take information that Man programs into it on standardized forms. Later, when it is activated by Man it can produce a printout that repeats or uses the information.

But Man's own brain does not acquire information in any such straight-forward, systematic way. It comes by way of smells, sight, sound, taste, and touch. Each person's life involves tens of thousands of unique experiences. And each person is subjected to millions of sensory impressions that aren't programmable on tape — a breeze blowing or a rooster crowing before dawn.

I can demonstrate to readers the incredible wonder of their own minds just by mentioning one word. Get ready. If you wish, get a pad and pencil to make a printout. The word is "Watergate." Out should tumble hundreds of facts, on reflection, about a cast of characters and a host of issues. Your brain has registered somewhere the gurgling sound of Senator Sam Ervin's voice. It has stored the image of Rosemary Wood, President Nixon's secretary, straining to demonstrate how she answered a telephone while transcribing a tape recorder. Your mind has preserved the novel use of words by the White House group in such comments as "let's stonewall it" or "that particular time frame." You picked these impressions up over a period of two years. At the start the word Watergate meant nothing to you. Certainly your brain has no single compartment that eventually became labeled "Watergate." It doesn't even have an ongoing compartment labeled "National Affairs, subsection Shenanigans."

For the most part, things we learn through our eyes and things we learn through our ears probably get stored in different areas. But that doesn't explain much. A particular memory apparently may be stored in a variety of areas. And it seems to move around. It may be in one part of the brain one day but doesn't seem to be there a few days later. If the person has a brain abnormality the memory seems more likely to stay in one area. The brain explorer Wilder Penfield of the Montreal Neurological Institute was electrically probing the brain of a woman suffering from epilepsy. He found that whenever he stimulated one particular area the woman heard a certain tune.[1]

We start with the ancestral knowledge built into our genes at birth. All that comes after is learned knowledge and acquired intelligence. Developing these involves five distinct processes. (Scientists sometimes use the words "learning" and "memory" interchangeably. The two actually are different though interdependent.) Here are the five processes:

1. The acquisition phase. This is pure initial learning. Sensory experiences are converted into neural impulses.

2. The temporary, unstable coding of what has just been acquired and considered to be worth storing. The names of many people we meet at parties escape within ten seconds. Or less.

3. The permanent coding and filing away of what was learned. They may occur as separate processes in subsequent hours.

4. The systematic recall (memory), days or years later, of what was learned.

5. The evaluation of many recalled facts, often seemingly unrelated, in

order to come up with a decision. This is called thinking, the most baffling of all to comprehend as a physiological process.

How all these five things happen is still just guessed at. Scientists call a stored bit of knowledge an engram, but they aren't sure what an engram is. At first they thought it was a new pathway or "trace" in the brain's nerve network. Now there is frequent mention of it as a coded chemical.

They have learned enough about brain function, though, to venture guesses about the kinds of substances that are deeply implicated in acquiring intelligence and knowledge. Their focus is on transmitters of nerve impulses, especially the transmitter acetylcholine. Its efficiency appears to affect stage number one of the learning process. An ideal amount of acetylcholine seems to promote swift learning. At least as important is ribonucleic acid (RNA), which is now believed to be crucial in "fixing" what is learned into permanent memory.

RNA molecules are critical components of all living matter. For example, they synthesize a great variety of proteins inside cells. Coded protein molecules may well be the building blocks of memory.

The first evidence that RNA was deeply involved in memory came in the 1950's from a Swedish neurobiologist at the University of Göteborg, Holger Hydén. He had performed the remarkable feat of isolating and analyzing single living brain cells. The brains belonged to rats. He put rats through a rigorous mental regime before they could obtain food. For example, he forced them to learn to walk a tight wire. He noticed that the brain cells of the rats that had been forced to learn in order to get food showed an abnormally high RNA content.[2]

BRAIN POWER THROUGH EXERCISE

A few years earlier a group at the University of California, Berkeley, had discovered that mind stretching seemed to build better minds. And this group eventually discovered a great deal more about enhancing brain function.

The central figure was David Krech, a witty scientist who, until his recent death, cruised the hills of Berkeley in a Citroën. Krech first aroused my interest after World War II, when it was disclosed that he had helped run a supersecret training and screening program for would-be United States spies. Some were European refugees. All were stripped of identity. They were subjected to fiendish psychological and physical stresses. And at the end they were required to develop cover stories that would withstand Krech's ingenious, slashing interrogation.

After government service, Krech went to Berkeley. Although he was at that time a social psychologist, he found himself again pondering the mystery of the brain. Early in his career he had apprenticed himself to the noted neurophysiologist Karl Lashley. Krech was well aware of the conventional wisdom that heredity shaped the brain. The brain was supposed to be fixed in size and performance, and that was that. It seemed logical to Krech, however, that there ought to be biochemical differences between the brains of bright people and those of dumb people. Change the chemistry of the brain and maybe you change its brightness.

He didn't know enough about chemistry to test such a presumptive notion but he was persuasive in winning support to put together a team of experts. He recruited, among others, Edward Bennett, a biochemist; Mark Rosensweig, a biological psychologist; and later, Marian Diamond, a neuro-anatomist. What would be the best way to test his concept? All agreed they should focus on acetylcholine because of its crucial role in transmitting information-bearing impulses. They guessed that a busy lively brain would have to be unusually well endowed with it.

But an obstacle loomed. You can't measure acetylcholine in an autopsy. It breaks down swiftly after use or death. The group finally decided that acetylcholine might be measured indirectly. The enzyme that breaks down acetylcholine persists, and can be weighed. If the amount of acetylcholine varies, the enzyme should vary proportionately. So they went after that enzyme in the brains of decapitated rats that were known to be very bright or very dumb. There was the evidence! The brains of the bright rats did have more of the enzyme than the brains of the dumb rats.

Then the Krech group, being unconventional, proposed that they turn the proposition around. If a brain with superior chemistry is better at learning, then would a lot of exposure to learning produce superior chemistry in the brain?

So they started sending a group of young rats to school in Berkeley. Each trial course lasted ninety days. A control group of other young rats continued to live normal lives. And a third group of young rats was put into solitary confinement in a dark, noiseless area.

The "enriched environment" of the young rats that went to school was a delight. They had big cages with lots of companions. Playthings were changed every day. There were spinning wheels, which young rats enjoy mastering. There were trapezes, slides — you name it. But it wasn't all play. Day after day there were tests to take. The rats confronted maze problems with rewards of sugar waiting at the end of the maze. They faced real brain-twisting discrimination problems. They had to decide which of two alleys — one lighted, one darkened — to go up in order to reach food. Once they learned the solution, Krech's group would make a systematic switch and watch how long it took each rat to unlearn the old habit and learn the new. Then, again and again, the correct alley would be switched, but each time in the same pattern of change. Krech explained to me: "A real bright rat will catch the patterns of change and stop bumping his nose, with a minimum of errors. A dull rat will find each change a brand-new problem."

After ninety days the rats of all three groups were decapitated and their brains immediately dissected. Meanwhile, new groups of young rats were put into the three environments.

The evidence mounted that the young rats who were sent to school in the enriched environment had more of the enzyme associated with acetylcholine than rats in the normal environment. And those in solitary confinement had less. The big surprise, however, was the changes that Marian Diamond, the neuroanatomist, began noticing. Rats from the enriched environments had developed greater thickness in the cortex layer, the main thinking part of the brain. The cortex covers the cerebrum. Even the brain cells of the rats who had gone to school had become larger. Learning tests made before they were all decapitated showed that the schooled rats became brighter than those left in ordinary or deprived environments.

This was the first time anyone had proved that *lasting changes* occur in the brain as a result of experience.

Krech sighed: "Thousands of rats gave their all for us to prove this point."

Since then many tests have confirmed the above findings, not only with rats but with kittens and dogs. Krech added that while an enriched environment works best with young animals, some changes in intelligence can be made in adults by a dramatic increase of enrichment in the environment.

No one has yet put a substantial group of humans through a rugged learning regimen and then decapitated them to check their enzymes or cortex thickness. It is probably not necessary. There is now mounting evidence that an enriched environment does accelerate human learning ability.

For example, there is the evidence of the worldwide Montessori movement. Maria Montessori took poor children from a big public housing development in Rome, and immersed them in a school setting rich in colorful, sensory stimulation. There was a great variety of interesting puzzles and objects to work with and to manipulate. The youngsters were reading eagerly by three or four. And by five or six they were deep in geometry.[3]

A study of a community of children in Jamaica by a group headed by Stephen Richardson also supports the critical role of a stimulating environment.[4] One finding, while not relevant to our immediate interest, certainly should be noted. The children who had suffered from malnutrition during the first two years of their lives were twice as likely in grammar school to be severely retarded mentally. However, some of the children studied had

had adequate nutrition but had come from homes impoverished as far as stimulation was concerned. (There are now inventories for measuring "home stimulation.") Healthy children who had come from stimulating home environments averaged 71.4 on an intelligence test, whereas healthy children from deprived environments averaged 60.5. Those from deprived environments who also suffered from malnutrition averaged 52.9. (Possibly heredity contributed to the differing home environments.)

Krech believes that the most important element for creating an enriched environment for humans is not mazes or colorful stimulation but language. Man is a language animal. Early immersion in talking, reading, and writing, he deduces, should improve brain power. The neurophysiologist Robert S. Morison has stated in *Science* that young children "who have books at home learn to read earlier than those who don't have these amenities."

Some investigators thought that if an enriched environment really produced more efficient brains, then the evidence should show up in graduates of the Headstart program. This program of the federal government has involved about 350,000 preschool children from the ghetto. It was supposed to broaden their horizons, improve their nutrition, and generally get them off to a better start in life. Early evaluations showed that as measured by school test scores, Headstart children did get off to a better start in school. But that advantage often persisted for only a year or so.

The brains of many Headstart children, it should be pointed out, had probably been stunted by malnutrition before or after birth. Also, the program did not put much emphasis on brain-stretching exercises or language or enriching the home environment. The focus was on increasing social competence. And in the beginning Headstart was largely a part-time summer program. Krech says: "Whereas our rats lived in an enriched environment twenty-four hours a day, the Headstart youngsters had such an environment only a few hours on weekdays, for part of the year."

Tests in a year or so may show more change in the Headstart children's brain efficiency. The program has been reorganized to last longer for each child. It now begins in infancy. Parents are more involved in teaching the children and reading to them, and are encouraged to create colorful, stimulating rooms for them at home. There is greater emphasis on language development.

A better test of the enrichment concept may come out of the Early Education Project in Brookline, Massachusetts. Within two years there will be at least one hundred youngsters ready for school who have been in that program. Since infancy they have been provided with an enriched environment, largely as a result of counseling with parents. The program sponsors believe that the first three years of life are the most important years in which to make an impact on later intelligence.

Perhaps the best precise evidence now available of the effects of enrichment on small American children has been developed in the slums of Milwaukee. In late 1976 the *Monitor* of the American Psychological Association cited reports that a decade-long study there was coming up with results that were "unique, exciting and spectacular." The sample is relatively small. The program, headed by Richard Heber of the University of Wisconsin, began with forty infants. They were born to retarded mothers who had IQ's of less than seventy-five.

Half the infants were put in a control group, half in the enrichment group. For the enriched group teachers first came to the home to work with the toddlers and to counsel the mothers. Later, the experimental children went to a special center five days a week, seven hours a day. The emphasis in enrichment was primarily on language and thinking problems.

By the time they enrolled in school at the age of six the children from the experimental group had a thirty-point edge on standard IQ tests over the control group. Now, after three years in school, the "enriched" youngsters still maintain an IQ lead of more than twenty points and are within the normal school-child range.

IQ PILLS?

When Krech and his colleagues found that they could produce both brain changes and learning ability by enriching the environment they asked the next logical question. As Krech put it, "We said, well now, let's see if we can find a drug that will do the same thing to the brain that the enriched environment will do."

One of Krech's student researchers, James McGaugh, now a psychologist and a university administrator, was the first to come up with such a drug. It was, of all things, strychnine, a deadly poison. One clue that led McGaugh and his colleague Lewis Petrinovich to try it was the fact that decades ago a very mild form of strychnine had been an ingredient in a spring tonic used to pep up youngsters after the winter. Also, a search of the literature showed that early in the twentieth century Krech's mentor, Karl Lashley, had listed strychnine as somehow making rats more eager learners.

In mild form it apparently stimulates the nervous system to greater activity. At any rate it made star pupils of McGaugh's and Petrinovich's rats. It could be administered before tests for brightness, or up to an hour after the tests. But the sooner it was administered after a test the better it worked. That suggests that a part of its stimulating effect is to improve memory fixing.

Later, after he became a full-fledged scientist, McGaugh and his colleagues discovered still other brain stimulants. All would be highly dangerous in large doses. There were nicotine, caffeine, amphetamine, and picrotoxin, a poison resembling strychnine. But the best seemed to be a convulsant called Metrazol, sometimes used by doctors in shock therapy.

McGaugh and Petrinovich tested Metrazol on genetically identical rats. Some got it. Some did not. The ones that got it could learn a maze in five attempts on the average. Those that did not get it needed twenty attempts![5]

More recently, getting closer to man, Metrazol has been tried on rhesus monkeys at the Wisconsin Regional Primate Research Center. Results were remarkably similar. Monkeys solved a problem in one fourth the time it took monkeys without Metrazol.[6]

A totally different chemical approach to enhancing memory has been taken by the Dutch pharmacology researcher David DeWied of the University of Utrecht. He has reported finding very significant improvements in the memories of rats after they have been injected with the synthetic form of a peptide. This peptide (ACTH 4-10) is found in the tiny pituitary gland — sometimes called the master gland — at the base of the brain.[7] Since it is a product found naturally in the human body it should not offer the toxic hazard that the longtime use of stimulants like strychnine might entail. It is being tested on humans. In 1975 DeWied announced finding that still another substance stored in the pituitary gland, the hormone vasopressin, had an even more dramatic effect in enhancing the memory of animals.

Several pharmaceutical companies are currently in a race to develop a safe memory-enhancing drug. Drugs that show promise of helping aged people retain their mental faculties involving memory are moving into the experimental stage.

A still-controversial nonchemical approach to giving the baby's brain a head start is the decompression chamber. Many decompression clinics have sprung up in England, Canada, and South Africa. There are also some in the United States. Thousands of pregnant mothers have been treated.

Under this procedure, an expectant mother in the last phases of pregnancy lies on a bed for a half hour or so a day with a plastic bubble over her abdomen. A partial vacuum is created to relieve the barometric pressure on her abdomen. Some women have climbed into decompression outfits resembling space suits. One aim is to promote easier childbirth. But many proponents claim it produces brighter children.

How? Why? In late pregnancy the mother's placenta virtually stops growing while the fetus continues to grow. More pumping is required by the heart to get oxygen-laden blood to the child-to-be's brain. A brain growing on a less-than-ample supply of oxygen presumably has more chance of becoming a less than marvelous brain. The bubble apparently permits more blood to flow. A pioneer in the technique, a South African obstetrician named Ockert S. Heyns, claimed that children who had undergone a program of prebirth decompression scored 18 percent higher than average on the Arnold Gesell test for infant development. Others have reported less impressive results.[8]

GROWING SUPER BRAINS

Joshua Lederberg, the geneticist, has pointed out that the shape of the human female pelvis has through the ages put a limit on the size of the infant skull. This in turn has put a limit on the size of the human brain cavity. If Caesarean delivery became routine the brain might in time grow larger. And perhaps the process could be speeded up by growth hormones.

Another Nobelist, the French biologist Jean Rostand, has pointed out a way that the human brain might be doubled in size fairly rapidly. It is based on evidence that prenatal growth in the human brain occurs as a result of a fixed number of cell divisions (or doublings). There are ways to speed up or slow down cell division. Rostand suggested that scientists might seek to develop a drug that would stimulate only the brain. The brain cells might then undergo an extra division — or doubling — before birth.

But is bigness in brains necessarily betterness? Up to a certain point in human evolution brain size was critical. However, many modern geniuses, among them Anatole France, have had smaller cranial capacity than the average Neanderthal man. Apparently, when it comes to brilliance, the quality of the brain counts most: lots of forebrain . . . a thick, highly wrinkled cortex (covering) on the cerebrum . . . an above-average ratio of working nerve cells to the jellylike glial cells surrounding the nerve cells . . . lots of impulse transmitter material . . . and perhaps most important, above-average intricacy of the network of interconnections between nerve cells.

Possibly by experimentation Man could achieve selective growth within the brain. Then you might have something really startling to reckon with. It has been achieved in rats. A group headed by the biochemist Stephen Zamenhoff at the University of California, Los Angeles, injected pregnant rats with growth hormones. The brains of the offspring weigh a third more than normal. More to the point, most of the growth was in the thinking part of the brain. There were more nerve cells in the cortex to every hundred glial cells. ("Glial" derives from the Greek word for glue, but recent evidence suggests that glial cells are not just packing material but may be involved in intellectual work.) As the rats grew, there was an unusually intricate proliferation of interconnections among cells. And most noteworthy, the big-brained experimental rats actually were brighter on standard maze tests.

BRIGHTNESS BY ELECTRICITY?

A group that included McGaugh at the University of California, Irvine, found in two tests that mild electric stimulation of an area of the hippocampus of rats facilitated learning. The hippocampus is suspected of being somehow involved with memory processing. It was surmised that the electric stimulation accelerated the process of memory storage.[9]

Karl Pribram, brain physiologist at Stanford, has stated: "I certainly *could* educate a child by putting an electrode in the lateral hypothalamus and then selecting the situations at which I stimulate it. In this way I can grossly change his behavior." He was just making a point and insisted that he would not do such a thing as a matter of ethics.

A number of researchers have produced what they thought were improvements in the speed of learning simply by passing a low-frequency current between plates placed on each side of the head. This has been tried in both monkeys and Man.

Meanwhile, a search is on to find some simple electrical way to measure a youngster's IQ. Consider how much time and bother could be saved at schools and colleges. A college applicant could be put on the doubtful list just by a tracing made by a brain-wave machine. A Czech psychologist has been testing the possibility that intelligence can be measured by the ratio of theta to alpha waves. And a Canadian psychologist is trying to prove that his electronic machine measures the speed of mental reactions (crucial to intelligence). The test involves tracking the speed with which the brain responds to flashes of light.

Such taps on the brain might eventually tell *something,* perhaps offer a crude culture-free measure of the potential for fast mental activity, and hence become popular with schools. But they probably would have little to say about the person's persistence, imagination, integrity, creativity, sanity — or potential for accurate thinking.

TRANSPLANTING OR MANUFACTURING MEMORIES

James V. McConnell, a colorful psychologist from the University of Michigan, came up with some discoveries about memory that seemed simply unbelievable. Some of his fellow scientists were at first aghast. Memories, he reported, were not necessarily personal. They could be transferred to other individuals, at least in the animal world. Some thought that such preposterous talk would give Science a bad name.

If memory can indeed be transferred, the finding will stand as one of the most astonishing discoveries of modern neurobiology. It will certainly support the concept that memory has a molecular basis.

McConnell started with only a skimpy knowledge of biology. But he had an enormous curiosity about behavior. And he got expert instruction on physiology while still a graduate student at the University of Texas. This came from his many sources, including his friend Robert Thompson, a physiological psychologist.

Memory molecules, McConnell told me, "float around anywhere and can theoretically wind up in a big toe. But you can't get any response from such a molecule until it gets back to the brain, in humans. Its chemicals will mean something only to brain cells." By floating around anywhere, he meant within the body's nervous system. He now believes that the central nervous system has a few sites where learning is converted into memory molecules. Quite probably these molecules get pumped out into appropriate areas of the brain during dream-sleep.

The professor in charge of the animal laboratory at the University of Texas was scornful of McConnell's proposals for studying memory. He was so scornful, in fact, that he refused McConnell and Thompson lab space. They therefore began their now historic experiments in McConnell's kitchen.

They decided to start with the simplest creature that could be considered to have a brain, the inch-long flatworm. Its "brain" is a network of about four hundred cells. They set up a water tank with electrodes at each end. As expected, they found that when they sent a very mild shock through the tank the worms would scrunch up. So then they began to turn on a light two seconds before the shock. At first the worms ignored the light. In fact it tended to relax them. But soon they began scrunching up the instant the light went on, just as Pavlov's dogs had salivated when the bell rang. The worms were capable of learning and remembering!

Later, when he was a psychologist at the University of Michigan, McConnell pushed on to bolder worm experiments. After educating worms to scrunch on the light cue he cut them in half. It seems that flatworms can be cut in half and in a few weeks each half will grow a new head or tail. The tails that grew back into worms had brains, and the brains had the old memories. This astonished even McConnell. It turned out that the educated tail-based worms remembered the light cue even better than educated worms that had not been cut at all! In worms, at least, it seems that memories are stored all over the body, not just in a "brain." He cut worms up into as many as five pieces, and all the pieces regenerated brains that "remembered."

McConnell had also learned that his worms were cannibalistic, and that gave him another bold idea. After conditioning worms to scrunch when a light went on, he chopped them up and fed them to hungry, untrained cannibal worms. He says that after letting the cannibals have a day or so to digest their meals, "we then gave them their first exposure to light-shock training. To our delight, the cannibals that ate 'educated victims' showed much faster learning [to scrunch at the sight of light] than did cannibals that ate 'uneducated' or 'untrained victims.' "

Well! All this was pretty unsettling to a skeptical scientific community. Psychologists began jokingly to envision the day when students would learn their lessons by eating their professors. A large number of laboratories tried testing the transferability of memory, not only on worms but on chicks, goldfish, mice, rats, hamsters. By 1973 at least twenty-two different laboratories had published positive findings of memory transfer with vertebrates alone. A substantial number of other investigators reported that they had been unsuccessful. Unfortunately for skeptics, the possibility of memory transfer seemed so highly unlikely that positive reports seemed much more impressive than negative reports. One group of researchers, who were perhaps trying to blow McConnell out of the water for good, wrote a letter to *Science* saying that each of them in separate studies had found no support for memory transfer. While the letter was going to press one investigator who had signed the letter came up with positive results. He recanted and became a believer.

Meanwhile, several of the positive-finders came up with something else that outraged those with conventional wisdom! They could transfer memory across species barriers. The brains of rats or mice that had been educated to shun darkness (which they normally prefer) were made into a soup and injected into hamsters. The injected hamsters soon began shunning darkness. It also worked vice versa. One of the cross-species pioneers, the neurochemist Georges Ungar, quite recently reported that with coded molecules he had isolated the specific substance in the brain of a rat that caused a conditioned rat to avoid darkness. He called it scotophobin, Greek for darkness fear. The brains of four thousand conditioned rats were pooled to obtain a sample large enough to permit the isolation of this substance. The material turned out to be a peptide with a complex amino-acid sequence. (Peptides are molecules formed by two or more amino acids.) It has since been synthesized.[10] Efforts to replicate the synthesis have proved to be a challenge.

Ungar's group has also reported discovering several other brain peptides that can evidently transfer learning from one animal to another.[11] If these peptides, too, can be synthesized, it is conceivable that a large number of "memories" can be manufactured.

If it develops that memories can indeed be transferred or manufactured for humans, what would be the implications? Some science writers have envisioned injections replacing grammar school textbooks. That seems unlikely, based on present knowledge.

The implanting of memories in humans, if it comes, will more likely involve situations than facts. It will more likely be perfected for purposes of indoctrination rather than education. McConnell believes that synthetic molecules could be manufactured that would make people want to help each other.

On the other hand, dictators might be more attracted to the possibility of supplying subjects with synthetic molecules that would make them fearful at the sound of a certain voice or when performing a disapproved act. Dictators might very well wish to invest heavily in the technique. Also, they might in their indoctrination wish to obliterate in the populace memories of popular democratic leaders, or what it had been like to enjoy liberty.

ERASING MEMORIES AND IMPEDING LEARNING

A number of drugs, including the antibiotic puromycin, have in animal tests prevented memory storage of very recently learned experiences. Why this happens is still in dispute, but it happens. Electrical stimulation of the amygdala also disrupts the memory-storage process, as does electroconvulsive shock.

If it develops that memory of events covering recent days can surely be wiped out in humans, a lot of people would be interested. Gangsters would no longer need to bump off witnesses to their crimes. Just give them a memory-blotting treatment. It would be useless to torture couriers for national intelligence agencies who were known to have delivered secret memorized information. Memory of the messages would be wiped out immediately upon delivery. A great boon for the mentally ill would come if medical research delivered chemical ways of wiping out memories of terrible events that still haunt them. But this would be more difficult since such memories are long-fixed.

As for impeding learning from occurring in the first place, we are discovering that this is all too easy. One procedure is to continue the present global nutrition pattern whereby millions of infants are severely undernourished before or after birth. They are destined to spend their lives with stunted brains. Shortage of protein in nutrients can be particularly damaging. A University of California team has estimated that the actual brain weight of a severely undernourished four-year-old is 10 percent below normal. An adequate network of fibers interconnecting the brain cells apparently fails to develop. Crash programs of feeding such children can overcome some of the brain stunting. But ability to remember what has just been learned remains impaired.[12]

In Huxley's *Brave New World* menial laborers for the future were created before birth at the state people-hatcheries. As noted, these laborers-to-be had their ration of oxygen cut while they were still fetuses. Semi-morons were guaranteed. Cutting back on protein would probably be just as effective.

For young and old alike, ability to learn could probably be impeded, at least for a while, by adding a bit of calcium to the brain fluid. It could also be inhibited by drugs that slow down the synthesis of RNA and proteins in the brain.

In general our society applauds brainpower. The more the better. Many educators, for example, are eager to get going on the use of drugs to enhance learning. Educational journals, as indicated, have discussed the possibilities for the use of these drugs in schools.

Since public schools are mandatory governmental institutions the public can reasonably expect that any drugs "introduced experimentally" will be introduced with parental consent. Psychoneurobiochemists take heed!

Techniques for improving memory may turn out to be a boon to long-living people who become senile. They may remember again how to play bridge and dial the telephone.

Efforts to improve intelligence at any early age by enriching the environment certainly should be encouraged, worldwide. Children raised in an environment with lots of talk with adults, and lots of chances to explore, will probably be brighter. Certainly they will be happier and more emotionally secure. And the effects will remind adults that being a good parent requires a lot more thought and patience than most of us realize. This may help reverse the trend of the early 1970's to downgrade the family, to assume that just anybody can be hired to take care of a small child.

Get-smart drugs, building bigger brains, manufacturing memories, and in general, crash programs to increase national IQ levels are another matter. Readers will recall David Krech's apprehension about any mass program to increase intelligence levels by, say, twenty points.

Krech assumes there will be a great rush to obtain get-smart drugs. They could become the basis of a multibillion-dollar industry, and the drug companies know it.

But who will decide who gets what? Will their distribution be left, as that of most drugs is, to private enterprise? Will they be pushed to the public through TV advertisements? And will they be promoted to doctors through the thousands of "detail" salesmen of the drug companies? Will the supply-and-demand mechanism set the price? If so, we might end up with two-dollar pills. Parents will presumably be willing to pay an enormous premium to obtain these wonder drugs for their children. Will those who can't afford the drugs suffer social damage as they see their children outdistanced by classmates whose parents are more affluent?

Krech suggests another possible scenario. Evidence indicates that drugs would help significantly only with duller pupils. In that case would we end up with a society where there is little spread in educability? If so, he asks, who would become the hewers of wood and drawers of water?

Still a third scenario, he says, might develop from the fact that there probably will be drugs that will selectively strengthen clusters of special abilities. Examples might be mathematics skills, verbal skills, artistic skills. In that case, who would decide how the talent pills would be allocated? Again, would it be decided by huckstering? Or by the decisions of parents, or pediatricians, or school boards, or national manpower resources boards? Krech thinks that a federal commission would be required to decide whether such drugs should be utilized at all, and if so, how.

Another thought. A younger generation made learned by chemicals might show condescension toward parents and teachers — who were endowed only with minds they built themselves.

McGaugh advises me that he, too, has reservations, specifically about memory enhancers. They could be of great benefit to people with serious memory difficulties. But the abusers, he says, would be "the rest of us — and therein lies the problem. While we would all like to be able to remember better, I'm not sure that it would be good for us if we could. Do we all wish to be like Mr. 'S' in Luria's *The Mind of a Mnemonist*?" (A. R. Luria, a Russian psychologist, studied "S," a stage entertainer who could perform astounding feats of memory. Offstage, "S" seemed quite a dull, disorganized fellow. Since he couldn't forget, he had trouble thinking coherently, apparently because of the clutter of details in his memory.) McGaugh adds:

"My own view is that forgetting is beneficial. It is only because we forget details that we can deal in the abstract." And only by forgetting can we ease the pain of old humiliations, shocks, grudges, defeats.

A person can be high in learning ability and memory and still remain a fool. The two do not add up to either brilliance or wisdom in thinking. Until someone comes along with a pill for wisdom, we might better aspire to become a more humane society rather than a more brainy one.

8

Keeping Track of—
and Controlling—
the Populace

Those who seek and hold power have found that technology can be a valuable ally.
— Stephan L. Chorover,
physiological psychologist

It was the worst quarter century in American history, if we judge the 1950–1975 period on the basis of increases in efforts to control our lives as private citizens and increases in the capacity to achieve such control. Right to privacy was under assault.

Partly it was the billions of dollars spent by United States security and law-enforcement agencies to develop more ingenious techniques for surveillance or espionage. Partly it was the surveillance capacity of microminiature devices coming out of the space-exploration programs. Partly it was the emergence of the Age of the Computer. Partly it was the obsessive concern with identifying and watching radicals. Partly it was the growing anonymity of life and the curiosity of employers, welfare inspectors, insurers, and credit agencies about our private lives. Partly it was the ambitions of politicians and police to develop better controls over millions of unruly or disenchanted citizens.

At any rate it was a period when scientific techniques for controlling the individual proliferated. They proliferated in the name of efficiency, order, security. The techniques continue to proliferate. They add to the pressures to create a more manipulatable society.

Back in 1967 *Daedalus* devoted an issue to forecasts for the year 2000. Its forecaster on problems of privacy, Harry Kalven, Jr., stated: "By 2000, man's technical inventiveness may . . . have turned the whole community into the equivalent of an army barracks."

From today's perspective that sounds fairly conservative. Samuel Dash, who served as chief counsel to the Senate committee investigating Watergate, has said of the late years of the Nixon presidency: "We almost had a police state." President Nixon, in the glow of his triumphant reelection in 1972, told an interviewer, in a fatherly manner, that in some ways the American people were like children.

So how are we treating children? The city of Providence, Rhode Island, in 1975 almost launched a program to fingerprint its school-age children. And the idea of making broad use of the technology of fingerprinting was not reserved for Providence children. The director of the Passport Division of the State Department urged that all Americans be fingerprinted. The prints would be put on a government identity card, which every citizen would be required to carry. This would protect us from "criminal impersonation," detect crime, and guarantee each citizen a "personal identity." A bill was pressed in Congress to require the fingerprinting of everyone. Some of the additional benefits mentioned were that universal fingerprinting might help identify amnesia victims and missing persons.

Here is a look at some of the programs for employing scientific ingenuity to keep track of, or control, people:

TRANSMITTERS ON HUMANS?

What a simple, modern idea! Attach tiny automatic radio transmitters to the bodies of people you want to keep track of. The transmitters would be locked onto the wrist, ankle, or waist, or be implanted in the body. Each would send a unique signal. And it would be a felony to remove them. Wild? A.D. 2500 stuff? Not really.

The technology has been pretty well developed. And many sober-minded people have been discussing it. A criminologist at New York State University spoke of it to an interested congress of the American Correctional Association.[1] A journal of the Institute of Electric and Electronic Engineers devoted twenty-one pages to exploring the possibilities.[2] The *American Psychologist* carried an article pondering its implications.[3] *Issues in Criminology* also had an extensive presentation on the use of electronics to guide the rehabilitation of ex-prisoners and to keep track of them.

Or consider two people closely identified with the idea. Ralph K. Schwitzgebel, for years a Harvard psychologist, calls himself a "social gadgeteer." He spent two years with a team of experts designing possible people-monitoring systems. J. A. Meyer has gone furthest in exploring the possibility of locking transponders on people (in the IEEE journal mentioned above). A transponder is a device that automatically transmits electrical signals when activated by a specific signal from an interrogator. Meyer is a computer specialist and has been working for the ultrasecret

U.S. National Security Agency. It is a reasonable guess that he knows what is possible.

Who would be asked or ordered to wear these tracking devices? Those most frequently nominated would be parolees. Want an early parole? Just clamp on this little gadget so that we can keep track of your movements. Or do you want any parole at all? No tracking device, no parole. The purpose would be benign. Schwitzgebel calls it an "electronic rehabilitation system." Meyer calls it an "externalized conscience — an electronic substitute for the social conditioning, group pressures and inner motivation which most of the society lives with." If a robbery occurred, everyone spotted in the general neighborhood of the robbery by transmitter signal would be picked up for questioning.

Other uses would include tagging people on probation and tagging juvenile delinquents. Schwitzgebel includes as possibilities suicide-prone people and released mental patients.

Meyer has broader ideas. He talks of including "bailees." (People on bail often haven't even been indicted for anything.) He also talks of "arrestees." An arrestee might be anyone hauled in for speeding, or for punching someone in the nose. Or it might include the tens of thousands of young people arrested for conducting unruly demonstrations. Further, Meyer tosses in as possibles certain types of people with criminal records who in the past have shown a tendency to violence. They might be pulled in and have transponders locked on them on a "retroactive" basis.

When Meyer adds up all his nominees he comes up with a figure of twenty-five million people in the United States alone. By a turn of logic that eludes this writer he calls them all potential "subscribers." The taxpayer shouldn't worry. Each subscriber would pay the cost of bearing a transponder on his body — say, $5 a week.

But Mr. Meyer is a gradualist. He would not try to put transponders on twenty-five million people right off the bat. At the outset, "law enforcement agencies" should be allowed to attach or implant only a "few million transponders." That apparently is not facing the challenge of eliminating inappropriate behavior head-on, but it is a start.

There have also been proposals that transponders be attached to the bodies of "aliens" and members of "political subgroups." Subgroups presumably are composed of people who are not in the mainstream of the American Way of Life. If the system gets going and we later come under authoritarian rule, it would be quite logical to expand the system to include known dissenters.

As a taxpayer I am concerned. (Even if dissenters aren't included.) I can see enormously swollen budgets if the police are called upon to monitor the random hourly movements of millions of people.

Meyer, however, is reassuring. The system he envisions would practically eliminate the need for prisons and jails. He does acknowledge that criminals always try to beat the system. But the system he visualizes would be so close to foolproof that he feels this problem of tampering is something that could be handled.

The simple electronic tagging of people is only one application of people-monitoring that has been proposed. Astronauts out in space were constantly monitored for such things as pulse and blood pressure. Why not perform such monitoring here on earth? A former sex offender could be picked up if sensors triggered a radio signal indicating that he was undergoing penile erection in an inappropriate place. The technology for such signaling is not formidable.

Or if the "subscriber" is fitted with a transceiver his behavior not only can be monitored but controlled. (A transceiver is a radio that both transmits and receives signals.)

Barton L. Ingraham and Gerald W. Smith, writing in *Issues in Criminology,* offer an example of how this would work:

"A parolee with a past record of burglaries is tracked to a downtown shopping district and the physiological data reveal an increased respiration rate, a tension in the musculature and an increased flow of adrenalin. It would be a safe guess, certainly, that he was up to no good. The computer in this case, weighing the probabilities, would come to a decision and alert the police or parole officer so that they could hasten to the scene; or, if the subject were equipped with a radiotelemeter, it could transmit an electric signal which could block further action by the subject by causing him to forget or abandon the project."

Ingraham and Smith also proposed that the electronic devices could be implanted right in the brains of certain parolees. A computer could send out an electrical zap to those who appeared to be about to misbehave. Electrical zaps delivered in or outside the brain also might well take the mind of the former sex offender off whatever had inspired penile erection.

In J. A. Meyer's bold explorations of the technically feasible, he proposed a network of alarm posts. These would warn of the approach of any person required to wear a transmitter of some sort. Owners of banks, stores, or other buildings would be encouraged to buy radio receivers that would spot the transmitter wearers. These receivers would sound the alarm to warn security personnel of a suspicious person's approach.

Imagine five million Americans setting off an alarm every time they tried to buy a pack of cigarettes or cash a check? This thought also occurred to Meyer. He asserts that no one should be stigmatized unreasonably. Only the transmitters of ex-cons who might be a "public danger" would set off the kind of loud alarm that might stigmatize the person. For the run-of-the-mill transmitter wearer, the store's surveillance would still operate, but would not make the person's presence "promiscuously known." Maybe a security officer would hear a quiet "pssst." Even if a transmitter wearer's presence was not promiscuously known the person might well have trouble paying by check or opening a charge account.

Schwitzgebel is quick to acknowledge that electronic rehabilitation systems raise questions of civil liberties that cause concern. But he is quoted as believing that "electronic and tracking devices would not seem to be directly prohibited by the cruel and unusual [punishment] clause [of the United States Constitution] within a broad view of the issue."[4]

It all sounds pretty unusual to me.

Meyer does note that such a tracking system could be misused. One misuse, he suggests, would be to arrest a person on a flimsy charge, put a transmitter on him, then stall the case in court. This would in effect keep the person under total surveillance while the authorities were trying to build a case against him. Another misuse that occurred to Meyer would be the arresting and belling of rioters or demonstrators as a means of intimidation.

He even suggests that such a system as he envisioned could lead to a "police state" and so should never be developed at all. But he seems to dismiss this when he states: "The same could be said about police, jails, courts, laws, taxes, and so on." Effective checks and balances will keep things in hand "without allowing a monstrosity to grow."

I doubt it.

THE BIG EYE

Americans like to think of themselves as openhearted, freedom-loving. In a number of their towns the first thing that happens to a newcomer is that his picture is taken surreptitiously. It is taken with a long-range camera. The film is filed at the police station. Or it may be sent away to be checked against some master file of known troublemakers.

In many towns and cities not only newcomers but all people who stroll into a central area are photographed by motion picture cameras. Some cameras can zoom and pan and take a close-up picture of a person hundreds of yards away. Mount Vernon, New York, installed two TV cameras at each end of its main downtown street. Hoboken, New Jersey, has TV cameras going in several public areas. So does San Francisco. The New York City Police Department has many TV cameras going. One installation is on Times Square. A small city in upstate New York had for some time more than twenty TV cameras on poles photographing its entire downtown area.

Initially, photographing people in public places was done to identify antiwar or civil rights demonstrators. Today, it is done more and more, at times with publicity, to try to deter street crime.

Technology has permitted a great leap forward in visual surveillance. Even in George Orwell's *1984* citizens were safe from the omnipresent telescreen Eye if they were in darkened areas of their homes. Today police have infrared light cameras that can take fairly good pictures in the dark.

The surveilling TV cameras have been going higher into the sky to cover a broader area of a town. First the cameras were put on top of tall buildings. In the University Circle area of Cleveland, heavily populated with collegians, cameras placed on two buildings during a period of student unrest were able to zoom in on anyone walking within an area of two square miles. New York City has at times installed TV cameras in hovering helicopters. Now, with the once-incredible expertise achieved by cameras on satellites, U.S. companies have been developing sky cameras for all-purpose surveillance. At least one of the cameras can hover in a stationary position without benefit of cables.[5] It is also considered feasible, through microminiaturization, to position an invisible sky-colored ball the size of a softball a couple hundred feet up in the air. It could both watch and listen to people below.

Of special interest to police photographers, often disguised as newspaper or television cameramen, are protest meetings, rallies, lectures, sit-ins, and the like. The "civil disobedience unit" of the Philadelphia police has covered thousands of such functions in the past few years. Policemen often find it more effective to stand in uniform taking pictures, or to pretend to take pictures.

Legally we are in a fuzzy area. The right to freedom of expression and to assembly in the United States is supposed to be protected under the First Amendment of the Constitution. Policemen standing around taking pictures is a form of intimidation and would seem to be an infringement of this right. I feel that such activity should be confined to situations where there is a clear threat to national security, which almost never has been the case in recent years.

It is not just the local police who have been taking pictures of assembled dissidents. As the anti–Vietnam war movement gathered force the U.S. Army Intelligence Command sent literally thousands of agents out to photograph and tape-record meetings. Some were in schools and churches. Files on more than one hundred thousand American civilians were developed. This was all very irregular. The military by law are supposed to confine their intelligence-gathering activities to matters of direct military concern. When this activity was first revealed the U.S. military issued a highly qualified order to stop most of it. Later it was revealed that the Central Intelligence Agency, which is specifically forbidden to engage in domestic intelligence, had in fact built up files on more than ten thousand individual Americans involved in domestic dissident activity.

Here are some other kinds of mechanical eyes:

• Eyes that can read letters inside sealed envelopes.
• "Eyes" that can duplicate printouts of a computer in another room by picking up electrical impulses.

THE BIG EAR

One of the yearnings of all authoritarian leaders is to control the information that reaches the populace. You can put the only available radios or TV sets in public squares, as some countries have done. Or in a big country with many individual set owners you can try to jam programs from other countries. But in affluent societies with millions of television sets the problem is complex. It would help authoritarian regimes greatly to identify any household in which a set is tuned to a program disapproved by the regime.

Technologists in telecommunications have now developed the tools for such monitoring. They did it at first for a benign reason: to survey viewing habits. But the tools can easily be used for surveillance. In one approach, a paneled truck equipped with a radarlike device roams the streets. The device identifies the TV channels being watched in each house. Apartments are harder to unscramble. The device has been demonstrated at conventions and used in several states in this country.

A second possibility will be developed as two-way cable television becomes widespread. And it will undoubtedly become so because it offers so many attractions to viewers and marketers. In times of unrest any authoritarian regime quite probably would arrange for the viewing habits of each subscriber to be recorded in a data bank.

As for listening to conversations, I was told by an official in research at the Department of Defense that to pick up conversations within a room no longer requires bugging. Laser beams can be bounced off the window of the room in which the conversation is being held.

Or if you seek confidentiality of conversation by taking a stroll in a park, don't sit down on a bench. A special gun can fire a spike mike into a nearby tree or bush.

Listening in by wiretapping has, of course, become commonplace in many countries, totalitarian and democratic. In France a senate report in a recent year estimated that the national government had several thousand taps on telephones on any given day, most of them illegal.

In the United States a great deal of wiretapping both legal and illegal is going on. No reliable estimate of wiretapping by the federal government is in the French range. There is, however, a great amount of wiretapping by state and local governments, much of it authorized by court order. And there is a considerable amount of private, illegal tapping. Twenty-odd states have authorized tapping by local and state agencies. Since court-authorized wiretaps became legal in 1968 the voices of at least two hundred thousand Americans have been legally recorded without their knowledge.

The problem is that each tap, on an average, picks up the voices of about twenty-five people. Many are using pay telephones they have happened to walk into. Federal court testimony indicates that less than one recorded conversation in ten is in any way incriminating.[6] Tapping of this kind, then, would seem to be a flagrant violation of the Fourth Amendment. The amendment is supposed to protect people against unreasonable search and seizure without evidence of "probable cause" of some specified criminal activity.

A prosecutor, in his application for a court order, must list the persons he would like to overhear and why. In most cases he cannot possibly know what people will actually use a public telephone that is to be tapped. The privacy of a great many innocent people is being invaded. And people not named who engage in suspicious conversation are being picked up. Prosecutors have been successful in writing into tap applications the phrase "others as yet unknown."

In 1974, the majority of the newly conservative United States Supreme Court upheld "others as yet unknown" as acceptable. They said it met the Constitutional requirement to describe *particularly* "the persons or things to be seized"! Speaking for the minority of three, Justice William O. Douglas suggested that this position was preposterous. He said: "Under today's decision, a wiretap warrant apparently need specify but one name and national dragnet becomes operative."

A new technological development that greatly enhances the powers of taps and bugs to incriminate is the voiceprint. A voiceprint offers a pattern of lines that comes close to being as precise for individual identification as a fingerprint. Not close enough yet to please the courts, but close.

The increase in the use of sophisticated surveillance devices — audio and visual — by both public and private organizations makes it imperative that the Congress enact a law that regulates strictly when and how the devices may be employed. They are becoming a threat to our free society.

THE BIG MOUTH

A fuss developed in the latter days of the Nixon administration with the exposure of a proposal to put a small government radio into every home in the land. By pushing a button in Washington, these millions of radios could be activated to broadcast administration messages twenty-four hours a day. The official purpose of this government network would be to improve the nation's disaster alarm system.

But the same three-hundred-page secret White House report outlining the proposal talked of using this technology to educate preschool children for world citizenship. Another need cited was to cope more effectively with "growing social unrest."

Congressman William Moorhead, as chairman of the House Subcomittee on Information, got his hands on a copy of the report. Every page was stamped "Administratively Confidential." He said it smelled of Big Brotherism. The report had been prepared for John D. Ehrlichman, chairman of the White House Domestic Council. As a close aide of Nixon's, he later became notorious during the Watergate exposures. The report had been prepared by the White House Office of Science and Technology, then headed by Edward E. David, Jr. Feasibility tests had been conducted.

When pressed by Congressman Moorhead, David insisted that the idea of the push-button network had already been abandoned on technical grounds. Moorhead contended that it was a "shocking document." He saw it as shocking because it gave "no consideration for the potential abuses of propaganda, no attention to the possible invasion of privacy."

THE BIG MEMORY

This same White House report proposed a "wired nation" system. It would store information from police and court records and health records in one great computerized memory bank. Apparently, information from the giant computer would flow back and forth between Washington and every corner of the country.

The idea of a national data bank was not new. I testified before Congress in 1966 to protest a plan very actively being pressed then by the U.S. Bureau of the Budget. One gigantic computer would pull together all the information on individual Americans held by twenty federal agencies, including the Internal Revenue Service, the Social Security Administration, and the Census Bureau. All three are supposed to treat information on individuals with strictest confidentiality.

The administration proponents argued that they wanted the information from such a central data bank only to develop statistics as an aid to planning. But they needed to keep individual identification on the data for updating and cost reasons. They could offer no convincing assurance that the bank would not be misused. The problem of safeguards against misuse, they acknowledged, was a baffling one. And it apparently had not occurred to them that under an authoritarian regime such a system could become a giant dossier bank on every citizen. The pooled information could contain every citizen's income, dependents, military service, employment record over the years, and many other private matters. The bank would become hazardous as more agencies were added. There was talk of assuring against abuse by setting up the proper administrative guidelines. But guidelines could be quietly changed by an authoritarian administration. Only a federal law setting forth privacy protections and requiring removal of individual identification could offer any hope at all that the system would not be misused. Otherwise it could be used ultimately to intimidate or discredit critics of an administration.

At this writing a single government data center linking federal agencies is still being pushed. One recent name for it was FEDNET. Beautiful! Now its main sponsor is the General Services Administration. When FEDNET ran into resistance there were reports that the GSA would just go ahead and, without any specific authorization, begin routinely pooling computer data. The idea was that it would be viewed as simply an improvement of the existing information-exchange systems in the government.

The federal government has at least five thousand computer installations for keeping records. And there are also many hundreds of large rooms full of individual records on old-fashioned file cards. The information still hasn't been programmed into computers. The Senate Judiciary Committee estimated in 1974 that the federal government had a billion personal files. That is five times as many files as there are people in the land. Added to that are perhaps another billion personal files in local, county, and state government agencies. Look at a few of the major federal filekeepers:

- The Department of Defense has at least 16,000,000 life histories.
- The Civil Service has at least 10,000,000 sizable files on individuals.
- The Internal Revenue Service has, of course, well over 100,000,000 files.
- The Social Security Administration has another 100,000,000 or so.
- The Federal Bureau of Investigation has at least 6,500,000 files on U.S. residents, including 600,000 computerized criminal histories. It also has files on more than 100,000 residents that at some time or another have been viewed as Communist sympathizers. Recently, while under criticism, it has been cutting back drastically on its political file keeping.
- The Secret Service has computerized files on hundreds of thousands of "persons of interest." You can become a "person of interest" by protesting welfare regulations, by being in a demonstration, by making harsh remarks about people in Washington, by insisting on personally seeing high government officials for the purpose of venting grievances.

The idea of having a "whole system" of pooled information on people in one data center is being adopted in numerous U.S. cities and counties. Santa Clara County, California, was a pioneer in "computer" government. Officials began putting into the central computer practically all official records that were available on more than a million residents. In went birth records, driver's license data, voter registration, jury status, juvenile delinquency records, welfare records, or any involvement with the sheriff or with the district attorney's office or the probation office. People who protested this countywide data bank were dismissed as dreamers. Access to the data was available through any of about one hundred teleprocessing stations. In some cases specific data could be retrieved only at specific stations.

New Haven, Connecticut, became International Business Machines' testing ground for programming everybody within a city. In went all the city's people files.

More recently, federal funding has been available for computerizing records on people in cities under a program labeled Integrated Municipal Information Systems. The acknowledged aim is to help cities create common, computerized information-file data that can be "accessed" by all departments of a city government. In this program Long Beach, California, and Wichita Falls, Texas, have been early plungers. Charlotte, North Carolina, has also moved toward the total-system approach.

As for states, Minnesota has been a pioneer. It is trying to pull together everything it can about people from files in about thirty-five hundred governmental jurisdictions. At a 1975 conference in Washington the state official in charge of this ambitious project, Daniel MaGraw, proclaimed: "We have everything about people that you can think of and it is in those state files." He was fascinated by the challenge. Minnesota patterned its system after the virtually universal data collection on individuals in Sweden. But recently Sweden has passed a law designed to protect the more important areas where personal privacy is involved. Too many people were getting hurt.

Some states in this country have disquieting special files. Maryland has been keeping a record of all people receiving psychiatric aid. With federal funding, Washington, D.C., has been asking applicants for driver's licenses a lot of highly personal questions about their life-style on a questionnaire. The subjects of the questions range from marital and money problems to feelings of frustration on the job, to a sense of loneliness or guilt, and even to night sweats. The questions touch on social variables that add up to a pattern suggesting that the applicant could be a problem drinker. So far, the questionnaires have been voluntary, and information on individuals is not made public. But consider the prospects if the questionnaire becomes required, institutionalized, and computerized in Washington, and in various states. It is easy to see how additional stigmatizing information on citizens could find its way into a central computer bank.

One hazard of the big memory banks is their image of infallibility. How do you argue with a vast computer that pours out derogatory information about you? The truth is that computers are more fallible than old-fashioned record-keeping systems. Most are ill equipped to correct errors or to bring records up to date.

Tens of millions of Americans are on record as having been arrested at some time during their lives. And this doesn't include traffic arrests. Local police and the local courts are notoriously careless about doing the paperwork needed to bring arrest records up to date. Many neglect to note "dismissed" or "innocent" or "acquitted" or "conviction reversed on appeal."

Yet most states send a record of all people arrested to the National Crime Information Center in Washington. This is a central exchange managed by the FBI. Thousands of retrieval centers around the nation can query the center for information. Included are not only law enforcement agencies but employers, credit agencies, insurance companies, licensing agencies. One estimate is that more than half the center's records are incomplete as to disposition of cases. In 1975 Massachusetts refused to forward its crime information files to the national center. The state officials knew that their records were grossly incomplete or full of inaccuracies.

In 1975, also, the FBI's plan to expand vastly its own role in computerizing the records fed into its crime information center by the states brought a sharp protest from its own sister agency, the Law Enforcement Assistance Administration. LEAA said that the plan might lead to federal control of the nation's police and could be perverted into a Big Brother type of arrangement.

And even if after there has been a conviction on a minor charge, should a person be haunted all his life by a petty youthful offense? Computers tend to petrify embarrassing information from an earlier era.

The hazard of an unforgetting, unrelenting computer was anticipated a century ago by Lewis Carroll in *Through the Looking-Glass:*

"The horror of that moment," the King went on, "I shall never, never forget!"

"You will, though," the Queen said, "if you don't make a memorandum of it."

The mere fact that the open-ended accretion of data on a person is done in a way that is beyond his control can clearly become a major threat to privacy. In April 1975, the National Bureau of Standards, which sets standards for the handling of computer data in the federal government, sponsored a symposium/workshop on "The Privacy Mandate." One speaker was Aryeh Neier, president of the American Civil Liberties Union. He cited — case by case — unwarranted damage or serious infringement of privacy that has occurred as a result of cumulative information maintained in school-system records . . . juvenile court records . . . armed service records . . . arrest records . . . medical records . . . mental hospital records.

The explosion of data collection on individuals, he said gloomily, has reached the point where "it is very difficult for us even to begin to establish any kind of control over our privacy. . . . Data collection, I think, is at odds with the very idea of individual rights." The person who somewhere, somehow, gets a "troublemaker" label on him, he said, "doesn't have much choice but to exist on the margins of our society."

The hazard is multiplied when information on individuals that is collected for one purpose turns up being used for some other purpose. This inevitably happens when you get a pooling of data from many agencies, with multiaccess terminals. Information you give on your marital problems when you apply for a federal loan to buy a house could turn up years later in a central data bank and be used against you at Civil Service if you apply for a job.

Beginnings are being made to try to keep the memory bank from becoming a tyrant. Thanks to Congressional action, by 1975 citizens were given the right to see all the information that certain federal agencies had about them. They were also given the right to demand correction of inaccurate information. Unwarrantedly, large exceptions were made for law-enforcement and intelligence agencies. And the law relates only to federal government records.

New federal laws have also given citizens the right to see and correct credit-bureau information. And parents of children now have the right to see records kept on their children in public schools.

These are modest starts.

The problem is grave. We are seeing a prime example of technology getting close to being beyond probable human control, except by the controllers. A judge of the state court of appeals in Cleveland — J. G. Day — states:

"Of all the great issues on the social agenda for the next thirty years — energy, peace, poverty, conservation, food, whatever — the protection of privacy . . . is one of the most important. Our success in protecting privacy may be the difference between a government that is bearable and one that is just not worth having."

Given the peril of that magnitude, it seems reasonable that Americans move to add an Information Bill of Rights to the U.S. Constitution.

There should certainly be developed a comprehensive Code of Fair Information Practices for all kinds of local, state, and federal agencies. It should set rules for personal automated data collection in a way that affords the citizen reasonable privacy and the right to redress.

And there should be established a National Privacy Board to enforce the code. It should have independence and broad enforcement powers.

Such a board should, I believe, be particularly alert to five problems:

1. Government information on private citizens should be safeguarded from anyone who does not have a reasonable need to know.

2. There should be public disclosure annually of all the personal data systems maintained, either by government or by private, interstate data-collection agencies. Clear, easy procedures should be set up whereby citizens can review and correct data about themselves (unless an individual is the subject of an active criminal investigation).

3. There should be a prohibition against commingling personal data collected for different purposes by different government agencies.

4. All obsolete personal data should be destroyed. Since bureaucrats are virtually incapable of considering anything they have collected as obsolete, procedures should be spelled out. In general, a review every five years would be reasonable. For arrest records there should be a review every three years when first arrests are involved.

5. There should be a clear prohibition on the organized collection by government agencies of any personal data unless there is an established need to know — and a fairness in knowing.

TAGS FOR LIFE?

There is now a movement to establish "universal identifiers" for all Americans.

The advent of memory banks has inspired pressure by bureaucratic types to tag every American with a number. Such a universal identifier would provide the crucial link for developing central memory banks. It would enable the pooling of all sorts of information from all sorts of sources, information that could be filed in a central memory bank under the universal identifier. And the bank could function, literally, from the cradle to the grave.

Some have suggested tattooing the universal identifier on each newborn baby's instep. But that is not really necessary. Several congressmen sought in the mid-1970's to pass a law requiring that all schoolchildren be assigned a universal identifier. The identifier proposed was a Social Security number (even though the children obviously were not employed). The American Civil Liberties Union helped defeat the move. But there may be others.

When the Social Security number was devised four decades ago there was great concern in the land. There was fear that the number might become some sort of cradle-to-grave number tied to you. Congressmen voiced these fears. The public finally received assurance that each citizen's number would be kept a closely guarded secret.

That secrecy began eroding within a few years as people were routinely asked to give their Social Security numbers when they changed jobs. The confidentiality really began crumbling with the advent of computers. The Internal Revenue Service got permission to ask each citizen to list his Social Security number on his form. And with computers, that number became the crucial key to keeping track of everyone's source of income. All parties who paid any taxpayer income last year were required to report the annual payment and in the reporting use the taxpayer's Social Security number. Reports on an individual coming in from a variety of sources all carried the same number.

The IRS could have set up its own numbering system and have notified each taxpayer to use that number. Each taxpayer would have been directed to give the number to employers paying him income, to banks paying him interest, and so on.

There was apprehension that taxpayers might react badly. How would they take notification that a brand-new number had been created to keep track of all payments made to them? Unfortunately (I believe) the IRS chose an easier route. It got the necessary permission from Congress to start using Social Security numbers. There was a problem. Millions of taxpayers didn't have a Social Security number. The Social Security Administration obligingly invented millions of new numbers for these taxpayers.[7]

Suddenly the Social Security number was becoming a universal identifier, a possible link between memory bank systems. The armed forces began using it instead of a service number in 1967. Here are some examples of how confidential the Social Security number has now become:

- The Retail Credit Company, which has files on the financial conditions of tens of millions of Americans, uses it.
- The United States Civil Service uses it.
- Massachusetts, to name one state, uses it on applications for driver's licenses.
- The Bureau of the Census uses it.
- The Federal Housing Administration uses it for loan applications.
- Virginia is one of the states that requires it for voter registration.
- Some states require that it be listed by all people applying for welfare and social services programs.
- The Medical Information Bureau, which maintains files on the medical condition of at least 12,000,000 citizens for the benefit of some 700 insurance companies, uses it.

This last use is particularly interesting. The bureau assembles information on the health of millions of Americans who are seeking health or life insurance. It manages to do this by assembling information on the examinations given applicants for life insurance. Or it assembles the information from claims for insurance benefits. Other data derives from information that insurance companies obtain from health-care providers, such as doctors and hospitals.

The bureau's headquarters is in Greenwich, Connecticut. But if a person wishes to check what information the bureau has on his health he must write to a post office box in Boston (at this writing, Box 105, Essex Station). He will eventually be requested to fill out a two-page application. Then he should either come in person to Boston to obtain the information or accept a collect telephone call. The bureau has seemed reluctant to put anything in writing. And he must waive all claims for liability against any misinformation the files may have about him.[8] In many instances he can apparently get the information only by having it explained to him by his physician.[9]

But back to the universal identifier. Before World War II, Holland had a superb identification system, one that included every citizen. On the other hand, France had been relatively indifferent to keeping records on its citizens. When the Nazi armies invaded Holland they seized the files and systematically tracked down the Jews of Holland. The records reportedly led them to arrest the father of Anne Frank, the girl who wrote the harrowing diary. In France the Nazis had a terrible time trying to find who was who.

We have reached the point where memory banks obviously are here to stay. There is probably no turning back. But a universal identifier that can provide linkage between different systems is dangerous and unnecessary. A "system identifier" is perfectly adequate. The IRS could have developed its own unique system. And the armed forces could have stayed with their perfectly adequate service numbers. It is the bureaucratic passion for consolidating the system, so that information on individuals can be swapped and pooled, that is ominous. Such centralizing requires the universal identifier.

The omnipotent potential of a universal identifier was anticipated by W. H. Auden even before memory banks were developed. In 1940, when governments were beginning in earnest to put numbers on people, he wrote a poem called "The Unknown Citizen (To JS/07/M/378 This Marble Monument Is Erected by the State)." Auden's unknown citizen was a curiosity because the government had no record of any complaints whatever involving him. The poem ended:

> *Was he free? Was he happy? The Question is absurd.*
> *Had anything been wrong, we should certainly have heard.*

The move to the universal identifier was slowed slightly at the federal level by the Privacy Act of 1974. It placed restrictions on further expansion of unapproved use of the Social Security number by other agencies. But there is no longer anything special about the Social Security number. What is needed is a flat, nationwide prohibition against the establishment of *any* kind of universal identifier. Let each organization — private or governmental — develop its own people-numbering system.

THE VETERAN'S SECRET LABEL

In 1956 the U.S. military began putting Separation Program Numbers (SPN's) on all military discharge papers, honorable or otherwise. The numbers were in the "For Official Use Only" area. Hardly noticeable. Many of these 530 "spin code" numbers were character ratings of the person discharged. Some covered undesirable habits or attitudes that some superior had reported. Remember, many of these citizens being labeled for life had been draftees ranging in mood from unenthusiastic to resentful.

Close to a million honorably discharged veterans were given discharge papers with spin code numbers that were derogatory, or partly so. Spin number 368 meant that the veteran had been labeled as having an "antisocial personality." The number 265 stigmatized him by alleging that he had a character disorder. Number 263 indicated that he had been reported for bed-wetting (perhaps under hostile fire?). Number 41A charged him with "lack of interest." Number 469 labeled him "unsuitable." In 1973 more than thirty-five thousand discharged veterans unknowingly carried this "unsuitable" label in their pockets as they headed back into civilian life, even though they had been given an honorable discharge.[10]

In modern America a recently discharged veteran seeking a job usually has had to show his discharge papers to personnel officers of the employing organization. The translation of these innocuous SPN numbers gradually was bootlegged to many major corporate employers, such as Firestone, Boeing, Chrysler, and Standard Oil of California.

After seventeen years the American Civil Liberties Union discovered and brought to public attention the meaning of SPN. It also brought suit, and within a year (1974) the Pentagon ordered an end to spin-labeling. Veterans were given the right to stand in line to get a new discharge paper without a spin number. But there are still thousands of veterans holding discharge papers who are unaware of the stigma that may be in them. And the Pentagon still has deep in its vast files the SPN numbers of every veteran discharged from 1956 through 1973. For most, the veteran's Social Security number is listed. If a comprehensive FEDNET memory bank ever gets humming, these records might become a part of it.

The decent thing to do, I believe, would be to issue new spinless discharge papers to all veterans holding spin-coded papers. The cost would be far less than building a single modern bomber plane. And the derogatory

Pentagon SPN listings on all honorably discharged personnel should be destroyed.

LEVERS OF MASS INTIMIDATION

During the late 1960's and early 1970's, observant people in the Land of the Free got a fair idea of how their society might slide into a semipolice state. The national mood was ugly. The White House, for most of the time, was dominated by people more preoccupied with mastering technologies of people control than with preserving citizens' rights.

The most common characteristic of all police states is intimidation by surveillance. Citizens know they are being watched and overheard. Their mail is being examined. Their homes can be invaded.

Richard Nixon, as President, was an ardent listener and intimidator. He wiretapped reporters, officials, his own brother. A former Nixon speech-writer, the columnist William Safire, wrote in 1974 that "the willingness to listen in . . . was second nature to Richard Nixon." Safire had reluctantly concluded that Nixon had "an addiction to eavesdropping." Only the bungled efforts to wiretap the headquarters of the opposition triggered the course of revelations that eventually brought his downfall. Long before then, in 1970, he personally approved the so-called Huston Plan for political surveillance. It involved such methods as burglary, wiretapping, maintaining mail "covers," and bugging. This plan purportedly was aborted, but soon after, the White House had its own personal secret agents who were using many of these methods.

Surveillance of critics and dissidents (as well as conceivable Communist sympathizers) was of course not new. But in this period it reached a new high, thanks to the new surveillance technologies. For example:

- The U.S. National Security Agency heard virtually every overseas telephone call during the six years ending in 1973. It also inspected virtually every cable.
- Central Intelligence Agency personnel stationed in the United States opened more than two hundred thousand letters illegally. And the FBI had mail-opening projects in eight states.
- A research organization, the Center for National Security Studies, estimated in late 1975 that more than two hundred thousand Americans had been the subjects of *active* surveillance in recent years.

I believe that as a general rule government agencies, whether federal, state, or local, should be barred by federal law from collecting and filing any data about a citizen's political activity, association, or expression, beyond voter registration. That is what the First Amendment is largely about.

The only exceptions should be when a person is being considered for appointment to high national office or when there is probable cause, confirmed by court order, that an individual has recently engaged in violence against other persons or may be engaged in espionage, sabotage, treason, or terrorist activities.

A few more words about the strategies of intimidation used by the Nixon White House. Nixon people tried to punish a *Newsday* reporter. The reporter had written a critical series on Mr. Nixon's good friend Bebe Rebozo, who was deeply involved in the Watergate scandal. A White House official requested that the reporter be subjected to an income-tax audit. Also, there was a memorandum proposing antitrust actions against critical news media.

The newer technologies of broadcasting news are more vulnerable to intimidation than printed media. They need to have their federal licenses renewed in order to function. In 1970 Mr. Nixon set up in the White House the Office of Telecommunications Policy. Among other things it began monitoring radio and television coverage of news involving the administration.

According to one memo uncovered by the Watergate investigation, White House aide Charles W. Colson had a meeting with the top executives of all three national networks. Colson reported, apparently triumphantly, to H. R. Haldeman, Mr. Nixon's chief aide, about how nervous the executives had been. They had been vastly impressed, he said, to discover how thoroughly their news programs were being monitored and analyzed. "In short, they are very much afraid of us and are trying hard to prove they are 'good guys.' "[11]

But apparently they were not afraid enough to please the White House. By 1972 the head of the Office of Telecommunications Policy, Clay T. Whitehead, gave a speech that some viewed as ominous. In citing it, Republican Senator Lowell P. Weicker, Jr., of the Watergate select committee, interpreted it as a threat to owners of individual television stations. The threat seen was that the individual stations might lose their licenses if they did not shape up by bringing pressure to force their networks to correct "imbalance or consistent bias."

This power to punish through licensing reportedly came up in a taped conversation Mr. Nixon had with his counsel, John W. Dean III, in September 1972. Mr. Nixon was angry that Watergate revelations were being published by the *Washington Post*. He was quoted as vowing vengeance and commenting, "They have a television station, you know." Within a few weeks two groups with close political ties to Mr. Nixon challenged the *Post*'s television licenses in Jacksonville and Miami, Florida.[12]

Another possible lever of control that seemed to fascinate many of the Nixon group was the Internal Revenue Service. Its memory banks contain year-by-year accumulations of data on people's financial affairs. This fact was seen as an ideal tool for intimidating or discrediting "enemies" or dissidents or political opponents.

In 1969, under prodding by the Nixon White House, the Internal Revenue Service set up a special staff to watch "activist" or "leftist" organizations. The staff compiled dossiers on 8,585 individuals and 2,873 organizations, according to records brought to light by the Tax Reform Research Group. Caught up in this dragnet were such solidly American organizations as the National Council of Churches, the Urban League, and Americans for Democratic Action. A great deal of information unrelated to tax status was assembled and in some cases was passed on to other law-enforcement agencies.

During the Watergate hearings, John Dean testified that several lists of people considered by the White House to be its "enemies" were drawn up. More than six hundred names were transmitted to the Internal Revenue Service with the suggestion that their tax returns get a good going-over. In 1975 the IRS began an investigation of charges that in the 1971–1973 period some of the "enemies" may indeed have been harassed by the IRS agents who investigated their sex habits, their drinking, and their social habits.

As it turned out, the White House was hampered in its campaign to use the IRS. The director, Johnnie M. Walters, and many of his subordinates resisted courageously. Walters took seriously his mandate to exercise fairness and to preserve the secrecy of tax returns. He resisted efforts to politicize the IRS. There is little evidence that the IRS unfairly harassed the "enemies" and "activists" it was asked to investigate.

At one point Mr. Nixon's top domestic aide, John Erhlichman, blasted Walters over the telephone for his "foot-dragging tactics" in pressing a case of special interest to the White House, according to Walter's testimony. Mr. Walters resigned. Soon thereafter Mr. Nixon was in such deep trouble because of the Watergate exposures that attempts to use the IRS were discontinued.

But taken together, the efforts of the Nixon White House to intimidate and control the people and institutions it viewed as opponents will go down as a close brush with authoritarianism. If a nightwatchman had not happened to notice an odd piece of tape stuck on the lock of a door at the Democratic headquarters in the Watergate, the history of the 1970's would make much more discouraging reading for those who cherish freedom.

On the broader theme of this chapter — the new technologies for keeping track of, and controlling, the populace — the evidence we have examined seems to reaffirm the validity of a question that was raised by Tom Wicker, the political columnist. Wicker asked:

"Can man preserve his liberty and individuality against his own genius?"

Molding Super Consumers, Super Athletes, Super Employees

9

To make people feel or act in a certain way is not manipulating them.
— Alvin A. Achenbaum, marketing expert

Mr. Achenbaum was testifying before the United States Federal Trade Commission. The subject of the hearings was "Can Advertising Manipulate Consumers?" A great variety of advertising practices were scrutinized. Federal Trade Commissioner Paul Rand Dixon treated with skepticism Mr. Achenbaum's disavowal that any manipulation was involved.

Whether specific advertising practices are manipulative can be argued. What cannot be argued is that advertising in general is a manipulative force. It has become a major instrument of social control in much of the Western world. Advertisers spend close to $33 billion a year to shape the consuming habits of Americans. That averages out to nearly $600 for each family in the land.

Some ads just announce the availability of a locksmith. But many of the messages are designed to modify life-style and attitudes. They seek to make people more hedonistic, more narcissistic, more status-conscious, more prone to live against the future. The goal is to create more insatiable consumers. And this is starting to be seen as functional in highly technological societies. Some years ago the sociologist Clark Vincent, in making his presidential address to the National Council on Family Relations, said that the family is no longer just a production unit. It was adjusting to a new challenge, that of being a "viable consuming unit."

The massiveness of the behavior-molding effort of advertising can be seen in the viewing of television commercials. By the time young people in the United States are eighteen they have seen and heard about eighteen hundred hours of commercial messages on TV. If you divide that up into thirty-five-hour workweeks you find that they have spent a full year of time listening just to commercials. Adding in radio commercials would take them well into a second year of time.

Since 1950 many hundreds of behavioral scientists have worked to enhance advertising effectiveness. Some have worked for consulting firms. These often are called motivation-research firms. A recent issue of *Bradford's Directory* included 124 of them. The list was not complete. Other social scientists have worked directly for advertising agencies. Still others have worked for companies trying to sell the products.

I have described in previous books a great many strategies devised by these consultants to modify consumer behavior.[1] A more recent and typical approach by one of the marketing-research pioneers was concerned with overcoming consumer resistance to a hair dressing for women. Sales had not risen in proportion to the massiveness of the advertising. This producer had been very successful in selling male hair dressings. The famed psychological consultant Ernest Dichter was called in. His interviewers talked with prospective users about the dressing. They seemed uneasy about its milky color and texture, and talked about the fact that the producer made products for men. Dichter advised his client that the male image of the producing company led women unconsciously to associate the feminine product with male semen. This caused them to feel that it was not for them. He recommended far more feminizing in the product's name and promotion.

Dichter advises clients that nowadays it is not enough to praise a product. First you must remove the customer's feeling that "I don't need it." Need, he explains, is a psychological phenomenon. As he sees it, the great majority of the services and products we use are not needed in a purely utilitarian sense. "Thus, in selling a man another suit you do some psychological selling 'on the pleasures of renewal of yourself.' "

Resistance now is being encountered by cosmetics manufacturers who have for so long been promising women an interesting sexual conquest. There have been so many disappointments that now, Dichter says, it often is better to focus on the fun and narcissistic pleasure of applying cosmetics, even if nothing happens.

In recent years the marketing journals have handled tough subjects with erudition. The *Journal of Marketing* carried a long presentation titled "Fear: The Potential of an Appeal Neglected by Marketing." (I didn't know fear had been neglected.) The principal author was a social psychologist who had been retained by a major advertising agency. He and his colleague reviewed ninety studies of fear to prove that advertisers had been too squeamish about boring in on it.

With an array of charts they demonstrated what levels of fear are most effective in selling such things as insurance, mouthwashes, dietetic foods, and safety features in automobiles to various groups. At the end he and his colleague raised the question of ethics. Was it possible that using the amount of fear necessary for optimum sale of the products might have "deleterious consequences for those high anxiety persons who happen to be in the message audience?" They immediately dismissed this by contending that the level of fear that is effective in marketing would not be high enough to be even remotely unethical.

The matter of sex in advertising received scholarly treatment by two other psychologists in the *Journal of Advertising Research*. The title: "Who Responds to Sex in Advertising?" They explained at the outset that their paper applied both Q-analysis and a heuristic clustering method to the same set of data to find out how various individuals interpret sex appeal in advertising. Heuristic clustering analysis proved to be superior in breaking down sex in advertising into its principal components and showed how "one group of individuals differ from another in their responses to sex in advertising." Three hundred advertisements were rated by several hundred young men and women.

A principal finding: the suggestiveness of the copy is a much more important variable in ads for women than for men, contrary to traditional thinking. And women proved to be more sexually aroused by nudity in ads than men. The study also identified groups who had sex fetishes concerning certain of the products seen.

ZEROING IN ON THE BEST PROSPECTS

In recent years the behavioral specialists who advise marketers have put the heaviest stress on identifying the best prospects for products. Vast amounts of research have been done and computers have been heavily used.

Identifying the likeliest prospects is achieved by "segmenting" the universe of consumers. First, you divide people up on the basis of "demographics." That is, you segment by age, income, education, occupation, ethnic background, size of family, and so on. But demographics have no *feeling* for people and their new life-styles. Today many ambitious marketers consider demographics just the beginning. They call next for "psychographics." Enter the behavioral specialists.

The word psychographics was coined by Emanuel Demby, the amiable, bearded head of Motivation Programmers, Inc. It made motivation research sound more precise — and new. It was more measurable, thus more scientific. With psychographics you build psychological profiles within the groups targeted by demographics. People are analyzed by interests, life-styles, status aspirations, self concepts and attitudes (including fears and biases). The aim is to arrive at groups with similar psychological profiles. One facet of a Demby study was to find, within the world of car buyers, the kinds of people who would consider buying a certain brand of car.

A rival of Demby's, the Commercial Analysts, studied at considerable length four thousand people on 360 psychographic dimensions. The computer came up with eight predominant life-styles for each sex.[2] Women readers may find some overlap. For women:

> The self-righteous social conformist
> The family-oriented churchgoer
> The downtrodden salvation seeker
> The happy materialist
> The blithe-spirited natural woman
> The romance and beauty seeker
> The fulfilled suburban matron
> The liberated career seeker

The eight categories for men:

> The inconspicuous social isolate
> The silent conservative
> The embittered resigned worker
> The highbrow puritan
> The rebellious pleasure seeker
> The work-hard, play-hard executive
> The masculine hero emulator
> The sophisticated cosmopolitan

A major advertising agency, Benton & Bowles, Inc., came up with somewhat different psychographic profiles from a panel of two thousand housewives it was studying. They were asked 214 questions about their attitudes. They were also asked about the products they bought. One clear area of interest to the researchers was concern about germs. Many product promoters thrive on germ anxiety. This study came up with six categories as "most meaningfully classifying" the housewives:

> Outgoing Optimists
> Conscientious Vigilants
> Apathetic Indifferents
> Self-Indulgents
> Contented Cows
> Worriers

It was discovered, for example, that the Conscientious Vigilants and the Worriers were more likely to be receptive to products that promised to kill germs. Contented Cows were described as "relaxed, not worried, relatively unconcerned about germs and cleanliness, not innovative or outgoing, strongly economy-oriented, not self-indulgent." (In short, not promising prospects.) The study produced clues on how to appeal to each category of housewife in selling messages for various kinds of products.

One enthusiast of psychographics, Alan R. Nelson, contends that psychographics is "light years" away from other methods in moving consumers into action. He heads Alan R. Nelson Research. Nelson conducted a research effort involving several thousand consumers to find the best way to sell decorated toilet paper, and similarly decorated facial tissue. His group tested thirty-two selling themes in relation to nineteen "life-style value systems." One finding: "Guests instinctively study bathrooms for messages about their hosts. . . . If we have any minor weaknesses, they are most likely to be betrayed by our bathrooms. . . . That's why we are so anxious to 'look good' in them." Decorated toilet paper was concluded to be one good solution to this dilemma of modern life.

Perhaps the most interesting psychographic study by Demby's Motivation Programmers, Inc., involved spotting people who are likely to be "aggressive" buyers of *new* kinds of products. Such people are called "creative" consumers. They are contrasted with drudges who are "passive" consumers. The research group found that lots of rich people with college educations are very slow in taking to new concepts in products. An example would be the electric blender. Other affluent people with identical demographics are three and a half times as likely to buy electric blenders, food liquefiers, and the like. And these others are seven times as likely to buy electric hot trays.

How do they differ? They differ in psychographics. For the kinds of products mentioned above, the heavy buyers lead a more outgoing, socialized way of life. They are joiners. Their life-style is directed outward, aimed toward contacts with other people. In contrast, the equally affluent but dreary prospects lead a life-style that is directed inward. Their important activities tend to be those that "involve the individual, the family and close friends."

MACHINES TO PROBE MOODS

Admen are constantly searching for a better way to get inside our minds. For one thing they want to know how much we are moved by their handiwork. Knowing our reaction beforehand to their persuasive efforts, they can make adjustments before launching a worldwide campaign that can make additional millions of dollars from sales for their clients.

A large part of their analysis involves our reaction to pictures and words: pictures and words in the media; pictures on cans and packages. In the early 1960's a German-born psychologist, Eckhard Hess, came up with what seemed to be a breakthrough. He developed a culture-free device that instantly revealed how much we were interested in a picture. The device was a pupillometer. It measures the pupil, that round spot in the middle of the eye. As the pupil expands (dilates) or contracts while looking at a picture, it purportedly reveals something. Hess had found that the degree of dilation is a measure of how intently we are examining a picture.

Many of the major advertising agencies plunged into pupillometrics, as did many university laboratories. The admen mistakenly leaped to the conclusion that their machines were giving them a yes-or-no answer on whether the beholding eye liked or disliked their ad. It turned out that the machine could not reveal likes and dislikes, and some admen became disgruntled. What the machine did reveal, though, was intensity of interest and concentration. And that was something. You can't either like or dislike a message until you get interested in it.

The machine also couldn't tell them what it was in the ad — the handsome man or the girl admiring the diamond ring being offered — that was causing the beholder's eyes to dilate. Another complication was that the pupil is more likely to dilate at the sight of dark colors than light colors.

In recent years both of these problems have been largely solved as the machines have become more sophisticated — and expensive (up to $20,000). Today's better pupillometers can show exactly what part of the picture is causing the pupil to react. Two psychologists who assessed the state of the art for *Psychology Today* have concluded: "Although the pupil is no panacea, it can play a substantial role in exploring the emotional and mental functions of man."[3]

Different kinds of machinery are used on a large scale to pretest television commercials and TV shows that advertisers have been invited to sponsor. People are brought in off the street. Their reactions are registered by turning dials or pushing buttons. For light entertainment, the previewers' fingers may be attached to electrodes measuring skin reaction. The skin of a person who is enjoying himself throws off less sweat than the skin of a fidgety person.

Viewer interest is also measured by the squirm test. The TV previewers sit in chairs wired to record movement. The assumption is that the more the buttocks move in a chair, the less the interest in what is being shown on the screen.

Recently the researchers have shifted from the buttocks to the brain in searching for clues to viewer interest and selling effectiveness. The *Journal of Advertising Research* reported a test of brain-wave response to advertising. One finding has been that people show much more mental alertness (beta waves) when they read an ad than when they listen to a commercial on TV. Here the more relaxed alpha waves predominate.

In 1975 a researcher at the Brain-Behavior Research Center at the Sonoma State Hospital in California reported that he had been testing reactions to a variety of TV shows. To his subjects' heads he affixed electrodes leading to a computer. He found that the brain-wave patterns of people really interested in what they were watching were clearly different from those of people who had "little interest" in the program being watched.

This brain-tapping presumably is the wave of the future in shaping messages for maximum behavior-shaping impact on the marketplace.

THE SCIENTIFIC SEDUCTION OF CHILDREN

In applying strategies and using machines to determine maximum sales impact the advertisers have focused in particular on one group: children. From the age of three up. Children consume at least $75 billion worth of goods and services in the U.S.A. alone. And they are influential in the adult purchase of many billions of dollars worth of additional items. An adman was quoted in *Advertising Age* as observing:

"If you truly want big sales, you will use the child as your assistant salesman. He sells, he nags, until he breaks down the sales resistance of his mother or father."

The average youngster sees more than twenty thousand television commercials a year. Corporations spend nearly a half-billion dollars on these commercials. Billions in profits are at stake. And most marketers think they are now learning how to get their money's worth. An economist for the Federal Communications Commission says: "Children's programming is the most profitable area of television programming."

Across the land there are dozens of motivation-oriented consulting firms that specialize in probing children's reactions to commercials, programs, and products. The findings are used to make modifications that will cause child viewers in general to be more eager users and hawkers of the sponsor's product.

Audience Studies, Inc., in Los Angeles runs a small theater for testing commercials and pilot programs. Children are tested apart from adults. Each year about four thousand children man the "interest machine." By turning a dial they can indicate five degrees of enthusiasm.

But the real probing occurs in the more intimate play-area laboratories. Usually these are inside motivation-research firms. Some years ago I visited one up the Hudson from New York where hundreds of children in the area were serving as reactors. About a dozen at a time were probed for their reactions. The playroom wall had a one-way mirror. Behind the mirror were cameras, tape recorders, and chairs for client observers.

Recently an editor for *Human Behavior* made a study of such laboratories under a pledge of confidentiality.⁴ One was in a West Coast suburb. The children's responses are tested by pupil-dilation measurement machines, finger sensors, and so on.

She reported that after the children had viewed a commercial a child psychologist questioned each one closely about his reactions. Each group of child-reactors were requested to make drawings about each part of the commercial. These would be analyzed later by specialists. The analysts also drew upon Stanislavski dramatic techniques. Teams of children were organized to act out how they thought their parents would respond to their request for the product. Also they acted out what pitches they themselves would use to sell it to adults. And they were probed on how their playmates would feel about the product.

This honing of sales appeals for maximum exploitation of the child market unquestionably is felt in the home. A survey at Michigan State University of several hundred preschoolers revealed that 80 percent of the children acknowledged that they had urged their parents to buy toys they had seen advertised on television. About the same proportion had appealed to their parents to buy cereals they had seen advertised on television. Many of these cereals are so loaded with sugar (the tooth-decay producer) that they could pass for breakfast candy. Until recently, at least, several drug companies have pushed vitamins to children with the appeal that they were wonderfully sweet.

In a book called *The Youth Market* two advertising executives reported a survey of mothers. How much more did the mothers buy in the supermarket as a result of urgings from their children to buy specific products or brands? The responses would indicate that for U.S. mothers as a whole the cost would come close to $4 billion a year in addition to their grocery bills.

SLIPPING MESSAGES INTO OUR MINDS

In the late 1950's there was a hullabaloo in much of the Western world when it was discovered that hidden messages were being tucked into TV, radio, and motion picture shows. Much of the tucking was being done by advertisers. The technique was called subliminal stimulation. It was based on findings by psychologists that the brain can receive quickly flashed images and whispered sounds below our level of conscious awareness.

In 1957 there was an announcement that a firm in New Orleans called Precon Process and Equipment Corporation was in the business of placing subliminal images in movies, on billboards, and in taverns. The firm was set up by a psychologist and a neurologist with engineering training. They said they had been experimenting for several years and had applied for patents. Later they claimed to have doubled the consumption of a beverage advertised subliminally on the premises where it was for sale.

In New York City, James Vicary, the head of a motivation-research firm, called a press conference. The purpose was to demonstrate subliminal stimulation and to announce that he had set up a separate firm, the Subliminal Projection Company. Clients were being solicited. Patents were being sought.

Soon after the unveiling of Precon and of the Vicary enterprise, television and radio stations across the country began trying out the technique. A Chicago radio station for four months broadcast "subaudible" messages at $1,000 for four hundred messages. At least two motion pictures were made using subliminal images of ghosts, blood, and skulls to enhance dramatic impact.

Meanwhile, a public uproar over this hidden seduction arose. (And my first book of social comment, *The Hidden Persuaders,* happily for me, was caught up in the uproar.) The *New Yorker* deplored the fact that minds were being "broken and entered." *Newsday* called it the most alarming invention since the atomic bomb. The *Saturday Review* devoted an editorial page to deploring it. Congressmen and senators joined in, and introduced a number of bills. (Nothing happened on them.) A few state legislatures passed laws outlawing subliminal stimulation. Great Britain banned it. Precon and the Subliminal Projection Company were having problems getting patent rights. (Precon finally got a patent in 1962.)

Broadcasters became nervous. The United States National Association of Television and Radio Broadcasters, which includes the three networks and a majority (but not all) of the nation's TV and radio stations, banned its use. Admen began calling it a terrible idea.

Subliminal stimulation disappeared from the news.

Several years later, in 1967, the political scientist Alan Westin speculated on what "probably would have happened had the protests not been so widespread." He said it was most likely that TV sponsors, display advertisers, theater owners, filmmakers and political advertisers would have tested the technique thoroughly and advertisers would have been arriving at a "costs against presumed return-in-sales" picture. Political users would have been reaching a similar judgment. "Subliminals would have then become part of the communication arsenal."[5]

As a matter of fact, interest in subliminal stimulation has continued, but much more quietly. I have reports on fourteen studies that have been made in recent years, and references to quite a few more. The psychologist James McConnell in his new, widely adopted textbook *Understanding Human Behavior* devotes a chapter to "Subliminal Perception." The *Journal of Marketing Research* has carried an article on the effects of subliminal stimulation on brand preferences.[6] The article told of experiments involving ninety-six people in which messages were subliminally flashed. The results indicated that "a simple subliminal stimulus can serve to arouse a basic drive such as thirst." The author concluded that "the field of marketing should maintain an *active* interest in this area."

Researchers have now developed more precise information on what works and what does not work, subliminally. Words with a high emotional overtone — such as "whore," "raped," "bitch," "penis" — can be flashed twice as fast as neutral words like "river" and still be remembered later. A person must be already motivated to make use of even the weakest hunches suddenly felt. (That is, you must be at least vaguely hungry to accept a subthreshold suggestion that you go buy popcorn.)

A few years ago *Advertising Age* reported matter-of-factly that Toyota Motor Sales U.S.A. was using subliminal images to enhance the impact of its commercials.

In the pre-Christmas season of 1973 the producer of a packaged family game called Husker Du launched a nationwide TV campaign involving subliminal stimulation. Starting on November 26, its minute-long commercial appeared on hundreds of television stations. Many appeared in hours when child-watching was heavy. Four times during the commercial the command "Get It!" was flashed subliminally.

A technician at one of the stations noticed the flickered commands. Within a week the Television Code Authority of the National Association of Broadcasters alerted its member stations that the commercial was in violation of the code. The code prohibits "any technique whereby an attempt is made to convey information to the viewer by transmitting messages below the threshold of normal awareness." Apparently, most members of the association snipped out the "Get It!" commands. But many major stations are not members of the code organization. The manager of the company distributing the commercial contended that the subliminal commands had slipped past him because of the Christmas rush and asked stations to delete them. Weeks later, a few days short of Christmas Eve, a number of stations in large cities were still carrying the subliminal messages.

There are reports that the flashing of messages subliminally is being outmoded by a new technique. The message stays on the screen throughout the movie or TV show, but it is so dim that you are not consciously aware of it. Since you are continuously exposed to the message, it is considered by some to be more effective. And it may be harder to detect than the flashed message.

In 1973 Wilson Bryan Key reported in his book *Subliminal Seduction* the results of inquiries made to commercial research firms in New York, Chicago, and Toronto. Thirteen of the firms were prepared to offer some sort of mechanically induced, subliminal message service to advertisers.

There are still no rules forbidding the use of subliminal images in motion pictures or in supermarkets or taverns. And there are very few legal prohibitions against any kind of subliminal stimulation. It is principally fear of wrath that limits its use.

I would guess that its greatest potential is not as a huckster's tool in democracies but rather as a conditioning tool for authoritarian regimes.

PUSH-BUTTON BUYING IN THE LIVING ROOM

The great dream of TV advertisers is to find a way to clinch sales immediately. Catch the prospect while he is relaxed in his own living room, perhaps with a drink in hand. Close the sale while his desire is aroused and the product is clearly in his mind. Then you won't have to wait until some future date when he happens to be near a shop where the product or service is available.

That is the dream. And the reality is being tested. The reality is two-way cable television. A push-button console sits near the TV set where you the listener obey the commands or appeals of the salesman on the screen. The device is easier to operate than a pocket calculator. Just push three or four buttons. The machine that makes ice cubes, the electric leg shaver, the mahogany backgammon set, or that ten-speed bike Junior is hollering for will be at your door next morning. Pay later.

Cable television installations are growing rapidly in the United States. As of late 1976 there were about eleven million installations. And by law those now installed in most metropolitan areas must have a two-way capability.

The Advertising Research Foundation has been carrying on an intensive analysis of the potential of push-button selling on two-way cable TV in El Segundo, California. Subscribers can order instantly products and services offered in commercials. They can also order sports, theater, or travel tickets. They can buy items from mail-order catalogues as the items are being demonstrated. They can even ask for a demonstrator to call.

In El Segundo the TV merchandisers have a pretty good idea of whom they are trying to sell to. They are provided with important information about each subscriber who is reached by that two-way cable TV system. This includes not only the standard demographic stuff but psychological characteristics, type of residence, reading and broadcast listening behavior, buying habits—and possessions already owned.

El Segundo is just one place where bush-button selling has been occurring. Two-way merchandising systems have been established in Akron, Ohio; Irving, Texas; Mesa, Arizona; Orlando, Florida; Overland Park, Kansas — to mention only a few places. The current costs of apparatus are still a factor causing hesitation.

Some enthusiasts are saying that two-way TV is the most significant development in social communications since Samuel Morse first transmitted his code over wires. Buying is just one thing a subscriber can do with two-way TV. He can see an instant printout of the news. He can see today's menu at a local restaurant. He can arrange that sensors report a fire or burglary while he is away. And political and marketing pollsters can get back his reactions instantly.

Most of the strategies cited above for building super consumers will of course intensify problems already besetting advanced societies. The pressures to keep sales soaring will aggravate the growing shortage of energy and many irreplaceable minerals. The pressures will also lead to increased desecration, pollution, and preoccupation with finding fulfillment through consumption.

SUPER ATHLETES

As big-time sport became a multibillion-dollar industry during the past decade, coaches became desperately eager for some extra edge. Their jobs were on the line if they didn't have winning teams or performers. They had exhausted most of the ways to achieve a physical advantage. Word got out that maybe psychologists and psychiatrists could give them that extra edge. Maybe these experts could tell them how to assess rookies, how to decide where each man would fit best on the team, how to get the players superpsyched up for important games. Coaches yearn for certitude. And a number of behavioral scientists were happy, usually for a fee, to help them feel they were achieving it.

I learned of this development by chance when a psychology professor at Bluffton College, in Ohio, offered to give me a lift to the Toledo Airport. This affable man was William J. Beausay. As we talked I learned that a major drain on his time was running the Academy for the Psychology of Sports International, in Toledo.

His academy had profiled the personalities of several hundred National Football League players. It had profiled more than a hundred automobile racers. Beausay had worked with three national hockey teams and three National Basketball Association teams. He had offered counsel to Ohio State and Notre Dame University football coaches in their selection of players and assignment of positions. Beausay was particularly enthusiastic about "maximizing" the performance of athletes selected to play in each position.

His academy does not assess and psych up athletes off the top of its organizational head. It is equipped with computers, cassette recorders, brain-wave measurement machines, oscilloscopes, biofeedback machines, equipment to enhance visual efficiency, and batteries of psychological tests, including the exhaustive Beausay Experimental Survey of Temperament Profile.

Beausay, I find, has had competition. The Institute for the Study of Athletic Motivation has tested the personalities of several hundred athletes. Its founders are Bruce Ogilvie and Thomas Tutko, psychologists at San Jose State College. The people they have tested include aspiring members of the Detroit Lions and the Chicago Bears. They have often been referred to by football players as "The Shrinks." Their mission, as they see it, is to tell coaches how emotionally tough a player or prospect is, whether he can stand up to rough handling. They forecast whether the prospect will do better on offense or defense. They also test the "coachability" of rookies under consideration.

Another kind of consultant is Arnold J. Mandell, a psychiatrist. He served for a couple of years as psychiatrist-in-residence to the San Diego Chargers, who were beset by problem players. Mainly he tried to get inside the players' minds and offer instruction on how they could be more effective. But he also developed profiles — with the help of a computer — for the kind of personality needed for every position on the team. Other developments:

• A clinical psychologist, Robert Nideffer, has been counseling the up-from-the-bottom Buffalo Bills football team.
• An international conference on "sports psychology," including behavior-modification techniques, was held at the University of Texas in 1976.
• At the 1976 Olympics several East European track teams were accompanied by staff psychologists.

What are the Success Traits to look for in athletes? Beausay analyzed, by sport and position, the personality patterns of a large number of successful performers. He found sharp and telling contrasts. For example, he tested thirty-five racing-car drivers who were trying to qualify for the Indianapolis 500. Of these thirty-five, the seventeen who qualified almost all had attitudes of blazing hostility toward the world. Surprisingly, they tended as a group to be impulsive. Also they were withdrawn, insensitive, domineering. All the traits but impulsiveness would seem functional for performers who continually put their lives on the line.

Top golfers tend to show up in tests as being very composed and self-disciplined. Good long-distance runners tend to be passive and tolerant, but extraordinarily self-disciplined and self-punishing.

In football, perhaps the world's foremost team sport with body contact, different positions call for different personality profiles. Good defensive linemen carry a heavy load of hostility into the game, and enjoy discharging it on opponents. (Mandell adds that the best defensive men are rebels who hate structure and tend to be rule-breakers.)

In total contrast, quarterbacks are high in self-control. They are perfectionists and cool, self-confident characters. Mandell points out that they are either naturally cocky, like the Joe Namaths and Sonny Jurgensens, or they get their assurance and calm certitude from On High. He notes that many of the best, such as Roger Staubach, John Unitas, and Fran Tarkenton, have had a strong religious orientation.

Beausay contends that offensive linemen should be selected according to personality profiles quite different from those of the defensive men they face. Offensive linemen are much more self-controlled. They need to be excellent team players, skillful executioners of plays. Precision is required. Impulsiveness would be counterproductive. The defensive lineman on the other hand has no idea what the guys opposite him are going to try to do or where the play is going. He has to wait for the snap of the ball and "have an impulsive reaction."

The profiles of the best linebackers, I gather, include ruthlessness combined with objectiveness and cleverness. Wide receivers tend to be as narcissistic as actors. Off the field they tend to be tactful but aloof. Mandell finds that they are rarely popular with their teammates.

In studying player personality Mandell discovered that he could distinguish between defensive players and offensive players just by looking at their lockers. "The offensive players keep their lockers clean and orderly, but the lockers of the defensive men are a mess. In fact the better the defensive player, the bigger the mess."[7]

Beausay feels that the psychology of defensive and offensive players is so different that they should be separated for most of the half-time pep talk. The defensive briefing should be emotional, the offensive cerebral.

Many pro-football players go into games under the influence of behavior-modifying drugs. Until the situation became scandalous the drugs were largely provided by the club. "Uppers" such as the amphetamines were favorites. A couple of years ago a young coach at the University of California, San Diego, surveyed some two hundred players of sixteen National Football League teams on the use of amphetamines in pro football. He wrote a graduate thesis on his findings. His figures indicated that on any given Sunday afternoon half the players were on stimulant drugs. One pro-football player told Mandell:

"Doc, I'm not about to go out there one-to-one against a guy who is grunting and drooling and coming at me with big dilated pupils unless I am in the same condition."[8]

In 1975 the NFL commissioner made a show of disapproval by fining the San Diego Chargers for its use of behavior-affecting drugs.

To superpsych football players up for games the consultants have been experimenting with autosuggestion. One approach is to instruct an offensive tackle to repeat his best move—perhaps "Charge! Throw! Shoot!" — several hundred times before a game. Other football teams have been experimenting with having certain of the players engage in Transcendental Meditation before games.

In baseball the Chicago White Sox have been counseled to lie on the floor, close their eyes, and think how much better they are getting. They are to imagine themselves at bat and try to achieve such total concentration that they can see the seams of the oncoming ball.

Beausay tries to get players into prime mood for a game by having them play a cassette tape he has developed. He says it is not hypnosis, but has a kind of hypnotic effect. It purportedly gets them into a mood where they can draw on their physical resources to the maximum. Beausay has also been having players whose main need is cool skill to sit taped to biofeedback machines before games. This is to help get them into a calm, alpha brain-wave state. He also prescribes it for pro golfers, pro bowlers, and tennis players.

Defensive linesmen in football he handles differently. He has them put on earphones and then subjects them to a prerecorded series of obnoxious impulses that are calculated to irritate and aggravate them—to put them on edge. A number of players think these techniques in mind conditioning have improved their performance.

Some of the psychological consultants in pro football have at times been more coach-oriented than player-oriented. Ogilvie and Tutko, for example, wrote a book that rankled some players. It was entitled: *Problem Athletes and How to Handle Them.*

Beausay insists that his main aim is to help the individual athlete improve his performance. And he said of one rival psychological consulting team that its members had aligned themselves "too closely with the coaches and owners, which puts the emphasis more on manipulation."[9]

The business of filling out long personality inventories has also annoyed some players and prospects. In fact, the whole business of submitting to psychological testing became such a touchy issue that it became part of general contract negotiations. One contract football players rejected specified that the clubs could require the players to submit to psychological tests. Management dropped the issue. Now, any psychological evaluating is done informally as each club sees fit. As recently as 1976 some individual National Football League teams were reported in the psychological newsletter *Behavior Today* to be giving new players psychological tests. Beausay would like to see the league adopt a standardized selection system based on some sort of psychological tests.

SUPER EMPLOYEES

As disciples of Skinnerian conditioning moved out of the rat and pigeon laboratories into prisons, schools, and mental hospitals, they began asking: why not go for the big apple? Industries have spent vast sums to try to find ways to make employees less poky. Why not use operant conditioning with schedules of reinforcement to create desired changes of behavior in company employees?

One problem with this idea is that operant conditioning works best with controlled populations. This means that reinforcements of the negative or aversive type would be risky in today's labor market. Employees can always quit. Only positive schedules of reinforcement should be considered, at least at the start. Another problem was that union leaders might object if laboratory findings with pigeons were applied too literally to union members. One of the first proposals to take Skinnerism to industry was offered in the 1960's. The psychologist Owen Aldis wrote an article titled "Of Pigeons and People," which was published in the *Harvard Business Review*. He asked whether the findings about what makes pigeons peck best and most consistently did not have implications for human employees. Pigeons, and in fact all organisms, he noted, like immediate rewards for work accomplished. The obvious implication was that industry should work to restore piece-rate pay in some new ingenious form. He acknowledged that in modern industry it is often the machine rather than its human operator that does most of the work. Still, some form of immediate reward per unit of production ought to be considered. He pointed out also that pigeons perform better if rewards are variable rather than fixed. This suggested to the writer that elements of chance should enter into the method of payment. People like to gamble. Why not, each week, pick a worker's number from a hat? How big a bonus the winner would get would be determined by how much he had exceeded some agreed-upon standard of performance.

Taking a tip from a comment made to them by Skinner, two behavior shapers in Ann Arbor, Michigan, used gambling as an "intervention strategy" to try to curb absenteeism. The sample consisted of 215 hourly workers at a manufacturing and distribution facility. The shapers used straight five-card-draw poker. Each day, each worker who showed up for work, and was on time, could draw one card. If he was on time every day he would at the end of the five-day workweek have five cards, or a five-card poker hand. The highest hand won $20. And there were seven smaller winnings. Over a four-month period absenteeism dropped 18 percent, while rising 14 percent in comparison groups at four adjoining plants. Perhaps it was not the reinforcement of money so much as fascination with poker that caused the improvement.

By the 1970's Skinner-inspired consulting firms, such as the Praxis Corporation in New York, were busy offering guidance on how to accelerate the performance of employees. Emery Air Freight is one of at least a dozen sizable corporations that went thoroughly to Skinner concepts. Emery instituted programmed training for its salesmen. The aim was to try to persuade more potential customers to use its services. It also provided daily feedback to employees on how well each of them was meeting company goals. Those doing well have been receiving huge doses of praise and recognition. Those doing poorly aren't punished; instead, the problem they face is discussed. First results were reported as gratifying in terms of output and profitability. Praise, as James McConnell pointed out to me, "doesn't cost anything." He himself is in management training. Currently he is teaching reinforcement strategies to U.S. Steel, Owens-Corning, and Du Pont.

By December 1973 representatives of forty major United States corporations were gathering in Atlanta, Georgia, to hear what "Behavior Modification in Industry" was all about. Some were numbed by lectures on "managing the reinforcement contingencies" until they gathered that it was mainly a matter of systematic rewards. Punishments were not much discussed. The real issue, one speaker said, is which premeditated control is superior to haphazard partially effective control?

The audience of officials learned that behavior modification in some form was being tried at B. F. Goodrich, Standard Oil of Ohio, General Electric, and General Motors, to mention a few. And it was being tried on everyone from truck drivers to vice-presidents. Some of the behaviors under modification were punctuality, accuracy, speed, and work attitudes.

There was a good deal of talk about how in daily feedback sessions it was important to praise whatever performance could decently be praised and either to ignore what couldn't or discuss it as a problem. By all means, a General Electric spokesman said, the worker's self-esteem should be protected.

The behavior-modification programs in force were more often described as "behavior modeling," since behavior modification was a phrase that was starting to make some people wary. In modeling, employees are shown a film on how the job is to be done, and daily feedbacks are based mostly on how well the model has been followed.

A number of the programs described did not embody pure, systematic, Skinnerian concepts for behavior control but rather tended to promote the wonders that could be achieved simply by being courteous. The sessions at least made participants more aware that there are alternatives — in modifying a subordinate's performance — to chewing him out in public or threatening to fire him.

At the end of the conference a behavioral scientist from Washington

University was asked, as an observer, for his comments. He said some of the programs presented seemed pretty fuzzy and short of documentation. The heavy emphasis on praise and preserving self-esteem suggested that the companies were starting to treat adult subordinates like children. He suggested that you will get still further if you treat subordinates honestly. Humanistic psychologists in general deplore human relationships that are not authentic.

Another recent scientific approach to improving the performance of employees has been to extend time-and-motion studies to white-collar workers. Years ago assembly-line workers became accustomed to having their hand movements fragmented and measured, but white-collar workers assumed that their services were more personal, creative, and brainy.

Experts with stopwatches have now been measuring how long it takes a secretary to open a letter (7.027 seconds). Microsecond values are assigned to a host of the daily activities of office employees.

More than a dozen consulting firms heavily armed with stopwatches will now for a fee of about $7,000 a month, show companies how to break each office task down to arm movements, eye movements, et cetera, and find out how long each task should take. Many depend on reports prepared by the Methods-Time Measurement Association of Fair Lawn, New Jersey. Its report R.R. 106 is titled "Short Reaches and Moves." Its report R.R. 108 is called "Arm Movements Involving Weight." At its fall conference in Reston, Virginia, in 1974, one of the numerous presentations made was "How Work Measurement Is Applied in Banking to Measure Teller Staffing Levels." Motion standards are also now being applied to white-collar workers in the offices of insurance companies.

Even top-management people have not escaped. They have been provided with computer systems that are supposed to improve their decision-making. The managers are being told by the machines what the proper decisions should be. Some contend that this imposed information has the same negative psychological effect on the managers as the findings of time-and-motion experts on lower-level employees.

Chris Argyris, an industrial psychologist at Harvard, has stated that these computer systems designed to program management behavior "do for man in organizations what Skinner hopes social scientists will do someday for man anywhere." He concedes that in some situations computers are more efficient than Man, but he contends that unless thoughtfully handled such systems leave managers with a sense of psychological failure.

While these efforts to build superior employees and managers were being made, companies began to hear ominous rumblings of discontent. Employees complained of the dullness of their jobs. Companies were hit by soaring absenteeism, high turnover of jobs, increases in the number of employees drunk or stoned on the job. And managers complained of what they felt were increased pressures to make them conform. A part of this reflected the general alienation of the times, and a deflation of the expectations of achievement. Millions had become overeducated for their jobs.

To counter this mood a great many companies have begun taking seriously the experimental findings of the 1960's made by such humanistically oriented psychologists as Abraham Maslow, Warren Bennis, and Chris Argyris. Their findings were that men and women work best when they are given considerable personal control over how the jobs will be done. The great majority of people, according to these findings, enjoy being given responsibility and a chance to rise to challenges. Their work suddenly becomes more interesting as they are trusted to choose the best way to do their jobs. This has proved true whether you are a manual laborer or a middle manager.

Engineering Voter Approval 10

Our Presidential campaign is really going to be waged between two television consultants nobody knows.
— Nicholas Johnson, former commissioner, Federal Communications Commission

Mr. Johnson was perhaps being a bit dramatic in the above statement in 1972. But if he had included some other people that "nobody knows," such as professional political consultants, computer experts, advertising agency representatives, demographic specialists, public relations specialists, and consultants on communications theory, he would have been close to the truth.

The year 1968 was the year in which professional imagemakers finally appeared on center stage in presidential politics. They came forward with their concepts for the total packaging of candidates. Central to their thinking was the new immense role of television, with what they viewed as its stern technical demands.

In that same year the executive of the advertising agency that handled advertising for the victorious Republican presidential candidate (Richard M. Nixon) made an odd complaint. He felt that his agency should have had more to say. (He was from Fuller, Smith and Ross.) In the future, he said, he would insist that advertising people be "included in the very highest councils of the candidate." Perhaps he wasn't in the very highest councils but advertising men from another, bigger agency, J. Walter Thompson, were indeed in the highest council of Mr. Nixon. One Thompson vice-president, H. R. (Bob) Haldeman, of later Watergate notoriety, was Nixon's chief of staff. And Thompson's "creative supervisor," Harry Treleaven, was on Nixon's national committee. *Advertising Age* indicated that one of the Democrats' handicaps was that they had changed ad agencies in midcampaign.

In the following year the American Association of Political Consultants was formed after an international conference in Paris. The AAPC holds seminars on such subjects as the use of computers in campaigns. At one recent session a presentation on the role of the professional consultant was made by an officer of Image Dynamics, Incorporated. Members of the association include not only full-fledged political mercenaries but television consultants, advertising agency representatives, pollsters, time buyers, and political scientists (many of them for hire).

One area of research for shaping voter behavior that intrigues the pros is what might be called biopsychopolitics. Political scientists have been asserting that simple polling can be deceptive. People often don't indicate their true attitudes in verbal answers. Now, in experiments, people are being polled while rigged up to instruments measuring heart rate, blood pressure, and electrical skin conductance. They are exposed to pictures, slogans, and phrases like "farm price supports." Verbal answers to questions on racial issues, for example, were found not to reflect accurately the intensity or direction of attitudes being reflected on the machines. This happened when subjects were exposed to stimuli with a racial content.[1]

A political scientist at the University of Pennsylvania has reported finding signs that people with high energy levels tend to be more reformist. People with low energy levels tend to be more conformist. He has received grants to look into such physiological variables as the effect of a voter's turning bald on his political attitudes.

In the view of many of the pros the new technology, with its emphasis on TV, calls for theatrical skills. In 1976, for the first time in United States history, a former actor and TV pitchman became a serious presidential candidate. That was Ronald Reagan, of course. He shook the Republican Party by almost ousting the incumbent President, something that had not been done for many decades. Reagan's handler during much of his rise was Spencer, Roberts and Associates, a California firm that has been in the forefront of the new industry of candidate packaging. At the peak of the 1976 nominating campaign Stewart Spencer went to work for hard-pressed President Ford. Mr. Ford was said to be having acute image problems in running for the nomination against Reagan. Spencer may have saved the day for Ford in the nomination race, and certainly helped him almost overcome Jimmy Carter's long lead in the presidential race. The Reagan handlers included two psychologists who specialized in in-depth studies of voter behavior. According to one analyst, the good-guy image that Reagan had been well trained to project in Western movies "has a soothing effect, like a warm bath."

On the Democratic side, in 1976 Jimmy Carter had six persons in his inner circle of advisors, plus his wife. Of the six, one was an advertising man, one was a public relations specialist, one was a professional political pollster, one was a psychiatrist. His ad manager spent more than $2,000,000 on TV spots placed in reruns of such well-stereotyped shows as those of Andy Griffith and Lawrence Welk. The group they appealed to were already thoroughly established by sponsors' market research.

In the new game of winning votes the firms producing TV and radio "spot" announcements became crucial. They were so crucial in fact that one of Mr. Ford's crises came when his media specialists fell into disagreement. The dominant group had begun showing a series of TV spots in which professional actors portrayed citizens discussing Mr. Ford's accomplishments. One of his eminent media advisors, Peter H. Dailey, who had helped put Nixon across in 1972, quit in protest. He said the technique was very effective in selling soap but still unproved in selling a man.

One of the acknowledged geniuses at producing political TV spots is Tony Schwartz. He built his career as an advertising man by creating "spots" for manufactured package goods. In election years he now also sells packaged politicians by the dozen. Joe Napolitan, one of the most eminent overall political consultants with international clients, says of Schwartz's prowess: "I know at least one man who would be President of the United States today if he had followed Schwartz's suggestions soon enough."

Schwartz doesn't waste much time explaining to the public how his candidate stands on specific issues. The big challenge, he contends, is to affect the inner feelings of the voter. (In 1976 Carter was receiving this same counsel.) Schwartz revealed his strategy in a journal of the AAPC. "The goal of a media advisor," he wrote, "is to tie up the voter and deliver him to the candidate." He sees TV spots as moving posters to create for each voter "auditory and visual stimuli that can evoke a voter's deeply held feelings. Indeed, the best political commercials are similar to Rorschach patterns. They don't tell the viewer anything. They surface his feelings and provide a context for him to express his feelings. The real question in political advertising is, how to surround the voter with the proper auditory and visual stimuli to evoke the reaction you want from him, i.e., his voting for a specific candidate." Schwartz then summed it up:

"So it is really the voter who is packaged, not the candidate. The voter is surrounded by media and is dependent on them in his every day life functioning. The stimuli a candidate uses on the media thus surrounds the voter."

Schwartz takes into consideration whether the message is being heard in a car or a kitchen, and at what time of day. "A person may use radio to cheer him up in the morning, to provide company in the afternoon and to tranquilize late at night." The images of Schwartz's candidates are geared to such variables. In the 1976 presidential campaign Tony Schwartz made two dozen television commercials for the victorious candidate, Jimmy Carter. And since the election Schwartz's views on image building via television have been carefuly studied by the President's media aides.

The kind of test sampling that business marketers pioneered before launching a new product is now widely used in politics. In Nixon's fantastic 1972 triumph, his managers again were mostly advertising and marketing people. The vice-president of Market Opinion Research, for example, guided the $600,000 polling campaign. Very specific questions were asked, such as "Should the President be smiling or somber in a poster directed at youth?"

Note that I referred to Nixon as President, not candidate. A study of the campaign will show that the name Nixon was rarely seen or heard in all the promotional efforts on his behalf. His campaign was directed by "The Committee to Re-elect the President." Its office director was a former marketing man, Jeb Stuart McGruder. After the campaign McGruder wrote that the Democratic chairman, Larry O'Brien, had protested angrily: "Why don't they use Nixon's name? Are they ashamed of him?" McGruder added, "But we knew what we were doing." Their polling showed that Nixon as a person was not seen as particularly likable. But he was seen as relatively strong on being "informed, experienced, competent, safe, trained and honest."

Two words are of particular interest here. One is "safe." The Republicans successfully depicted Nixon's opponent, George McGovern, as unpredictable. They did this largely by the heavy use of a TV spot comparing McGovern to a weather vane. In this they were exploiting what Tony Schwartz calls the LOP factor. Schwartz says that often "the logical task of the media specialist is to make his candidate the Least Objectionable Politician in the race."

The other word of interest is "honest." Nixon, with expert counsel, had overcome his earlier image of being "tricky." *Advertising Age* reported that Nixon's promoters had at one point considered the idea of basing his whole campaign on the theme "A good, honest President." In the campaign his image managers did encourage substantial emphasis on morality and law-and-order. Some months after his electoral sweep he became the first President of the United States ever forced to resign. The resignation came because of revelations and charges questioning his honesty, his interest in law-and-order, and his morality.

In that same year, the year of Nixon's Sweep, the *Journal of Applied Psychology* carried a long article titled "The Image of Political Candidates."[2] The author stressed the importance of projecting an image of benevolence and humility. He suggested that when polls were being conducted political pros use personality inventories to get a better reading on how potential candidates stack up. Another possibility, he wrote, would be to send prospective candidates to workshops in benevolence and humility training. Or, he added, the political managers could, for candidates, "select those individuals who already have this skill such as actors who have characteristically played the role of the unassuming good guy."

Victory in modern U.S. politics requires consummate skill in putting together winning combinations of demographic blocs. Special attention is given to ethnic and religious combinations. Nixon was so aware of this that in at least one case he personally decided in advance what voting blocs would be represented among the little girls who would hand flowers to Mrs. Nixon.

Playing the blocs like an organ became possible with computer technology. You punch keys on a computer, which rolls out the names of potential voters in the blocs desired.

One Democratic state chairman attributed Nixon's 1972 victory less to image building than the adroitness of his aides in computer politicking. Jeb McGruder boasted in his memoir, *An American Life,* that his Nixon committee had made use of "an extensive, sophisticated, entirely computerized operation" in singling out prospects for mailings. The mailing list was broken down into many dozens of groups. He said that his experts could "by pushing the right buttons, mail to middle-aged black dentists, or to the presidents of small Midwestern colleges or whatever." The message would be tailored to that group's interests.

During the 1976 campaign I received a seemingly personal letter from President and Candidate Gerald Ford, who was seeking political funds. My name appeared four times in the letter, which is laying it on a bit thick. Microscopic examination showed that only the portions carrying my name were typed — presumably by automation-controlled typewriters. The rest was expertly printed. I assume I was sent the letter because my house in Connecticut is on the border of a census tract that is high in affluent conservatives. The names of people by census tract are for sale.

To return to 1972, in theory Nixon's Democratic opponent, Mr. McGovern, should have had the best of it in computer politicking. At the time, the biggest political data bank on voters ever assembled was owned by a company catering only to Democrats. This data bank had useful information on forty million Democrats and Independents in many states. The firm that owned it, however, had not taken on McGovern as a customer. This omission was quite probably due to the firm's close ties to Hubert Humphrey. Senator Humphrey had been outmaneuvered by McGovern enthusiasts in the bitterly fought 1972 Democratic convention. This data bank firm, Valentine, Sherman and Associates, is located in Minneapolis, Humphrey territory. Valentine was trained in political science. Sherman is Humphrey's former press secretary.

This company's computer doesn't just send out letters tailored to target blocs. It can send out individually typed and personally addressed letters. It can make thousands of telephone calls saying, "Hello, this is Hubert Humphrey on a recorded message. I'd just like a moment of your time to talk about . . ." What the candidate talks about depends on the bloc identity of the person being called. The computer can also send out standard letters with different paragraphs for women and men. It can print out a "walking list" of a district that will enable a canvasser to reach each house and in the proper order. It can run off a list of people in a district who are undecided or who haven't registered.

With the help of the Valentine, Sherman and Associates' computer, Senator Howard Cannon of Nevada won a lopsided victory in 1972. In his previous election he had won by a frighteningly thin margin. For Cannon, the computer was programmed with information about every chunk of federal money and every federal project that the senator had managed to get for each town, neighborhood, and county in Nevada. Also programmed was information on the location of every voter in the state. These two categories of information were put together, by the computer, to determine what message went to which voter. All together about twelve thousand types of mailing came out of the computer. If a letter included a paragraph reminding the voter that Cannon had gotten funds for the local airport, the paragraph was deleted if the recipient lived in a house that was under a flight pattern![3]

By 1976 a large number of computer firms were competing for political clients. Carter called himself a peanut farmer but he and his aides made astute use of computers. As governor of Georgia he had many times been asked to address national or regional meetings held in the state. His only "price" for a speech was a list of all the people who would hear him. And their addresses. Thus, as a start toward running for the presidency he developed a computerized list of fifty thousand people he could write to as a possible friend. These names were added to the more conventional mailing lists he had had programmed into the computer. These included carefully chosen lists of subscribers to magazines or former magazines. In soliciting advertising, the magazines had already determined quite precisely the demographic and life-style characteristics of their readers.

In a society, like ours, where anonymity is growing and where presidential candidates are advised to wear bulletproof vests, it is probably inevitable that the political process will become more mechanized and juiceless. And this may help account for the growing absenteeism at the polls. It may be depressing to see so much preoccupation with image-making, packaging, and computerized bloc building. But it is not necessarily all deplorable, considering the modern condition. The computer that gives a canvasser a "walking list" makes sense. And the TV Eye, even in the hands of imagemakers, can be revealing of character.

Perhaps it is degrading the political process to put personality above issues. Certainly we want to know a candidate's general orientation on issues that concern us. But in a society of strangers we are, in truth, primarily trying to get a feel for the candidate. We are trying to decide if he is competent, honest, compassionate, and steadfast enough to handle whatever new kinds of problems will be arising. As they do.

But with the new expertise in packaging candidates, we need more than ever to be alert to manipulation and phoniness.

THE NEW SCIENCE OF ENGINEERING JURY VOTES

Trial lawyers have long sought by hunch to spot and challenge prospective jurors who might take a dim view of their client. And of course they have sought any juror who might be sympathetic. In the 1970's lawyers for the first time began bringing in sociologists, psychologists, market researchers, computer specialists, medical hypnotists, and other people-experts to help them assess jury prospects. Jury stacking came closer to becoming a science.

Scientific jury stacking got its start, interestingly, in celebrated cases involving people who felt oppressed. The cases involved protestors, civil liberties, minorities, the right to a fair trial. It was done almost entirely at first by the defense. The defendants were known to the public as the Harrisburg Seven, the Camden Twenty-Eight, Joan Little, the Wounded Knee Indian Rebels, the Gainesville Eight, Angela Davis. In the Davis case, handwriting experts analyzed the signatures of prospective jurors for relevant character traits.

The first major effort at scientific jury selection occurred in 1971 in the Harrisburg Seven case. Perhaps the most famous of the seven Catholic radicals was Daniel Berrigan. The Seven were charged with conspiracy because of their actions in protesting the Vietnam war. A social psychologist from Columbia University, Richard Christie, and Jay Schulman, a New York sociologist, assisted the defense.

Their first move was to force a redrawing of the entire jury list. They produced evidence, after making a survey, that the list was overloaded with older people, and thus probably was not a cross section of Harrisburg, Pennsylvania. Elders, they surmised, might be more inclined to have fixed law-and-order views, and so be unsympathetic to protestors in general.

Then they set out to identify the kinds of people in Harrisburg who might be sympathetic to the defense. They interviewed in depth 252 people scientifically selected as representatives of Harrisburg. First they noted a large number of demographic facts about each individual. They also inquired about reading habits. Then they asked questions about attitudes, such as trust in government, acceptance of antiwar protests. With this mass of data they were able to make a number of correlations between demographic characteristics and the attitudes of people in the Harrisburg area. This provided the lawyers, at jury-selection time, with guidelines for questioning prospective jurors. The idea was to determine if they would probably be hostile or sympathetic.

The jury chosen voted 8 to 2 for acquittal on the principal charge of conspiracy, and the case was dropped as a mistrial.

Soon Schulman and Christie found themselves being called in as consultants on other protest trials. They discovered that they could not assume the same correlations they had found in Harrisburg. For example, women as a group proved to be less dependably sympathetic at a protest trial in Gainesville, Florida, than they had been in Harrisburg.

At the trial in St. Paul, Minnesota, of Indian militants Russell Banks and Dennis Means, the two scientists used a computer to work out the correlations. The two defendants were leaders of the American Indian Movement that had occupied Wounded Knee, South Dakota. The computer came up with the best "predictor variables" for people in St. Paul, where the trial was being held. People of Norwegian or Germanic backgrounds, the computer indicated, were more likely than average to feel that the Indian defendants should be dealt with severely. College-educated people in general were likely to be more lenient toward the Indians than were people with less education.

Before the trial, researchers also circulated in the neighborhoods of people in the jury pool. The aim was to pick up information from neighbors, co-workers, acquaintances. They discovered that one prospective juror, who looked acceptable otherwise, had assaulted an Indian several years earlier. He was challenged and dismissed.[4]

At the voir dire to select the jury the defense lawyers had a team of ten advisors, including a body-language specialist, an Indian psychologist, and a tribal medicine man. Their task was to watch each prospective juror and observe on how he or she chatted, dressed, answered probing questions.

In the celebrated, inflammatory case of Joan Little the defense had a $325,000 defense fund. They used much of the fund to develop a panel of scientific experts that again included Christie and Schulman. A lot of things were felt to be on trial besides Joan Little: sexism, racism, Southern juries, prison conditions in rural eastern North Carolina.

Prisoner Joan Little, a black, had killed her night jailer, a white. An ice pick was the weapon. Eleven blows. She disappeared. The jailer's body was found naked from the waist down with traces of recent sexual activity. The state concluded that she had tempted him with sex, then killed him and escaped. The defense claimed that she killed him in self-defense against rape and ran for her life. It would claim that the jailer could not have been in her cell that night for any legitimate reason.

The defense demonstrated with scientific polls and surveys that she probably couldn't get a fair trial in Beauford County, where the episode occurred. Most of the residents there already considered her guilty. And the jury system did not produce a cross section of the county. The trial was shifted to Raleigh.

Here Schulman and Christie went to work. Their computer came up with twenty-three factors that would incline people in the Raleigh area to be for or against the defendant. Schulman and Christie were learning all the time. Three factors in addition to demographics seemed to take on increased significance:[5]

1. The body language of prospective jurors. A body-language expert from the University of Nebraska, David Suggs, was sizing up each prospective juror. Did the prospect fidget more when questioned by the prosecution or by the defense? From which did the prospect avert his eyes? Did he seem to be relaxed or stiff when questioned?

2. The personality of the prospective juror. For the kind of trial at hand, Christie and Schulman concluded that a person who had a compassionate personality was a good prospect. A person with an authoritarian personality was seen as a bad apple, regardless of other traits. People with authoritarian personalities are not usually powerful persons themselves but have a strong respect for authority and seek to please those in authority and to agree with their contentions.

3. Reading habits. Christie was quoted by Edwin Tivnan as expressing surprise at discovering, in the Joan Little case, how important magazine-reading habits were. "People can start out well," he explained. "Their education is right — some college. Their age is right — under forty-five. Their residence is right — urban." But, he said, it turns out they read the wrong magazines. In the Little case, *Sports Illustrated* seemed to be an example of a wrong magazine. On the other hand, he said, everything could be wrong — low level of education and living in the boondocks — but if the prospect read national newsmagazines or *Harper's* or the *Atlantic* the evidence indicated that he would be a better juror for the defense.

Miss Little was acquitted. Whether the extensive scientific jury screening was a factor is a matter for speculation.

An altogether different approach to getting the kind of jury you want occurred in a case related to the Watergate scandal. There were two defendants. One was the formidable John Mitchell, former U.S. Attorney General and also former chairman of the Committee to Re-elect the President (Nixon). The other was Maurice Stans, former U.S. Secretary of Commerce and the treasurer of the same Committee to Re-elect the President. Mitchell and Stans were charged with conspiring to impede a federal investigation of a fugitive financier, Robert L. Vesco. They did so, it was alleged, in return for a secret $200,000 contribution by Vesco to Mr. Nixon's campaign fund.

For guidance in jury selection, the Mitchell forces hired not social psychologists but an advertising market researcher, Marty Herbst. His specialty was media analysis. This involves studying the readership and broadcast listening habits of people that advertisers want to reach.

To start, Herbst, like Christie and Schulman, made an extensive survey to find the correlations between demographics and states of mind that might favor a defense in New York City. His ideal jurors turned out to be drastically different from those sought by Christie and Schulman. Herbst's ideal jurors would respect authority figures, such as Mitchell and Stans. They would be relatively low in education. And they would not be particularly interested in social or national issues.

Herbst is reported to have set up a weighted point system from 0 to 6, plus or minus, for assessing prospective jurors.[6] A college graduate was −6. A Catholic was +4. A Jew was −5. A clerical worker earning between $8,000 and $10,000 a year was +4. In the report cited, Herbst explained:

"Limousine liberals were to be avoided. We wanted people who were home-established, to the right, more concerned with inflation than Watergate. We looked for jurors who did not read *The New York Times* but who did read *The Daily News* (a mass newspaper)." Herbst had the defense lawyers probe not only the prospective jurors' reading habits but their TV-viewing habits. Herbst was able to surmise, from the kinds of programs they watched, whether they fitted within his profile of a desirable juror. The ideal juror loved old John Wayne movies. John Wayne was authoritative, strong, a preserver of property rights, a right-winger. Wayne in short was a man like Mitchell.

After the trial the *New York Times* carried a report that eleven of the twelve jurors fitted within the mosaic Herbst had sought to create. It might have been a total fit if one of the jurors had not gotten sick. Her alternate was a college-educated, widely read international banker. But other factors still made him a plus. He was a longtime Nixon supporter and an emphatic right-winger. Although he read the *New York Times,* he had strongly objected to its editorial policy on the Vietnam war.

Even though Herbst got the kind of jury he hoped to get, the verdict was still a chancy matter. The first sense of the jury, according to the foreman, was 8 to 4 for conviction.[7] And the banker was said to be one of the four favoring acquittal. As they argued further, questions of law and particularly what evidence was relevant became central issues. There was substantial bafflement among the jurors. The banker with his superior grasp of judicial procedures, perhaps unwittingly became the dominant force. The final verdict was all twelve votes for acquittal.

In the dozen cases I am familiar with in which large-scale research efforts have gone into jury selection all the verdicts favored the side employing the research. Conceivably, verdicts would have been the same without the efforts. Really fair-minded people of course will often vote against their inclinations.

The noted Columbia University sociologist Amitai Etzioni believes that a skilled team of social scientists can exercise up to 80 percent control over jury selection. Is this new power to manipulate a hazard? He says he can't imagine a more effective way to undermine the impartiality of juries. Certainly there is a clear potential for distorting justice.

Thus far, in the United States, scientific jury research has been used mostly to aid individual defendants. In a civil suit in West Virginia it was used by plaintiffs suing to collect damages for 650 survivors after the Buffalo Creek dam broke. They won $13,500,000.

But prosecutors can use the same techniques. As the practice becomes more commonplace, jury selection will become a more dragged-out process than it already is. And prosecutors will have access to even more sophisticated computers in government law-enforcement agencies. In 1976 the *New York Times* disclosed that in some states prosecutors have routinely been obtaining information about prospective jurors from the Internal Revenue Service. The main aim is to find out if the prospect has had any run-ins with the IRS that might leave him in an antigovernment mood. An IRS spokesman, in acknowledging that in some instances such cooperation had occurred, explained that while federal regulations forbid making tax returns available they do not prohibit answering an inquiry on whether a specific prospective juror has been investigated.

Meanwhile, in many cities detective firms are now routinely selling information on prospective jurors to trial lawyers. In San Francisco, for example, there is a firm that sells information only to defense lawyers, and another that sells only to plaintiffs. The amount of digging done varies with the significance of the case. A California businessman who has served on three juries has sued the state for $50,000, claiming that such investigations violate his basic constitutional rights.[8]

Corporations involved in lawsuits have started showing an interest in the scientific approach to jury selection. They get involved in cases where vast sums can be won or lost. Some have been offering sociologists up to $35,000 for guidance in jury screening.

Scientific jury stacking requires a great deal of money or volunteer effort. And computers can be expensive. The only clients who can afford to indulge in it are the wealthy, and groups, companies, or government agencies that can raise tens of thousands of dollars. It is not for the average poor or middle-class defendant unless she or he has a celebrated cause.

Thus it would appear that we will end up with less equality before the

Bar of Justice. This brings us back to the problem we have noted in other contexts: that of well-intended scientists who are developing new powers to manipulate Man and his institutions in ways that are not altogether fortunate.

The only situation in which justice is improved by this innovation, I believe, is in supplying data to force a change of trial site. This can obtain a fairer environment for a trial. Otherwise scientific jury selection subtracts rather than adds to the process of justice.

Now that the genie is out of the scientific vat, what can we do about it? Probably nothing can be done to prevent lawyers from bringing body-language specialists into the courtroom. And probably nothing can be done to prevent general populace surveys from obtaining clues on the types of people to seek and avoid for a specific trial.

Some overhauling of trial procedures, however, can and should be made to reduce the hazard of this new potential for stacking juries. Etzioni proposes that the traditional ban on tampering with the jury be extended to cover all out-of-courtroom investigation of individual jurors. He would make it a crime to collect data about prospective jurors, analyze their handwriting, or interview their neighbors. These seem to me to be excellent ideas.

Another corrective action, I believe, would be to reduce the number of prospective jurors who can be challenged and dismissed without cause. In the Mitchell-Stans case the defense was allowed twenty such preemptory challenges. That number certainly made it easier for the defense to get a jury that fitted a predetermined mosaic.

Perhaps the best approach of all would be to require that all factual information on prospective jurors possessed by one side be made available to the other. That might discourage the whole practice of juror analysis.

Behavior Control by the New Hypnotechnicians

11

More than 10,000 physicians and dentists, according to a Chicago investigation, have received professional instruction in the use of the technique [of hypnosis].
— *Journal of the American Medical Association*
(1966)

One of the oldest mysteries of Man is being tapped today on a large scale to induce people to behave as they would not ordinarily do. I refer to hypnosis. It is being used by the police, by advertisers, by physicians, by educators, by attorneys, by athletic coaches, by psychotherapists, by military personnel. Many of these users have taken extensive courses in hypnotism and are themselves practitioners. More often, they have been given three-day briefings and then employ professional hypnotechnicians to do the actual hypnosis.

Why and how hypnotism works is still not fully understood. But that it does work in a great many (but not all) situations is no longer seriously questioned by scientists.

There are eleven national or international associations that I know of promoting the serious use of hypnosis. Membership in most of them is limited to people with advanced degrees. Two are international centers: one in Milan, Italy, one in Uppsala, Sweden. Hypnotism is now being studied as a subject in many universities in the United States.

This is quite a rise for a black art that until recent years was left almost entirely to stage entertainers and charlatans.

When the phenomenon of hypnotism was discovered is lost in the mists of history. Many primitive peoples fell into a kind of hypnosis during their prolonged ceremonies, when drums were pounded rhythmically and chants were repeated over and over. Presumably they slept it off.

In the late 1700's Fritz Mesmer, a Viennese physician, began "mes-merizing" patients intentionally. He thought some sort of magnetic fluids created the hypnotic behavior. A half century later a Scottish physician, James Braid, dismissed magnetic fluids or currents and decided that the phenomenon was an abnormal form of sleep. He called it "hypnosis," the Greek word for sleep. He was wrong. Hypnotists usually tell subjects they are going to sleep. But the subjects don't go to sleep. They go into a wak-ing trance. Their brain waves are not those of a sleeping person but rather those of a very, very relaxed person.

A series of noted French doctors brought hypnotism, for a while, to medical respectability. Sigmund Freud toyed with it as a way to probe the unconscious. Perhaps he wasn't very good at it. Anyhow, he dis-carded it for free association.

By the twentieth century hypnotism had become pretty much left to entertainers. The new shift to serious attention started around 1958, when the American Medical Association formally approved hypnosis as a valid tool for doctors to use.

At a three-day workshop which I covered at the Statler Hilton Hotel in New York and which was sponsored by the Ethical Hypnosis Training Center (South Orange, New Jersey), sixty-nine people attended. There were fifteen psychologists, thirty doctors and dentists, seventeen nurses or doctor's assistants, and seven hypnotechnicians (some did most of the instructing). There were demonstrations. Much of the discussion was quite technical.

How can people be induced to do such strange things under hypnosis? What is it? Hypnosis involves an increased susceptibility to suggestion. In its milder form a TV commercial in which an admired person repeats key words over and over can induce some people to hasten to buy the ad-vertised product. A spellbinding orator usually employs a repetition of phrasing that builds to a crescendo. Martin Luther King's powerful speech in which he kept repeating "I have a dream . . ." may have created at the least a hypnoidal state in the minds of many listeners.

True hypnosis, however, involves trance. The subject becomes pliable because he has been persuaded to reduce greatly his organized brain activity. Input from the five senses is also greatly reduced. Awareness is narrowly focused on the voice of the self-assured hypnotist. To a large extent the subject's ego comes under the hypnotist's control. A person susceptible to hypnosis has, under hypnosis, a strong desire to please the hypnotist.

Erika Fromm, a behavioral scientist at the University of Chicago, esti-mates that 10 percent of us are highly hypnotizable, about 10 percent are highly unlikely prospects, and the rest of us are in between. People susceptible to hypnosis are by no means necessarily lame brains. Quite the contrary. They are strong in imaginative powers. Ernest Hilgard, a Stanford University psychologist and a leading investigator of hypnosis, cites evidence gathered by Josephine Hilgard that readily hypnotizable people often had imaginary playmates as children.[1]

A fairly reliable way to determine whether a person is readily hypnotizable is to note which way he turns his eyes when he is asked a question requiring thought. A good, hypnotizable subject will usually turn his eyes leftward.[2] Test yourself. Here we are back to our two contrasting brain hemispheres and which one is dominant. A person highly susceptible to hypnosis has the imagination to accept unconventional instructions. Good prospects also tend to have an above-average number of alpha (relatively slow) brain waves when they are awake.

Herbert Spiegel, a psychiatrist at the Columbia University College of Physicians and Surgeons, has developed the Hypnotic Induction Profile for rating a prospect's hypnotizability. Prospects are rated from zero to five. A high-five person will be high on readiness to trust. He can suspend critical judgment. He is high in tractability and has a high ability to concentrate. It is also easy for him to roll his eyes toward the top of his head and gradually close the lids over them.

Different hypnotists use different techniques to put a person into a trance. Some count slowly as they tell the person he is becoming more and more relaxed. Some, mostly entertainers, do the eyeball-to-eyeball thing. Some ask the subject to concentrate on a light above eye level. This quickens the time the subject's eyes will begin to feel heavy, and close. Some use deep breathing.

None of these are necessary, however. Harry Arons, who conducted the three-day workshop mentioned earlier, puts people into a trance in fairly quick stages. First, he gently suggests that the subject pick out a spot on the wall and watch it. Then he explains that the subject will soon start feeling relaxed. His arms and legs will feel heavy. His eyelids will start to droop. Gradually he will fall into a deep sleep. By the time Arons gets through telling the subject what is going to happen, the subject is already slumping and his eyelids are narrowing. When Arons sees these signs he goes into the next phase. In a monotonous tone he goes through the whole thing again, but in the present tense. When the eyelids are closed, Arons's voice shifts to a more take-charge tone. He instructs the subject to do whatever he is told to do. A hypnotized person will, if directed, see words on a blackboard where there are no words. A fully hypnotized person, if told there is no feeling in his cheek, will report that he feels nothing when his cheek is pricked by a pin.

HYPNOSIS: WHAT GOES ON?

Many psychologists have decided that hypnosis involves a special kind of social influence in the same explainable category as brainwashing. They explain the lack of response to the pinprick in the context of what they term role playing. The subject is eager to do what he is told. This desire may cause him to ignore pain.

Some years ago Martin Orne, professor of psychiatry at the University of Pennsylvania who is also a hypnotist, demonstrated that role playing is indeed a factor in hypnosis. He divided a class into two groups and offered to demonstrate hypnotism. To one group he explained that when a person becomes truly hypnotized his dominant hand becomes immobilized. The other group was given the identical explanation about the hypnotized state except that no mention was made of an immobilized hand.

Each student was then subjected to hypnosis by an outside hypnotist who did not know which students had been told about the immobilized hand. A clear pattern emerged. All the students who had been told about the dominant hand becoming immobilized did manifest an immobile hand. Those who had not been told did not.[3]

Hilgard's studies at Stanford, however, convinced him that more than role playing is involved with highly hypnotizable subjects. He had twenty of these subjects thrust their arms into ice water up to their elbows to find out what their natural level of endurance of pain was. Then he and his associates hypnotized half the subjects and told them under hypnosis that the next time they would not feel pain. The other half were told the same thing but *while they were not hypnotized.* Normal suggestion did produce some increased tolerance for pain, but not half as much as when the suggestion was made while the subjects were under hypnosis. Hilgard concluded:

"There is overwhelming evidence that some people get relief from pain by hypnotic suggestion; they are not merely withholding reports of suffering to please the hypnotist."

Studies were made at the biophysical stress section of the Wright-Patterson Aerospace Medical Laboratories, in Ohio. The goal was to find how to keep aerospace pilots functioning during emergencies of extreme heat or cold. Students were hypnotized, put in hot boxes, and told to work diligently on a task. They learned to perform up to an hour in temperatures of 140 degrees. One student said he felt cool. He saw himself as a lifeguard at a swimming pool.

In investigating this study, the psychologist Perry London found it significant that tolerance of extreme heat and extreme cold apparently was not all in the mind. There was some evidence that hypnosis actually affects bodily functions. Hypnosis helped diminish the debilitating effects of freezing temperatures on the heartbeat and shivering. London and a colleague conducted "cold box" experiments in California on subjects who were taught to hypnotize themselves. The results were generally similar.

In a study of pain tolerance made under a grant from the National Institutes of Health, hypnosis easily outperformed Valium and aspirin, narrowly outperformed morphine and acupuncture.

Hypnosis is often used to ease the pain of childbirth. In a few cases mothers have even been delivered by Caesarean section under hypnosis, without any anesthesia.

Consider another situation: How do you explain that people under hypnosis can perform feats of strength far beyond their normal capacity? This has been demonstrated hundreds of times in such tasks as weight lifting and making the body rigid.

I myself was once, under hypnosis, stretched out on three chairs and told that I was becoming as stiff as a plank. The middle chair was removed. I remained rigid. People were invited to sit or bounce on my midsection. Still I remained rigid. I recall feeling pleased with myself. It was certainly something I could not have done ordinarily.

POSTHYPNOTIC SUGGESTION

Another, and potentially dangerous, aspect of hypnosis is the ability of an expert hypnotist to influence a person's actions after the trance is passed. This smacks of the brainwashed man in the fictional *Manchurian Candidate*. On receiving a certain signal he would return to a zombie-like state.

In practice, posthypnotic suggestion is used primarily by therapists to reinforce a subject in what he already wants to do. The subject wants to be calm when he goes into a dreaded scholastic examination. He wants to have a good night's sleep. And so on. The instruction given in trance will serve as a reinforcer in producing a future conditioned reflex. Some of these reinforcements are reported to have lasted for months. But with the average subject the reinforcement is apt to wear off in a few days.

AGE REGRESSION

Highly suggestible people under hypnosis have substantially better recall of their early childhood than similar people in a nonhypnotized state. They are more likely to remember the name of their first-grade teacher. They can often add new details to traumatic, repressed events.

Some people in deep hypnosis seem able, on signal, to regress back in age until they talk and behave as five-year-olds. Actual age regression apparently can occur in highly suggestible people. However, tests made on hypnotized subjects whose early backgrounds had been investigated independently, raise questions about normal subjects. In the tests the hypnotized people enthusiastically acted as five-year-olds might and babbled out details of events, many of which had never occurred. One German-born subject relived his sixth birthday in English although he did not learn English until he was in his teens.

HYPNOSIS BY REMOTE CONTROL

The hypnotist does not need to be in the same room with the subject in order to put him into a trance. He can in fact be miles away. One way to hypnotize by remote control is by television.

The psychiatrist Herbert Spiegel demonstrated this in an experiment at Columbia. A subject known to be hypnotizable sat in a lounge chair before a TV set. Spiegel was seated before a closed-circuit TV camera four stories below. He talked to the subject just as if the subject were in the same room and put him into a trance. On another case it was a thirty-year-old male stranger. While the man was in trance Spiegel told him that his hands were locked together, and the subject found this to be true. Then Spiegel told the man that he, the subject, was now going out of the trance but that his hands would remain locked until Spiegel came up and tapped him on the head. After waiting a while for an elevator, Spiegel entered the room and found the subject normally composed except that his hands were locked firmly together. Spiegel tapped his head and the hands parted.[4]

Spiegel suggests that televised hypnosis could have a number of uses. It could be used in group therapy and mass education. (There is some evidence that simple rote learning can be enhanced by the mental relaxation that goes with hypnosis.) But Spiegel warns that the technique could have dangerous consequences if used in any way in public broadcasts. He has called for stringent controls as a safety measure.

The warning is in order. Several years ago a radio performer in England inadvertently hypnotized a portion of his listening audience. There was a furor, and a ban was placed on any demonstration of hypnosis on British radio or TV. In the United States the National Association of Broadcasters (representing the majority) has a similar ban. But a danger persists. With all the current, almost obsessive interest in relaxation and meditation, a person skilled in hypnosis could profess to show a television audience ways to achieve deep relaxation. In the process he might immobilize for several minutes a large portion of an audience numbering in the millions. Maybe it wouldn't matter. But conceivably it could be designed to occur at the outset of a national crisis. The people might even be admonished to ignore all distractions (such as interruptions to announce an emergency). How effective the admonition would be on large numbers of people is still unclear.

HIDDEN INDUCTION TECHNIQUES

It has often been said that no one can be hypnotized against his will. That, technically, is correct. But a person — or group of persons — can be hypnotized without their being aware of it.

The *Journal of the American Medical Association* has carried an article on hypnosis in which one example involved the use of unannounced hypnosis. (Parental consent may have been obtained.) The patient was a child frightened by the prospect of a tonsillectomy. The day before the operation the anesthesiologist, who had been trained in hypnotism, talked the child into a mild trance and played a game with him. Before releasing the child from the trance he mentioned that the child would feel no pain during the operation and promised to play the same game just before the operation. In the operating room the next day the child was eager to repeat the game. After putting the child into full hypnosis the anesthesiologist duly carried through and played the game. During the operation he supplemented the hypnosis with a small amount of gas.

A manual on the use of hypnosis in criminal investigation has a chapter called "An Indirect (Disguised) Technique for Inducing Hypnosis."[5]

In many situations where hypnosis seems indicated but the subject is nervous and might balk, a euphemism is used. Some doctors and dentists tell patients they are about to use "sedation" or "psychoanalgesia" or "narcosynthesis" or "controlled relaxation." Thus the patient's "consent" is obtained.

What they really use is the form of hypnosis known as "progressive relaxation." Harry Arons tells how it is induced. Care is taken not to use the words "hypnosis," "sleep," or "drowsy." (This is because, as noted previously, most people mistakenly assume that hypnosis involves sleep.) The conversation turns to the wonders that a few minutes of total relaxation can achieve in coping with the strains of life. Or the doctor may stress the importance of relaxation for the medical procedure he is about to perform. The subject is asked if he would be interested in trying it. Usually he will say yes.

Casually the physician (or a trained aide) suggests that the subject get in a comfortable position, close his eyes, and make sure his jaws are relaxed. Then the subject is taken on a verbal tour of his body, from his scalp to the tip of his toes. Is every inch of his body relaxed? By the end of the tour he is "completely immobile." The physician then proceeds to deepen the trance. At the same time the subject is assured that he is so relaxed he couldn't open his eyes even if he wanted to. Where he is taken after that depends on the objective.

"To awaken him," Arons explains, "you simply tell him that upon the count of five (or any other signal) he will open his eyes and come out of the state of progressive relaxation. . . . Your subject may never suspect for a moment that he was in hypnosis."

SELF-HYPNOSIS

Do you want to increase your self-control? Endure pain more easily? Relax your tensions? Keep your temper? Stay off cigarettes? Avoid fattening foods? Ordinary hypnosis can help many people reach such goals by posthypnotic suggestion (at least briefly) or with frequent repetitions. For long-term success do-it-yourself hypnosis may be the answer, according to a number of clinical people who are working with hypnosis.

There are a number of ways to achieve self-hypnosis. You can, for example, get yourself in the right frame of mind by practicing scalp-to-toe progressive relaxation every night for a week. Next you are taught, while under hypnosis, how to hypnotize yourself fully. Arons, for example, gives subjects the posthypnotic suggestion that they hypnotize themselves three times a day. To accomplish this they are told to get themselves into a relaxed mood and position, look at a spot on the wall, and take five deep breaths. On the fifth deep breath they are to start counting backward slowly from five to one and exhale. The counting is the signal that puts them into a trance. They maintain the trance by counting backward from fifty to one. Then they release themselves from hypnosis by counting forward from one to five. It all takes about five minutes. (Self-hypnosis is hard to sustain for more than a few minutes except by counting backward.) Once the subject learns how to do all this, Arons instructs him, under hypnosis, to write out his resolution for improvement on a card. Then just before the subject hypnotizes himself each day, he reads the resolution slowly five times. Arons offers a course in self-hypnosis for $200.

The trance of self-hypnosis seems in a number of ways similar to the transcending stage of Transcendental Meditation, which uses a secret word as a trigger. Some hypnotherapists also contend that autohypnosis contributes to the success of acupuncture, the Chinese pain-killing technique.

USES BY MEDICAL PEOPLE AND BEHAVIOR THERAPISTS

Despite the wonders of modern medicine, about six thousand Americans die each year from complications related to the use of general anesthesia during operations.[6] The evidence that hypnosis can permit milder anesthesia has attracted a great deal of medical interest. There is startling evidence, incidentally, that at least some patients who have been anesthetized with drugs unconsciously *hear* the conversations of the personnel around the operating table, especially if what the patient hears is alarming to him. Under hypnosis after their operations, a number of these patients have become agitated as they repeated, accurately, what they had heard.[7]

Physicians have been experimenting with hypnosis as an adjunct to the use of drugs in controlling the pain of cancer or arthritis. Subjects are taught self-hypnosis to help them through difficult periods. Hypnosis has also been found helpful in easing or eliminating the inevitable pain of such medical procedures as removing sutures and probing the body with instruments. Examples: proctoscopy, cystoscopy, urethroscopy, laryngoscopy.

Perhaps the most compelling argument for the use of hypnosis is its value in childbirth. The expectant mother who is trained in self-hypnosis learns to think of contractions as warm, positive experiences rather than dreadful ones. And when the pain does intrude she minimizes it by counting backward from, say, sixty to one, repeatedly. Pain reduction is not the only gain in childbirth. The first stage of labor can be shortened by a couple of hours. And the fully conscious mother is better able to cooperate with the doctor in the final stage, thereby reducing the need to manipulate the emerging infant with instruments. Also, since little or no anesthesia is required, some doctors are convinced that there is less chance of respiratory or circulatory infection in both mother and infant.

Dentists who employ hypnosis are sometimes called hypnodontists. They use mild hypnosis to relieve needle phobias or to raise the pain threshold for minor drilling or extraction.

The report of a remarkable use of hypnosis in extracting teeth from forty-nine hemophiliac patients appeared in 1973.[8] (Hemophiliacs can bleed dangerously.) Each of the patients was hypnotized and told that he had ice cubes in his mouth. Could he feel them? Yes, he could. Then he was told that when he "woke" after the extraction his gums would feel cold but there would be little or no bleeding. "Little or no bleeding" proved to be true. None of the patients needed to be hospitalized and given massive blood transfusions as they ordinarily would have.

Behavior therapists still are of several minds about using hypnosis in their work with emotionally troubled people. Hypnosis is claimed by some to have helped in relaxing patients and in reinforcing suggested alterations of behavior. Two psychologists, the authors of *A Handbook of Hypno-operant Therapy,* cite a number of case histories in which hypnosis seemed to be helpful.[9] In treating a person who wanted to stop smoking they were able to so alter the subject's sense perceptions that after he came out of trance he vomited at the odor of tobacco. They cautioned, however, that hypnosis was contraindicated in a number of situations. They advised against trying to use it with psychotics, homosexuals, or people sexually frigid because of marital discord.

POLICE, LEGAL, MILITARY USES

In writing of the notorious Moscow trials in the 1930's, the author-historian Aleksandr I. Solzhenitsyn stated: "It is reliably known that in the Thirties a school of hypnotists existed in the N.K.V.D. [Russian secret police]."

In the United States of the 1970's, police are using hypnosis, but only sparingly on the *prime* suspects of crimes. For one thing the defense might be able to make a damaging issue of it at the trial. Also, if suspects know that hypnosis is about to be tried on them, they normally resist, unless they have been led to accept the need for "progressive relaxation."

In recent years police departments have shown considerable interest in hypnosis as an investigatory tool. It is used to help witnesses, victims, and policemen recall accurately what occurred. In some cases it is used to clarify the tales of possible suspects or accessories.

The Sheriff's Office of Sedgewick County, Kansas, created an official position of hypnotist on its staff. A licensed ethical hypnotist has served as a guest instructor at the J. C. Stone Memorial Police Academy in Orlando, Florida. The Police Department of Ridgefield, New Jersey, has set up an officially approved course in hypnosis for policemen. The Los Angeles Police Department now has a special unit — known jokingly as the Svengali Squad — consisting of more than a dozen persons trained in hypnosis. As of the end of 1976, these special interrogators have worked on several dozen cases.

A leading force for training law-enforcement officers in hypnosis has been Harry Arons, already mentioned. For several years he conducted two courses a year for New Jersey detectives. In his travels to conduct his regular seminars for physicians, psychologists, and hypnotechnicians he has also taught small groups of law-enforcement officers in many cities he has visited. His classes have included police chiefs, sheriffs, detectives, prose-cutors, and many dozens of lawyers.[10] One of his assistant instructors was a former detective from the Somerville, New Jersey, police force.

In late 1974 the *New York Times* reported that hypnosis had been used on a policeman in Lakewood, New Jersey. He had seen but had been unable to recall the license number of a car in a hit-and-run accident. Under hypnosis by a physician he was able to recall four of the six num-bers. That was enough, along with the other information the police had, to locate the car.

Hypnotism apparently brought about one of the big breaks leading to the arrest of three suspects in the celebrated kidnapping of a busload of

California schoolchildren in 1976. It was reported that the bus driver, while under hypnosis, was able to recall most of the digits on the license plate of the van involved in the kidnapping.

Some suspects have confessed rather than undergo hypnosis. Others who volunteered to undergo hypnosis have been cleared. A person can lie under hypnosis, but he may give himself away by his slowness in answering questions. Under hypnosis, a subject can supply answers at a more rapid rate than when he is not in a trance. Thus, if he never hesitates in giving his replies, the police can presume that he probably is telling the truth.

When a hypnotist is brought in to chat with someone suspected of involvement in a crime, he often is not identified as such. He may simply be referred to as "Doc" or described as a memory expert, a psychic investigator, or a relaxation expert.

Arons tells of a gangling seventeen-year-old boy who was charged with a brutal mugging. He was sullen and totally uncommunicative. The police had given up on him. Then one of them remembered that the boy's mother was about to pay him a visit. In the boy's presence it was suggested that it would be a shame if she saw her son in such a nasty, up-tight mood. Why not get "Doc" Arons down to see if he could relax the kid enough so that the mother would not suffer needlessly during the visit.

Arons chatted with the boy a while, talked about the wonders of progressive relaxation. It would help the boy be more normal and cheerful with his mother. But, he said, such relaxation could be achieved only if the boy cooperated. Then he told the boy not to answer yes or no but simply make a hand motion if he was agreeable. The hand moved. Arons suggested he would relax better if he shut his eyes. He continued chatting in a relaxed tone and started "a procedure which is familiar to most hypnotists." After about ten minutes Arons began making suggestions on how the boy should behave when his mother arrived. She did deserve a little consideration. Arons recalls:

"As soon as I started talking about his mother, I noticed a change in his breathing pattern. Gradually his breathing became more excitable, his hands begin to tremble and his face becomes flushed. Suddenly he starts to sob violently, and blurts out, 'I can't stand it any more! I'll talk — I'll tell them everything.' " Apparently he did.

Arons asks: "Had I placed this boy in a hypnotic trance? If I had, he certainly was not aware of it. Whether it was hypnosis or not is not the point. I had cajoled him into a 'subjective' mental state. It helped 'break' his mental set."

Evidence obtained under hypnosis is not admissible in court. In the above case the boy supplied the evidence *after* he came out of what may have been hypnosis.

As for the interrogation of spies and prisoners of war, they are not protected by the same legal restrictions that protect criminal suspects. Some have yielded up useful information when beguiled by disguised techniques into a trance state.

Military planners have long been intrigued by the possibility of getting warriors, via some form of hypnosis, to perform with extraordinary strength and endurance in times of battle. The American military have experimented successfully with using "hypnotic couriers." The psychologist G. H. Estabrooks, a Rhodes Scholar who obtained his Ph.D. from Harvard, revealed that he was involved in preparing many such couriers during World War II.[11] Codes can be broken. Captured couriers can be tortured into revealing their messages. But a hypnotized courier is virtually unbreakable.

The way it typically worked was to tell a hypnotizable courier that he was to be sent on a routine mission to pick up a report. Then Estabrooks would put him under hypnosis and tell him that only two people in the world could hypnotize him: Estabrooks and the man at the other end to whom the courier was to report, say a Colonel Brown. Both would use the signal phrase "the moon is clear" to unlock the courier's mind and to permit him to be hypnotized again. Then, Estabrooks would give the courier the secret message verbally. By posthypnotic suggestion the secret message would be washed from the courier's memory. When he reached Colonel Brown in the Orient or wherever, he would ask for the report he was supposed to pick up. Colonel Brown would give it to him, then hypnotize him after saying "the moon is clear." Under hypnosis, the courier would deliver the secret message, and receive a secret reply. Again, by posthypnotic suggestion the message would be wiped from his memory. And he would be sent on his way home.

For at least twenty years the CIA has been testing and using many types of behavior control. Hypnosis apparently has been included, sometimes in combination with drugs. *The Control of Candy Jones* by Donald Bain, which was published in 1976, is based on the CIA's alleged combining of hypnosis and drugs. Herbert Spiegel wrote a favorable introduction.

The beautiful Candy Jones, a former model who is now a radio personality, apparently served without her conscious knowledge as a CIA courier to various nations for a number of years. Spiegel ranks her as extraordinarily high in hypnotizability, so much so that she inadvertently goes into a trance on cue, such as seeing a flickering light. And her trances can be so deep that amnesia results.

According to Bain's account, she was friendly with a CIA agent whom she had known as a medic during the war. He became an expert in mind control. During a chat with him she complained of certain ailments. He gave her shots of "vitamins." While she was under the influence of drugs and hypnosis he reportedly split her personality. The second personality,

Arlene, was a much tougher person than Candy. She was named after a childhood playmate. It was Arlene who served as courier, complete with wigs and passport. This second personality, according to the account, was discovered accidentally when Candy's husband, who was trained in hypnosis, tried to ease her acute insomnia by subjecting her to hypnosis himself.

USES IN SELLING, MINISTERING, ATHLETICS

A publication of the Advertising Research Foundation a couple of decades ago reported indications that hypnosis helped in getting opinions from people. These candid opinions could be about proposed copy, or true brand references. At least one advertising agency in the past has employed a hypnotist as a researcher.

There has been speculation that the use of adored hosts and hero figures to sell products on TV to children might be having a hypnoidal effect. This apparently is especially likely if the hero-host talks for more than a minute and looks directly at the children. In 1973 a group of women who formed Action for Children's Television succeeded in forcing the National Association of Broadcasters to ban the use of program hosts in the delivery of commercials aimed at children.

Can hypnosis make salesmen more self-confident, enthusiastic? Some organizations believe so. Personality Guidance, Inc., has used hypnosis in training insurance salesmen.

Skeptics have long assumed that there is some element of hypnosis in fundamentalist preaching, and in the techniques used by faith healers. Calvert Stein, a neurologist and a student of hypnotism, has commented: "Like ancient priest-healers, modern clergymen can accomplish similar anesthesia and healing via holy water, sacred springs, laying on of hands and other rituals — especially at renowned religious shrines." Power Publishers of South Orange, New Jersey, advertises two relevant books: *Pastoral Use of Hypnotic Techniques* and *Religious Aspects of Hypnosis*.

The same company also offers *Hypnosis in Athletics*. According to the psychologist Perry London of the University of Southern California, there are recorded cases of athletes being hypnotized before major contests. The aim was to improve their performance, and some provided evidence that performance did improve.

A paper presented in New Orleans at the fourteenth annual conference of the Association to Advance Ethical Hypnosis cited the value of hypnosis in improving athletic performance. It was asserted that hypnosis can improve concentration, coordination, motivation. Arons contends that hypnosis can profitably be applied to all sports. He dismisses the argument of some psychiatrists that there is the hazard that the hypnotized athlete will exceed safe limits. Arons says no, that we all have built-in "safety valves" that protect us. That is at least arguable.

The aspect of hypnosis that most intrigues psychologists for the future is its seeming ability to enhance motivation. London is inclined to believe that hypnosis can have a booster effect on whatever instructions are given.

Hypnosis is as yet far from being a science. Still, it is a fascinating and potentially profitable area for exploration in some fields. Its use by qualified personnel, with permission and full explanation, in many areas of medicine seems clearly to be an advance.

On the other hand its use, overtly or covertly, with criminal suspects should be banned. Professional associations should tighten their codes to outlaw any use of disguised techniques (except when there is parental permission). Posthypnotic suggestion should be confined strictly to suggestion and should not leave the subject under future control by the hypnotist. (Again, military couriers might be an exception.) Any radio or television program that potentially could have a hypnotic effect should be banned from the public airways. And hypnosis should be permitted on closed-circuit television only after an explicit announcement.

TECHNIQUES FOR
RESHAPING MAN

II

Thus far we have been exploring new methods for the direct control of human behavior. We proceed now to inspect new approaches for reshaping human development. The next four chapters will examine technologies for handling human seed and altering its nature. Then, in Chapters 16 through 20, we will inspect current efforts to engineer super people. And in Chapters 21 and 22 we will take a look at strategies for lengthening the human life-span by substituting new body parts for worn-out or malfunctioning old ones and by resetting the body's biological clocks.

Donald Fleming, a historian at Harvard, has suggested that these developments are major ingredients of a "Biological Revolution likely to be as decisive for the history of the next 150 years as the Industrial Revolution has been for the period since 1750."

12

Selling and Storing
the Seed of Man

*Artificial Insemination with Donor (A.I.D.) is
only the first breach of what has until recently been
understood to be human parenthood as a basic form
of humanity.*

— Paul Ramsey, ethicist

For many hundreds of years the male human has been fascinated by his
ejaculated semen. It was the human seed and was to be cherished. More
than a hundred years ago an Italian researcher proved that male semen
could withstand freezing. It was theoretically possible, he announced, for
men to put their semen on deposit before going into battle.

Only in recent times has the male become aware of the deflating fact
that it takes two seeds to start a baby — a male sperm and a female egg.
(Scientists refer to them as a spermatozoon and an ovum.) The female
egg had not been noticed in humans. A further deflating fact is that a single
sperm is pretty insignificant compared with the vastly larger egg. It is as
though a wee minnow were trying to enter an egg the size of a pumpkin.

Nature works in profligate, peculiar ways. A man must deposit several
hundred million sperms in a female to have a fair chance that one of them
will reach and fertilize the egg. The egg, at monthly ovulation, starts down
one of the woman's Fallopian tubes toward the womb. After male ejaculation occurs in intercourse, millions of sperms race to reach that egg. If the
timing is off — that is, if it is the wrong time of the month — there won't
be any egg in the tube and the race will have been in vain. It is a long trip
for the tiny sperms, equivalent to at least a ten-mile jog for humans. And a
jogger knows where he is going. Some tens of millions of the sperms don't
seem to know which way to go. Other millions don't bother to try to make
the trip. Those that do must swim through fluid in the female's vagina and
cervix into the womb (uterus). From there they must find the entrance to a

Fallopian tube, and take a chance that it contains an egg. If a *single* sperm reaches the egg and manages to penetrate it, the egg becomes fertilized. A membrane quickly forms over the egg to prevent the entry of any late-arriving sperms.

The odds are that even one sperm will not make it in the couple's first try at conception. For the average not-too-well informed couple pregnancy occurs after nearly half a year of unimpeded intercourse. Intercourse is not likely to result in conception if it does not take place within two days before ovulation or one day after.

For some couples years of attempt at conception produce no results. The problem may lie with the female. She may have, for example, a clogged Fallopian tube. Or the problem may lie with the male. His ejaculated semen may be inadequate. It may not contain a sufficiently high density of mobile, upward-striving sperms. In Western nations the percentage of married couples who decide they can't conceive babies by normal sexual inter-course runs between 15 and 20 percent. In about half the cases the cause of the difficulty resides with the male, and in the other half with the female.

LESSONS FROM ANIMAL HUSBANDRY

In recent years Man has begun solving infertility problems, in part by taking advantage of technologies pioneered in animal husbandry.

It is a rare purebred bull these days that has the pleasure of mounting a cow in the usual way. A compliant, no-account cow may be kept around for him to mount; but he is deceived into spilling his semen into an artificial vagina attached to the cow. More commonly, the purebred bull rarely even sees a cow of any kind. And few purebred cows see bulls. (Whether this unnatural state of separation is disappointing to the bull and cow can only be speculated upon.) The bull is induced by the stimulation of an electro-ejaculator to deposit his semen in a container. A single ejaculation contains a few billion sperms, enough to impregnate more than one hundred cows in estrus. The quality of the sperms, however, begins deteriorating rapidly within an hour. And it is inconvenient to have a hundred cows in estrus waiting nearby for hasty artificial insemination with a syringe.

Attempts were made to store the semen by freezing, but icing was a serious problem. The breakthrough came in 1949, when it was discovered that coating the sperms with glycerol prevented icing very effectively. Researchers at the National Institute of Medical Research in England made this discovery.[1] The quick freezing of semen was perfected some years later, when it was discovered that liquid nitrogen as a freezing agent could virtually halt all molecular action.

By the 1960's sperm banking had transformed both the dairy-cattle and beef-cattle industries. I was born on a dairy farm in Bradford County, Pennsylvania, U.S.A. We had a bull. Everybody did. I was therefore startled when in the 1960's I visited one of my cousins who runs a dairy farm there. It was dark outside when I arrived. The barn was lit up. I went in. The milking machines, invented since my farmboy days, were going. The old bull pen that I remembered was gone. My cousin stood by a desk studying a thick manual. It was an ordering catalogue for frozen sperm. Prices in the catalogue seemed to run from $20 to $60 per unit. He pointed out to me bulls that he had had good luck with. His milk production per cow was definitely up.

Many millions of cattle have now been procreated by breeding with frozen semen. The American Breeders Service cites a study made of four generations of dairy cows bred with frozen semen from one bull. In the thousands of offspring there were no abnormalities that could be laid to the use of frozen semen.

SEMEN FROM HUMAN DONORS

The banking of frozen human sperm did not get under way on any substantial scale until the early 1970's. But hundreds of thousands of human females had been artificially impregnated by male sperm before the first sperm banks began operating. The aim of artificial insemination had been to solve specific fertility problems rather than to improve human breeding. The technique was practiced in secrecy but is fairly simple. The male who is to provide the semen delivers the freshly masturbated specimen to the doctor just before the ovulating female arrives at the doctor's office. She places herself with her legs in stirrups just as in a routine examination. The gynecologist mechanically spreads her vaginal walls. He inserts a long, thin tube to the entrance of the cervix, squeezes a bulb, and in go two hundred million sperms. They have a somewhat shorter distance to travel than sperms in normal intercourse.

The semen may come from the husband of the woman. Perhaps he has been unable to deliver it in the regular way because of impotence, premature ejaculation, or anatomical problems. Or perhaps he has been making deposits at a bank which freezes the semen in order to build up a good supply of sperms. This makes it possible for his wife's reproductive tract to be teeming with sperm during the two days nearest to ovulation. That at least improves the odds. If the husband's ejaculates are small but of normal sperm density, several units from storage can be combined for one assault. More commonly, the husband's semen is of normal quantity but has a low count of *mobile* sperms. It is technically possible to combine several of his ejaculates and by centrifugation produce a more thickly concentrated batch. This improves the odds, but not much. Such sperms have less chance of surviving well in freezing.

In practice, because of the problems indicated, most semen used successfully in artificial insemination comes from paid donors (AID's). A donor may be used because of the inadequacy of the husband's sperm or he may be used to avoid a known risk of creating a genetically defective child.

Married couples may consider a donor as an alternative to adoption. And as an alternative, the AID is increasingly attractive. Legalized abortion has greatly reduced the supply of adoptable white babies. There is a long wait at the adoption agencies, and the couple's suitability is intensively investigated. An alternative way to adopt a child is to make a flat purchase of a child on the black market for several thousand dollars.

Psychologically, the use of an AID may seem more attractive than adoption. The wife actually gives birth to the child. There is no nagging fear that the natural mother will change her mind and somehow show up to claim the adopted child. Also, there is less reason to struggle over the decision whether to tell the child he is not the couple's natural child. The child does have 50 percent of their genes. And in some states, New York is one, the child is legally 100 percent legitimate.

If the infertility is due to the wife's reproductive apparatus, it is possible to collect an ovulated egg from a donor, insert it in the wife's Fallopian tube, and have it fertilized through normal intercourse. This is difficult, but possible. Also, some progress is being made in freezing eggs as well as sperm. Freezing eggs successfully is much more difficult.

The practice of inseminating the woman with semen bought from a donor is becoming more and more common. College students can work part of their way through school by masturbating for the AID market. As sperm banking becomes more common, one human male may be siring hundreds of children.

Newsweek reports a case in which a doctor urged two young people to call off their plans to marry. He knew they had both been conceived by artificial insemination and that the same man had supplied the semen for both. Thus they were half sister and half brother, and their marriage would embody a new kind of incest. Wedding plans were canceled.

GROWTH OF THE HUMAN SEED BANKS

In many countries the successful use of frozen semen in animal breeding inspired research, very quietly carried on, with stored human semen. Records indicate that the first human offspring conceived by the use of stored semen was born, either in Japan or in the United States, in 1953.

By 1970 a number of the world's universities and hospitals had facilities for freezing human semen. The first commercial human sperm banks in the United States were getting into operation in 1971: Genetics Laboratories, Incorporated, with headquarters in Minneapolis; and Idant (Greek for fertilized egg), with headquarters in Manhattan. One of the founders of Genetics Laboratories was Arthur Beisang, a biologist at the University of Minnesota who had been a pioneer in working with frozen bull semen. One of Idant's advisors was J. K. Sherman, of the University of Arkansas, a pioneer in freezing human semen.

Both firms began to branch out in the early 1970's, by establishing offices in several cities. Idant announced that it was in the process of opening eight branches and had plans for twelve more. Several other enterprises entered the competition. Within three years all were in financial difficulty. Several things had gone wrong.

The people assumed to be the prime prospects did not show up in significant numbers. These were men about to undergo vasectomies. In 1971 nearly a million American men were undergoing surgery to prevent sperm from being present in their seminal fluid when they ejaculated during intercourse. It was assumed that if only 20 percent were uneasy about the possibility that they might change their minds about fatherhood later, then there were two hundred thousand excellent prospects each year for the sperm banks. Semen would be deposited "just in case." Just in case the couple decided it wanted a second daughter. Just in case the man found himself with a new wife who wanted her own children by him.

Idant promoted sperm banking as "fertility insurance." It put out a brochure entitled *Considering a Vasectomy? Consider Sperm Banking First.*

The sperm-bank operators had not researched their market. It turned out that men undergoing vasectomies weren't particularly concerned about changing their minds. Furthermore, within two years microsurgeons were reporting frequent success in reversing vasectomies.

Another thing that went wrong was that some of the commercial sperm banks made claims that certain national health organizations were not ready to accept. There was still argument about how long sperm could be frozen and remain viable. There was still skepticism that semen with a low sperm count could be made potent by combining samples through centrifuge. A third and major problem was that many gynecologists were content to keep on using their favorite hungry students as suppliers of fresh sperm. Why change? They paid a student about $25 per ejaculate. Frozen sperm by the time it was delivered from some distant bank might cost $50 per unit.

By 1976 Idant was the only surviving U.S. firm. It is expanding its clientele of doctors and has recently opened branches in Atlanta and Buffalo. Its extensive laboratory on Madison Avenue in Manhattan now operates partly as a conventional medical testing center. It does urinalyses, Pap smears, blood chemistries, and so on. The profits from these other activities have permitted the sperm banking to continue. They also underwrite research in advancing and refining sperm banking. The new medical director of Idant is an internist, Joseph Feldschuh. He and the medical advisors have all been serving without salary. His wife, Roxanne Feldschuh, supervises the day-to-day operations of the bank.

He advised me that the use of frozen sperm is now on the increase. The deposits are made not only at Idant's offices but at banks maintained by university and hospital clinics. He was confident that Idant's bank would be self-sufficient soon.

"The potential for sperm banking is enormous," he said. Now that the arguments and misjudgments are behind, there are a number of reasons to believe he may be right. To date at least fourteen hundred human babies have been fully certified as having been conceived with sperm that was previously frozen. The use of stored sperm in artificial insemination by donor offers certain clear advantages over the use of fresh sperm hastily procured to coincide with the woman's time of ovulation. For example:

- A good sperm bank can offer a vastly broader selection of donors to choose from. Idant now has more than ten thousand samples in storage. It has many more donors and far more details about the background of each donor than an individual doctor is likely to have. Thus there is an improved chance of getting a good specimen and a reasonably plausible match of child to father.
- A good sperm bank maintains a far higher level of quality control than a doctor can because it uses sophisticated microscopic methods of checking the sperm it will accept. It will not accept sperm from a student who had intercourse with his girl friend the night before. (The best sperm emerges in the first contractions of ejaculation.) And it will not accept sperm — as has happened at least once with fresh-sperm donors — from a young man who unknowingly had gonorrhea. Tests are made.
- A good sperm bank with its coding builds in far greater assurance of anonymity between the donor and recipient than a doctor using fresh sperm can normally assure. After all, the doctor and perhaps the nurses on his staff know the identity of both donor and recipient.
- If samples of sperm have been deposited at the bank repeatedly by a husband or a donor, the chances of impregnation are improved. This is not because the stored sperms are more lively. Actually, they tend to be a few percentage points less lively. But a far greater amount of sperm can be introduced into the woman around her period of ovulation. In its

research, furthermore, Idant has finally been having success in concentrating many samples from a normally infertile male to achieve a vial of semen that has a good chance of producing fertilization.

A VISIT TO A SPERM BANK

Because of the unrealistic expectations of Idant's early financial promoters, the company has a fashionable address in midtown Manhattan. The sixty-foot-long reception area is grandiose and futuristic, with avant-garde wall coverings. Through a glass partition the laboratory area can be seen, where a number of white-robed technicians are at work.

Roxanne Feldschuh, who greeted me, is a youngish, attractive, matter-of-fact woman. First, she showed me the room that happened to be nearest to us. It is an ejaculatorium, known to some of the staff as the playroom. Some donors prefer to masturbate at home into a little brown bottle and bring it to the laboratory. Idant prefers that they use the ejaculatorium in order to assure the freshest possible sample. The room is soundproof and has comfortable lounge chairs and an assortment of erotic picture magazines such as *Playboy* and *Penthouse* to stimulate the imagination of the masturbators. The average donor is in the room for ten to fifteen minutes. If the strange surroundings make him self-conscious (and unsuccessful), he is casually told to collect the sample at home.

Many of the prospective donors coming to the laboratory leave unpaid. Their sperm is examined and found lacking. In that case Idant gently advises them that Idant deals only with "the superfertile range." Interviews knock out many more prospective donors. Perhaps a fifth of the prospective donors make the Idant list of acceptables. And even they get paid only if each sample of sperm submitted is up to Idant's level of quality. Sperm that is subjected to marked changes of temperature can become sluggish. If the donor is addicted to taking hot showers or wears tight underwear that pulls the testicles up against the warmer body, quality is likely to fall off. If the donor hasn't been obeying the rules about abstinence before delivery, this fact will show up in the sperm check and the sample will be rejected.

While a healthy well-endowed male is likely to have a billion sperms in his body, he ordinarily ejaculates only a portion of his supply. The average ejaculate can be contained in a teaspoon. Some men, much pleased with themselves, turn in a supersized specimen. Many are surprised when they are told that their semen is unacceptable because of its low concentration of sperms. There appears to be no correlation, incidentally, between penis size and the amount or quality of the ejaculate.

A high proportion of the donors at Idant are graduate students, many from Columbia University, some with Ph.D.'s. Idant occasionally runs ads for donors in the Columbia University newspaper.

"The average male," Roxanne Feldschuh said, "is not a good prospect for a one-shot or two-shot insemination. So many college males are on the border line. Our cut-off rate for consideration is 100,000,000 sperms per cubic centimeter."

To explain what she was talking about she led me to a television screen which showed, magnified four hundred times, a drop of semen that had been turned in forty-five minutes earlier. It was considered to be a good superfertile sample, concentrated and active.

The sperms were visible all right, shaped like wiggly long-tailed tadpoles. Most were scooting about at a great rate. Some moved very little. Others swam aimlessly. But a lot of them were swimming purposefully in a fairly straight line in various directions across the TV screen. A full 65 percent were actively moving forward. They are the kind that are esteemed.

Idant grades each sample for the percent of forward movers, and has an automatic sperm-counting machine that determines if the sample is above the cutoff rate.

If a prospective donor passes the semen quality test he is interviewed quite intensively about his medical and genetic history, going back to all four grandparents if possible. His blood type is noted, along with such physical characteristics as height, weight, hair color, eye color, complexion, body type. Racial and ethnic background is noted also, as well as religion, education, occupation, and special talents. Any drug habit or any visual signs of possible neurosis, such as hand-wringing, are watched for.

If the man is tentatively accepted for future deposits, he is assigned a code number. His name and code number go into a locked file in the medical director's office. A final test — and this is made also with future deposits — is for cryo-survival. For reasons not clearly understood, about one sample in four does not survive freezing well. Thus a small sample of every deposit is put into the freezer, frozen, and then thawed to check how well it has survived.

Semen to be frozen is drawn up into plastic straws that will be used in the insemination. These are put in tubes resembling aluminum cigar containers. The sperms are first frozen slowly to $-100°$ F., to ease the shock and then they are put in the liquid nitrogen freezers, which take the temperature down to $-321°$ F., a state so frigid that metabolic action stops and no oxygen is needed.

When orders come in from doctors, the specimens are shipped off in liquid-oxygen containers. Orders come from such distant points as Venezuela, Canada, Alaska, Arizona, and Nicaragua.

How long will frozen sperm remain potentially fertile while frozen? Human babies have been born from semen frozen for thirteen years. However, there is still an argument about how long frozen sperm remain superfertile. Idant at present keeps most of its samples for only a couple of years.

THE BEST AGE FOR DEPOSITING SEED

Some argue that taking semen from people who are in their twenties —
even if they are graduate students — is not logical. A New York psychiatrist, Willard Gaylin, has pointed out that animal inseminators wouldn't
dream of trying to get top dollar for the frozen sperm of young bulls. Too
little is known about them. It would be more logical with humans, he
suggests, to "find a sixty-year-old proven stud. First, he has lived to be
sixty, so we know he doesn't have certain congenital diseases. Second, he
has had children and grandchildren so we can see how they have come
out."

A substantial number of males retain their capacity for fertilization into
their eighties. Idant has an age cutoff of thirty-nine, simply because in
human males any possible effect of aging on the process of generation of
new sperm has not been sufficiently researched.

On the other hand, young men and women looking forward to procreation by their own seed might put their seed in storage at a relatively
early age. This was suggested by the geneticist H. Bentley Glass, and has
been spelled out by the embryologist Robert T. Francoeur. Such a procedure would offer a number of advantages. For one, it would, theoretically
anyway, reduce the incidence of certain genetic defects. It is well known
that the older a woman is, the greater the chance that she will bear a
genetically defective child. Because all the eggs a woman ever produces
are present from birth, her eggs are the same age she is. Apparently, the
older the egg, the greater the risk. Therefore, banking the younger eggs for
future use could be advantageous. In contrast, the male's sperms are
created by a manufacturing process that ordinarily continues for most of
a man's life, though the quantity and thus the fertility tend to lessen with
age.

Today more and more wives have careers. The age at which a woman
bears her first child is being pushed forward, and first childbirth may soon
commonly occur when the husband and wife are in their thirties. In addition, with more people marrying a second time, births are often occurring
when couples are in their forties.

If people at the time of marriage could put a number of samples of their
sperm and eggs in a seed bank they could face late parenthood with more
assurance. When they were ready to have their first child, they would
withdraw their banked deposits. One of the wife's eggs would be inserted
in her Fallopian tube, and the husband's freshly melted sperms would be
put into a position where they would have an easy race to fertilize that egg.

Francoeur makes the more dramatic proposal that as a regular ceremony
teenagers after puberty put a supply of their seed on deposit. Then they
could have their respective tubes for passage of egg and sperm tied off.[2]

If it develops that frozen seed banks can indeed keep the frozen seed of both males and females viable for fifteen or twenty years this would be a most provocative possibility. The girl could be induced by hormones to superovulate a dozen or more of her eggs at once. They could be recovered and frozen.

The majority of young men and girls in most Western nations now start having sexual intercourse, with or without marriage, before the age of twenty. Having one's tubes tied off does not affect the pleasures of intercourse. Some males might develop psychological hangups, at least until the practice became commonplace. But the advantage, it is suggested, would be that people would have intercourse purely for fun and to express affection. The anxiety and planning involved in contraception would be eliminated. Also, the possible physiological disturbance of constantly interrupting the woman's hormonal cycle by the use of the Pill — still not fully investigated — would no longer be a factor.

From a moral standpoint, intercourse without fear of pregnancy between unwedded people would, in such a proposed arrangement, become a matter of personal standards. There would probably be an overall increase in intercourse of, say, seventeen-year-olds. But in the past ten short years intercourse among teenagers has already become pretty common. One might even speculate that by removing the hazard of pregnancy, premarital intercourse might become less of a show-off thing — especially among boys bragging of conquest — and more a means of expressing true affection.

PSYCHOLOGICAL EFFECTS OF DONOR INSEMINATION

A number of scientists have expressed concern that artificial insemination by donor leaves scars on parent or child or both. Some have asserted that the child will grow up without any distinct sense of identity because it does not know who its father is.

A number of psychoanalysts in the 1960's saw insemination of a wife by another man's semen as striking a sharp blow at the husband's psychosexual development. One wrote, after observing five disturbed patients who had been involved in AID, that the experience could revive repressed castration anxieties. The husband is clearly singled out as the sterile one, whereas in adoption the responsibility can be left blurred. The analyst offered the sweeping conclusion: "Finally, I believe that a decision to participate in artificial donor insemination, in itself, is indicative of an emotional disturbance."[3]

Still others have assumed that "the produced child" will share many of the same stigma problems of the adopted child. One asked how parents would introduce their child created by donor insemination. Instead of saying, "This is my adopted child," would they say, "This is my artificially conceived child"? (Why not be lighthearted and call it "our AID child"?)

Granted that the child is created by scientific manipulation, I think the above concerns may be somewhat overblown. Take the "stigma" problem. Many doctors involved in donor insemination deliberately arrange that paternity be left an open question. They may urge that the married couple engage in intercourse several times during the week that artificial insemination occurs. Or if that presents difficulties, they may mix the husband's semen with the donor's semen.

In a world of increased mobility and second marriages there has — for better or worse — been a sharp drop in concern over preserving the "purity" of family bloodlines. It has in fact become fashionable for young couples to announce that they plan to adopt any children they may have because the world is already overpopulated. (Adopting also permits the young wife to continue a career without much interruption.)

As for the claim that the husband is likely to be traumatized when semen from another man is artificially introduced into his wife, several studies indicate otherwise. The husbands recognize the need of an outside contribution. And they usually see true fatherhood as involving a lot more than providing a successful specimen of semen.

The Dutch sexologist L. H. Levie investigated fifty-eight couples (or 116 spouses) who had had children by donor insertion. The reports of the couples tended to be enthusiastic. Many husbands expressed delight in their AID children. On the question, "Did the coming of the children influence your marital relations?" not a single respondent reported that the relationship had become "less good." On the other hand, forty-six reported the relationship was "closer." Levie concluded that the often-stated view that AID may endanger a marriage "is completely without foundation."[4] He qualified this slightly by saying that couples encouraged to go the AID route should be "carefully selected" in the first place. If a husband is inclined to be insecure to start with he might have a bad reaction to AID. But a perceptive gynecologist should sense this and offer the couple consolation rather than insemination. Such a husband would probably have problems with being an adequate father, no matter how the child was conceived.

ARTIFICIAL INSEMINATION AND THE LAW

No serious legal problems arise in most countries if the husband's semen is used in artificial insemination. But both in the United States and abroad the legal situation is in flux regarding donor insemination. At this writing, Russia has refused recognition of children born in this manner. In Germany, AID was legalized, but donors were given the right to know the

names of the offspring they sired and to claim to be their fathers.[5] In Great
Britain the Feversham Commission expressed severe disapproval of AID
but stopped short of recommending that it be prohibited. In the years since,
AID in Britain has increased.

A few U.S. states, notably California, New York, Georgia, and Oklahoma, recognize that children born in wedlock are presumed to be legitimate. This is especially so if the couple has been regularly engaging in
intercourse. Oklahoma, however, has insisted that papers be filed in court
in the same manner as in adoption proceedings. Other states have rejected
legislation that would legitimize the AID process. And in some states AID
is illegal. Many states still view AID children as illegitimate and thus not
entitled to inheritance.

There have been court decisions in Cook County, Illinois, and in On-
tario, Canada, declaring that AID constitutes adultery. In the Illinois case,
back in 1954, the judge came down hard. He said that artificial insemination with or without the consent of the husband was "immoral, the wife an
adulteress and the child born was a bastard." It was suggested that he was
reflecting, as sternly as he could, a church viewpoint.

In some states doctors performing AID are apprehensive about filling
in that part of the birth certificate that calls for the full name of the father.
Some duck this by referring the mother-to-be to another doctor for actual
delivery.

If the couple with an AID child divorce and get into quarrels about
support and visitation, the wrangle can become more than normally
complicated.

Until the legal questions are resolved, couples trying AID make sure to
blur the situation. They arrange to have some of the husband's sperms
swimming in the wife's reproductive tract at the same time as the donor's.

DONOR INSEMINATION AND PUBLIC POLICY

Sperm banking has in two decades revolutionized livestock breeding.
Cattle breeds in particular have been measurably improved. Thus far,
people involved in AID have not been particularly concerned with improving the human breed. The overall effect might gradually be in the
direction of improvement since virtually all donors have at least some
college education. As a group they are well above average in intelligence.
Sperm banks and other new developments that we shall shortly explore
will in a few years put a new focus on the concept of improving the human
breed. The State and powerful lobbying groups may even want to define
"desirable" or "ideal" humans to breed toward. For the moment we will
confine our attention to the use of donor seed to assist human procreation,
without reference to quality of seed.

The goal of the couple is to have a baby. For many people personal procreation remains a profound aspiration. Donor insemination offers the wife procreation and offers the husband social fatherhood. For both of them this advance in the technology of human reproduction can well be seen as a gain.

But what about society's interest? Is infertility a human defect that demands correction? One might say "Hardly," since overpopulation has become one of the world's most desperate problems.

Will society benefit by institutionalizing a brand-new form of human life, a citizen whose progeny is 50 percent anonymous? Even with illegitimate children the mother, and often close friends, know or have a good idea who the father is. Now that doctors can manipulate impregnation without lust or sexual contact being involved, should we give legal sanction to this new type of human creation?

In an era when human relationships are becoming more and more impersonal should we add to the impersonalization by creating a large scale technology for no-touch human impregnation by the seed of a stranger? A number of adults who had been told that they were adopted have been suing for the right to know the identity of their real parents. One woman said plaintively on national television: "I would like to know who I am, who I look like." At present, many of our social ills can be traced clearly to the growing dissolution of the natural family. Should we formally institutionalize techniques that further decrease the importance of the traditional family?

These questions suggest to me that society should treat donor insemination with caution. It may be good for individuals, but it could be regressive in its total social impact.

Miracles always can happen. If donor insemination is invariably accompanied by father insemination the question of legitimacy need never arise. Both the individuals' happiness and society's interest would seem to be best served by letting it go at that. In short, we need not, and should not, institutionalize a new kind of human family based on artificial insemination.

Starting Man in a Test Tube 13

All hell will break loose, politically and morally, all over the world.
— James Watson, biologist, in forecasting the reaction to the placement of laboratory-grown embryos in women's wombs

The development Watson warned us against is best known to scientists as *"in vitro* fertilization" followed by "embryo transplant." What this means is that the human egg is fertilized by a sperm *in vitro* ("in glass") in a laboratory. Then after the tiny embryo has grown for several days it is implanted in the womb of a female. There it continues growing in natural gestation.

As Watson predicted, hell has indeed broken loose:

- A distinguished obstetrician lost his job at the Columbia-Presbyterian Medical Center in New York City for going too far in helping a woman have a baby fertilized in his laboratory. A $1½ million lawsuit was brought against an official of the hospital by the woman and her husband.
- An Italian scientist who claimed to have fertilized a human embryo and to have let it keep growing in his laboratory for some weeks was asked by religious authorities to discontinue his work. He did so, at least inside Italy. Some wanted to prosecute him for murder for terminating the life of an embryo.
- A world-famous British physiologist has been severely criticized by some of his colleagues for his alleged insensitivity to the implications of what he has been doing.
- A number of experimenters, to escape the pressures to stop experimenting, have reportedly been going underground in order to continue their research.

In early July 1974, while I was attending a conference on biomedical ethics at Manhattanville College in Purchase, New York, I heard the startling news from a noted Chicago anesthesiologist that Russian scientists had brought to natural birth three infants who had begun life as a result of laboratory manipulation of human seed. This was the first actual birth I had heard of that had started in a laboratory dish. He added that according to a report he had heard previously, the Russians at one point had 218 fetuses growing which had been fertilized *in vitro*.

Two weeks later I read that England was in something of an uproar because of a seemingly casual announcement by a noted doctor. He made it in press-release form before presenting a formal paper to the British Medical Association on the general aspects of embryo research.

The doctor was Douglas Bevis, professor of obstetrics and gynecology at Leeds University Hospital. Until then he was best known as an expert on the biochemistry of the female reproductive organs during the first weeks of pregnancy.

Without amplification, the gaunt, mild-mannered Bevis announced that three babies conceived in a laboratory had been born in the natural way. The infants were then between a year and eighteen months old. They appeared to be normal. Under further probing he indicated that at least one was in England. He also intimated that he had had a hand in managing their conception and the transfer of the embryos.

Although he reported success with these three, he said that thirty others had been failures. What happened to them? Were any born as monsters or aborted as monsters? Or didn't they survive the transfer to the womb? The answers to these questions were not forthcoming.

Within a week of his announcement the Medical Research Council in Great Britain called for a halt to research with human embryos until the serious ethical problems involved could be studied. Meanwhile the editors of the *Journal of the American Medical Association* had asked for a moratorium on experimentation with *in vitro* babies-to-be. A British Sunday newspaper reportedly offered Bevis approximately $70,000 for an exclusive story on the "test-tube babies." Bevis complained that all the publicity had made it impossible for him to continue his work with embryo transfer. Many of his British colleagues berated him for not making a proper scientific report to permit evaluation of his claim. Some expressed skepticism, others outrage. Bevis said he felt impelled to withhold details in order to protect the privacy of the three mothers, but added that in due course he would publish an official report of the experiments.

The announcement by Bevis came as quite a shock to scientists in several countries who had been working on test-tube fertilization and embryo transfer. Bevis had not been counted among the front runners, although it was known that he had been working in the field. Two of his countrymen

— R. G. Edwards, a physiologist at Cambridge University, and his working partner, Patrick Steptoe, an obstetrician at Oldham General Hospital — were far better known. They had had massive funding. They had worked with hundreds of women volunteers. And their efforts had received wide attention. Steptoe made a critical and skeptical statement when Bevis announced his feat.

Edwards, Steptoe, and their associates were in a position to know the magnitude of the problem. Achieving *in vitro* fertilization and then successfully transferring the embryo to the womb is an extremely formidable enterprise. Here are some of the problems:

1. The first is to obtain the human egg that will be used. The egg must be just ready to leave the ovary and start the slow trip down a Fallopian tube. In early human experiments the egg was taken from a fresh cadaver or during a gynecological operation in which the ovary was opened. A two-step process developed by Steptoe and Edwards in the 1960's solved the problem more simply. Sex hormones were introduced into the woman. As a result a bumper crop of a dozen or more near-ripe eggs were produced in the ovary. Then Steptoe went in after the eggs by making two tiny slits through the woman's abdomen to an ovary. Into one slit went a long, slim illuminating telescope. Into the other went a slender suction device. Several eggs could be siphoned out and put in tiny laboratory dishes.

2. Even though you surround the egg with thousands of sperms, it will not be fertilized until the sperm destined to succeed has been exposed to a chemical found in the Fallopian tube of the female's reproductive tract. (This was a finding of M. C. Chang's at the Worcester Foundation for Experimental Biology in Massachusetts.)

3. Once fertilization occurs, the cells will cleave and multiply only if the nutrients are precisely right. They must be precisely right hour by hour as the biochemistry of the developing embryo changes.

4. Before the new, growing embryo can be transferred to the woman's womb with any chance of success it must be checked for chromosome normality. This requires manipulation that can cause damage.

5. The transfer itself, which also obviously involves manipulation, must be geared precisely to the time the fertilized egg would reach the womb if it had been able to go down the Fallopian tube. Normally the egg takes five days to make the trip.

6. The embryo must attach itself securely to the life-giving lining of the womb before the flushing out of menstruation occurs. This has proved to be a major hurdle.

The question arises: why take on all these challenges and hazards when Nature's Way has worked fairly well for millions of years?

Some investigators explain that they are simply searching for more knowledge about conception and that in the process they hope to obtain clues to simpler, less troublesome contraception. An ideal contraception might be an injection of antibodies that would surround the ovulated egg and prevent penetration by the sperm. The July 1976 *Proceedings of the National Academy of Sciences* indicated that animal studies support the feasibility of this approach. Protection might last a year. And apparently there would be no side effects. Investigators say that they hope also to learn how and why monsters may be created if the mother absorbs radiation or ingests certain drugs during the early weeks of embryo growth.

But some investigators assert, flat out, that they also want to grow the babies conceived in their test tubes. Edwards states: "For the first time in human history we are deliberately initiating embryonic development outside the mother with the intention of her eventually bringing the baby to full term."

One reason he cites is to relieve the suffering of infertile couples. About 5 percent of all women are unable to conceive because their Fallopian tubes are blocked. A common cause is a pelvic infection such as gonorrhea. Edwards also mentioned that the test tube may overcome the problem of the infertile husband with too low a sperm count. The long race of sperms up through the female's reproductive system is eliminated by putting with an eye dropper thousands of sperm directly in the vicinity of the egg.

Less often mentioned but often in the minds of experimenters are predetermining sex in the test tube and adding donor cells at conception to improve the genetic makeup of offspring. The latter is currently most often mentioned in relation to improving the quality of livestock.

The research in test-tube fertilization and embryo transfer is more advanced with animals than with humans.

ANIMALS CREATED IN TEST TUBES

During the 1950's and 1960's it took a long time for animal researchers to work out the nutritive substances that permitted test-tube fertilization in *some* animals. Their findings were later helpful to the experimenters working on human seed.

Much of the important animal work was done in the United States at the Jackson Laboratory in Bar Harbor, Maine; at Johns Hopkins University; and — perhaps most importantly — at the University of Pennsylvania School of Medicine. Apparently the first animal to be fertilized under glass was the rabbit. Then came the hamster and the mouse, and later the cat, guinea pig, gerbil, and rat, according to B. G. Brackett, a biochemist at the University of Pennsylvania.[1]

Brackett and his colleagues, in what they call their Womb Room, have been taking many dozens of rabbits from test-tube fertilization on through implantation into the womb. And from there the embryos have gone on to normal, natural birth. They say this has happened with rabbits and mice. Still, for every rabbit or mouse egg introduced for fertilization, only one in ten eventually ends up producing a live offspring.[2]

The eggs of large farm animals have proved much more difficult to fertilize under glass. But as we will see later, multicelled embryos taken live from farm animals can now readily be cultured artificially for two or three days and then transplanted into host mothers.

What is most distressing to critics, however, is that virtually no success has been achieved with Man's nearer biological relatives, not even with such primates as monkeys. And very little work of any kind has been done with the great apes, biologically the closest of all animals to Man. Chimpanzees are expensive. And the technique of superovulating and harvesting their eggs without damaging the animals for further experiments has proved to be exceptionally frustrating. As for other apes, there has been no success with them either.

PIONEERS IN THE LABORATORY CREATION OF POTENTIAL HUMANS

The British team of Edwards, Steptoe, and their associates is generally given the credit for achieving *in vitro* fertilization in 1969. They are given the credit primarily because of their massive documentation of the early stages of fertilization. But a number of possible or probable successes occurred before then.

As early as 1944 John Rock, the American codeveloper of the birth-control pill, attempted test-tube fertilization as a part of his research on sperm and egg behavior. In a few cases he reported observing a sperm apparently fertilizing an egg, and later seeing the egg actually dividing. M. C. Chang raised a question about this. He pointed out that if you take a human egg out of the body and leave it alone in a culture dish it will, as a part of the process of dying, show signs of dividing.[3]

In 1959, a confusing series of events began in Italy and continued in Russia. A physiologist at the University of Bologna, Daniele Petrucci, reported that after forty failures he had fertilized a human egg *in vitro* and had kept the embryo alive for several weeks artificially. (He apparently did not try to implant his creation in a mother's womb.) Many scientists were skeptical of Petrucci's claim to success. But he was taken seriously enough to be called into conference at the Vatican and requested as a good Catholic to stop his experiments. He did announce a halt, a few months later. But still later he was invited twice to Moscow to confer at length with officials of the Institute of Experimental Biology. They too were working on the test-tube fertilization of babies.[4]

The most solid claim to have beaten the British team of Edwards, Steptoe, and their colleagues could be posted by the obstetrician Landrum B. Shettles in New York. Shettles is one of the giant figures of reproductive biology. Physically he is no giant, but a wispy, amiable man of modest origins and blunt speech. In conversation he calls gonorrhea by its earthy, more familiar name. And he speaks of Edwards with something less than reverence.

Shettles quite certainly beat the British team in the follow-up of transferring to the lining of a mother's womb a human embryo that had been fertilized in a test tube. But what about the actual fertilization?

For twenty-seven years Shettles was a professor in the College of Physicians and Surgeons of Columbia University. He told me that by 1951 he had achieved *in vitro* fertilization. During the 1950's he took phase microphotographs of laboratory-created embryos. They convinced many scientists that he had succeeded in growing a multicelled human embryo in his laboratory. His pictures have been included in some seventy-five medical textbooks and were displayed, greatly enlarged, at the American Museum of Natural History.[5] The pictures have remained on display there and can be seen at a number of other museums, including the Academy of Science in Moscow. They certainly look impressive to my naïve eyes.

But some specialists were not convinced. There was the possibility that the egg could have been stimulated into cleaving and multiplying by some chance factor, without the aid of a male sperm. This is called parthenogenesis, or virgin birth. It is not miraculous in nature.

According to one report, as early as 1963 Shettles was experimentally implanting test-tube–grown embryos in women.[6] He mentioned to me that around 1968 he felt he had fully mastered implanting. He said he took a picture of the implant in the woman's womb but "kept it quiet." Implanting was already a touchy subject.

In January 1971 the British journal *Nature,* one of the world's most respected scientific journals, carried a detailed report, with photographs, by Shettles on his experience in collecting and fertilizing a human egg *in vitro* and growing it in culture until it was ready for implant. (It is ready after five days, when it has become a 100-cell cluster.) An embryo at this critical point is ball-shaped and known as a blastocyst. He said that the reason implantation did not follow was that the woman involved, from whom the egg had been taken, was recovering from an extensive gynecological operation. (He may have deliberately set up the project so that it would stop short of implantation. There was controversy among his own colleagues on the wisdom of attempting it.)

Later that year an American publication reported that Shettles had repeated the feat and then had plunged on and implanted an embryo in a woman. The implant was reportedly placed in a woman (not the natural mother) who was scheduled for a hysterectomy two days later. The operation was duly performed and the embryo appeared to be growing normally.[7] Later Shettles described the identical implanting during a symposium. An account of the symposium was published in the November 1973 issue of the *Journal of Reproductive Medicine.* He said that the embryo was "nidating properly. It consisted of several hundred cells at this point and no contraindications for continued development were discernible." The menstrual cycles of the two women involved had been synchronized with hormones. Two days, of course, do not constitute a full-blown pregnancy.

All the documentation I have assembled indicates that regardless of who gets credit for the first *in vitro* fertilization, Shettles had at least one "first." He apparently was the first researcher in the Western world to report in detail the implant in a womb of an embryo created by *in vitro* fertilization. (What the Russians were doing is perhaps another matter.)

In 1973 Shettles's pioneering got him into a mess of trouble. While still at the Columbia-Presbyterian Medical Center he sought to help an infertile woman from Florida have a baby. Her Fallopian tubes were blocked. He arranged with a surgeon to obtain one of her eggs and exposed it *in vitro* to her husband's sperm. Soon thereafter, according to Shettles, his superior at the hospital confiscated the specimen. Implantation became out of the question. The superior's explanation was that while it was all right to experiment with fertilization it was much too early to try implantation of an embryo that could grow to term and be born naturally. There had been too little work on primates to establish what the risk of malformation was. And further, he added, how do you get informed consent from a fetus? (The code of medical experimenters requires — though the wording is imprecise — that permission must be given by a human subject before experimentation can begin. Is a fetus human? If it is, at what point did it become so? Obviously, with a fetus consent can come only from the parents.)

Shettles felt impelled to resign. Subsequently, the couple involved brought suit against the hospital official on the grounds that without their consent he had deprived them of the chance to have a baby. At this writing the case is scheduled for trial soon, with Shettles as a key witness.[8]

Shettles then became director of research at the New York Fertility Research Foundation. Since July 1975 he has been continuing his experimentation at the Gifford Memorial Hospital in Randolph, Vermont, where he is chief of obstetrics and gynecology.

I heard a doctor publicly state at a conference that Shettles had "gone underground" in embryo research. When I raised the question of Shettles's activities in embryo research while he was still at the Fertility Research Foundation, the doctor murmured, "I'm not supposed to go into that." Now, Shettles advises me that he is very much involved in embryo transplant research as well as in presexing babies.

Word of his presence at the Vermont hospital has spread and the hospital has become a minor mecca for infertile women. He has been assured that he can proceed with work on *in vitro* fertilization and embryo transplant without constraint. Shortly before this book went to press, Shettles told me that he was just about ready to try to help an infertile young wife have a baby by the revolutionary new method.

What did he think of the Bevis announcement in 1974 about those three apparently normal test-tube babies? He said that from what he had seen from his own work with more than two thousand human eggs he had no doubt that Bevis, and possibly others in the world, had now "done it." Then he mused:

"If I had gone ahead back in the 1950's and 1960's I would have had the cat skinned before the current uproar."

Meanwhile, in 1973 an eight-person team at the Queen Victoria Hospital in Melbourne, Australia, implanted a very young embryo in a woman and it apparently remained viable for nine days.[9]

By that same year, the Edwards group in England had succeeded in getting eggs fertilized *in vitro* to grow in wombs, but for no longer than twenty-one days. Four months before Bevis startled the scientific world by announcing the three natural human births, the *Quarterly Review of Biology* carried this discouraged statement from the usually bold, and coolly confident Edwards:

"At present there seems no immediate chance of any serious attempt to grow human embryos well beyond the blastocyst stage through implantation *in vitro,* and any claims to have done so can clearly be dismissed."

Of course it is possible that the babies claimed by Bevis will not withstand scientific scrutiny if that becomes possible. One cautious speculation is that he may himself have been deceived. Women are sterile for reasons other than blocked tubes. One suggestion is that if he obtained his eggs by using hormones to grow a crop of near-ripe eggs in the ovaries, the hormones may somehow have overcome the three women's sterile condition. In that case, just possibly, and despite the implants, their babies could be the result of normal pregnancies following intercourse. That explanation seems to require a lot of coincidence. But only time will tell.

Two years after the unsubstantiated announcement by Bevis, the *Ob. Gyn. News,* a journal for obstetricians and gynecologists, carried this interesting headline, in June 1976: "FIRST DOCUMENTED PREGNANCY FROM IN VITRO FERTILIZATION IS REPORTED."

Guess who? Steptoe and Edwards made the announcement. They stated that after forty attempts they finally had achieved a successful implantation. The embryo lived for nine weeks before perishing. It had moved somehow from the back of the womb, where they had placed it, up into the opening of the partially clogged Fallopian tube. There, the environment might be more favorable. From now on, they said, they would implant farther down in the womb. They stated that everything was "very well documented," and that they were then working with several other women in an effort to achieve a lasting pregnancy.

TRANSPLANTING LIVESTOCK EMBRYOS

While Edwards of Cambridge was having trouble keeping an implanted human embryo alive, a nearby scientist, L. E. A. Rowson, was revolutionizing animal husbandry. To Rowson, embryo transplant (but not *in vitro–*fertilized implant) is easily mastered. A quarter century ago he began putting together a team of specialists at the estate of the Agricultural Research Council, near Cambridge.

By 1957 his group had transplanted their first *naturally fertilized* sheep's eggs from one ewe to another. This was not the first successful livestock transplant. That apparently had been done with a naturally fertilized cow egg by E. L. Willett and his associates at the University of Wisconsin in 1953.[10] Rowson's group, however, went on to achieve a number of spectacular firsts.

As I have indicated, *in vitro* fertilization of livestock eggs has proved to be extremely difficult. But to animal breeders, test-tube fertilization is of little interest. The medical justification of correcting sterility in humans is not much on the minds of livestock researchers. Rather, they are seeking ways to improve production and the breed by the straight transplant of eggs that were fertilized normally. (Nowadays, that means by the bull's semen squirted by hand into the cow.)

One purebred bull can by artificial insemination impregnate thousands of cows. But that is transmitting only 50 percent of a pure breed. If purebred cows could mass-produce eggs that were then fertilized by artificial insemination, extremely interesting possibilities would open up. Ordinarily, a cow, no matter how purebred, can produce only seven or eight calves in her lifetime. A lot of her lifetime is spent being an incubator. If her fertilized eggs, 100 percent purebred, could be transplanted to ordinary cows for incubation, she could spend her time more productively in turning out eggs. And if instead of turning out one egg at a time, as is customary, she could turn out thirty at a time, you could have a still more intriguing situation. Superovulation by hormones permits such an outpouring of eggs. One purebred cow could engender hundreds of offspring in her lifetime.

That is what is being accomplished with cows, and also with sheep and pigs, by breeders in many countries. The fertilized eggs (or embryos) are washed out and transplanted.

A major problem has been how to transport the purebred fertilized eggs for long distances; for instance, to regions where the breed may be rare. Rowson and his Cambridge group solved this in a way that startled the world of reproductive biology. First, they established in experiments with several kinds of animals that the substitute mother did not have to be the same breed as the mother supplying the fertilized eggs. Then, in 1962, they shipped eggs of purebred Border Leicester sheep by jet to South Africa. The container for shipment? The Fallopian tubes of a live rabbit. The lower ends of the tubes were plugged with cotton.

These fertilized eggs continued to grow normally during the journey. In South Africa they were removed from the rabbit and transferred surgically to the wombs of two female sheep of a quite different breed, the Dorpers. The Dorpers had been prepared by hormones to be ready for pregnancy. In due course two normal lambs of the purebred Border Leicesters were born from the Dorpers!

In 1964 an important improvement came from Japan. Experimenters at the Livestock Industry Experimental Station there developed a technique to eliminate surgery in making the transplants. They used carbon dioxide gas to distend the womb while the transplant was being made. This was done with cows.

At about the same time, an American named E. S. E. Hafez, who was carrying on research at Washington State University, reported experiments in which cows were superovulated by hormones and the large number of fertilized eggs were stored in a single rabbit. Soon Hafez was predicting that before long a whole herd of purebred cattle, in fertilized-egg form, would be flown across the Atlantic within one rabbit and would be transplanted to cows.

It remained for Rowson and his Cambridge group to bring this forecast fairly close to reality. In a 1971 experiment they washed forty-eight fertilized cow eggs out of seven heifers that had been treated with hormones. The fertilized eggs were packed in rabbits for several days. Of those eggs recovered from the rabbits, thirty-four appeared to be normal. Selected eggs from this batch were transplanted to twelve heifers whose onset of estrus would be fairly close to that of the donor heifers. Eleven calves were born.[11] Rowson has concluded that when the onset of estrus in the two heifers is exactly synchronized successful pregnancy rates can run close to 90 percent.

The idea of storing cattle eggs in rabbits is a charming idea, but is short on commercial practicality. Storage is limited to a few days. Scientists, despite frustrations, have been pushing hard to freeze eggs or embryos. In the frozen state they could be held for use indefinitely — even until after the donor had died.

In freezing embryos, as with freezing single eggs, the big problem has been to prevent the formation of ice crystals. Two major breakthroughs occurred in 1973, both at Cambridge University. The two principal investigators were Ian Wilmut, a member of Rowson's government team, and D. G. Whittingham, an associate of Edwards's at the Cambridge University Physiological Laboratory. They were able to boost the survival rate of frozen mouse embryos to almost 90 percent by using an antifreeze, dimethyl sulfoxide. They achieved this percentage on very early embryos (of eight cells or less). The freezing and thawing had to be done very slowly. More recently, Whittingham has upped the survival rate to nearly 100 percent, but has had only a moderate success in getting the unthawed embryos to grow inside wombs as far as the fetal stage (twelve weeks). The results are better when he puts the unthawed embryos into culture for a day before attempting implant.

The other breakthrough was the successful use of embryos that had been frozen for six days before implantation. In the experiment thirteen defrosted embryos were transplanted to as many cows. Two grew to term and were born — frisky little bull calves. While mouse embryos seem to survive only if a few cells are involved, the successful calf transplants were done with embryos that had grown into spheres containing several hundred cells. Long-term studies were begun to see if the use of antifreeze has any serious side effects.

So far, investigators have had little luck in freezing human embryos. But both Landrum Shettles and Joseph Feldschuh have assured me that it is feasible. Others forecast the freezing of human embryos within a very few years. Meanwhile, scientists may, for practice in keeping embryos alive, begin storing small human embryos in the Fallopian tubes of rabbits. (In the next chapter we will come to the startling possibilities that open up if human embryos can be transplanted from one woman to another.)

LEGAL ASPECTS OF STARTING HUMAN LIFE IN A TEST TUBE

The Vatican has taken the position that human life starts with the fertilization of the one-cell human egg. Anything a scientist did to terminate a test-tube experiment on a viable embryo thus would be illicit and theoretically leave the scientist open to charges of homicide. It is doubtful that any civil prosecutor would ever bring a scientist to court for terminating the sixteen-cell speck he had grown in his laboratory, even though the speck contains the blueprint for a human life.

After three months of gestation, when the embryo becomes known as a fetus and has taken on visible human form, it might well be entitled to some legal protection.

There are laws in some states against fetal experimentation. If a defective human — not to mention a monster — is born as the result of an experiment, the scientist might also be vulnerable to a damage suit from the parents. And possibly the institution funding the experiment might be vulnerable.

Edwards, who has understandably looked into the matter carefully, seems to feel moderately secure under British law. The parents bringing the suit would have to prove that something the doctor-scientist did was the direct cause of the defect. Expert witnesses presumably would come forward to testify that any risk assumed was not unreasonable. Edwards mentioned that in some countries, including the United States, the burden of proof shifts to the defendant if the case involves "ultra-hazardous activity."[12]

He also mentioned as a conceivable hazard that a plaintiff might charge that the experiment was performed without sufficient prior experiments on animals. But he discounted this by saying that "the work already carried out in animals would appear to make this argument irrelevant."

Since virtually no work has been done on nonhuman primates, that seems an astonishing position. If the customary scientific approach had been followed, at least several dozen monkeys and twenty chimpanzees would have been born as a result of test-tube fertilization before experiments with human embryos were undertaken. And once born, the monkeys and chimpanzees — and their offspring — would have been studied rigorously for abnormalities, including mental retardation.

A French commentator on the legal aspects of human-embryo transfer offered this equally astonishing opinion: "At present, the risk to the foetus is neither known nor capable of evaluation; therefore we believe that the doctor is not professionally responsible if the child is born malformed."[13]

This seems to hold that the doctor is innocent because he is ignorant.

In the United States a proposed guideline for institutions seeking federal funds from the National Institutes of Health had this to say about "Products of *in vitro* fertilization":

"No research involving implantation of human ova which have been fertilized *in vitro* shall be approved until the safety of the technique has been demonstrated as far as possible in sub-human primates, and the responsibilities of the donor and recipient 'parents' and of research institutions and personnel have been established."[14]

That is not law. But since many United States institutions involved in early-life research seek financial help from the NIH it has stimulated some prudence.

THE CONSEQUENCES OF MANIPULATING EMBRYOS

The principal argument for growing babies from test-tube fertilization in humans is the possibility that it will permit certain childless couples to have children. Some of these couples suffer acute depression.

Edwards contends that such a process "does not pose any moral problems, and the right of couples to have their own children should not be challenged." Some scientists and laymen have tended to agree. A leading Methodist official in England some years ago said this process for creating babies should be "welcomed," provided that no harm is done to the offspring.

Strong support in America has come from a leading ethicist, pink-cheeked, robust Joseph Fletcher. He is a pioneer of the situational ethics movement and believes that by and large people should make their own decisions in novel situations. "My own conclusion," he says of test-tube babies, "is that given the unknown quantities . . . the most human thing to do [that is, the best thing morally] is to explain things as they are in embryo transplant to patients and let them choose or reject the treatment, as we do in all surgical care. Medicine's forward movement is due fundamentally to investigative or experimental treatment."

Others defend test-tube experimenting and *temporary* implanting (with termination before a fetus can develop) as an important research procedure on early human gestation. Under a microscope scientists can study, in laboratory-made embryos, the beginning of genetic defects.

A major incentive influencing many of the experimenters, I suspect, is the challenge of helping create the world's first documented test-tube baby. It stands to them as a stirring goal, like being the first to transplant a heart, or climb Mount Everest, or land on the moon.

The argument most usually advanced — that of helping barren couples have children — is losing force each year. There are other ways. Japanese investigators at Nihon University are, in animal experiments, increasing the friskiness of listless male sperms by simple hormonal injection.[15] And sperm banks are learning to increase the concentration of sperms where there is a low density. As for the wife's trouble with tube blockage, some surgeons are achieving close to 60 percent success in opening blocked tubes by reconstructive surgery.

And then there is the broader viewpoint, well put by *Nature* when it editorialized: "in an overcrowded world infertility is an individual problem, not a great social issue."

Great social issues, however, do indeed arise in the mere effort to start fabricating babies. Are we to launch upon creating a new kind of human — a human incubated in a man-made broth?

And what do we create when we start freezing these embryos so that they can be stored until they are implanted in someone's wife or other suitable female? The British science writer Gerald Leach points out: "People are born, people die, people are killed; but 'people' have not yet had their life cycles suspended in deep freeze."

The imaginative and respected animal biologist E. S. E. Hafez (who forecasts herds of cattle flying across an ocean in a single rabbit) has a forecast for people. He envisions that within a matter of years "a woman will be able to buy a tiny frozen embryo, take it to the doctor, have it implanted in her uterus, carry it for nine months, and then give birth to it as though it had been conceived in her own body. The embryo would, in effect, be sold with a guarantee that the resulting baby would be free of genetic defect. The purchaser would also be told in advance the color of the baby's eyes and hair, its sex, its probable size at maturity, and its probable IQ."

Neat! But what happens to the concept of the sanctity of the individual? What happens to the concept of the family as the basic unit of civilization?

To a couple longing for a child, laboratory manipulation to achieve fertilization seems highly reasonable. It looks like progress. But it is precisely this kind of experimentation, if encouraged, that will open the gates to a host of other new technologies in human reproduction. That is why this moment of decision about *in vitro* fertilization and embryo implanting looms as so important in the history and evolution of Man.

In subsequent chapters I will deal with some of these *Brave New World* technologies that would open up shortly. I refer to the use of surrogate mothers as gestators; growing a human fetus to term entirely outside the womb; predetermining sex; prefabricating humans genetically to order; attempting to breed "superior" humans; producing exact copies of "desirable" humans; developing man-animal mixtures; and so on.

The geneticist Bentley Glass, former president of the American Association for the Advancement of Science, says of the *in vitro* technique: "It should be obvious that the technique can be quickly and widely extended." He is an enthusiast for genetic manipulation.

Test-tube experimentation with human seed is, from the scientific viewpoint, a wedge. More picturesquely, it is the camel's nose pushing under the tent. If it is permitted to go unsupervised, it can put us on a slippery slope toward the dehumanizing of humans. The biochemist Leon Kass of

Washington, D.C., calls it "a step in the long series of steps yet to be taken." He believes that the technique of test-tube fertilization of babies has a considerable potential for abuse, and perhaps challenges the very nature of human reproduction.

At the same time, it must be stressed, many of these same techniques offer the world great benefits in increasing livestock production. We can applaud work to achieve superior breeds or standardized breeds of quality or greater quantity in animals. To me, humans would seem to be quite another matter. And can we hope that scientists will experiment vigorously with the seeds of animals but will be very cautious with human seed?

A number of people who would call a halt to *in vitro* human reproductive research on ethical grounds focus on the risks. Scientists often speak of the possible "teratogenic effects" that may occur during the manipulation that goes with the fertilization, inspection, and implant of embryos. "Teratogenic effects" is a much more comfortable phrase to use than the more familiar translation: "monster-producing effects."

Some gross defects can be detected before implant or during fetal growth. But many cannot. The human body has a tendency to abort monsters, but monstrously deformed fetuses often get born naturally. If we permit the implant of *in vitro* human embryos in women, are we prepared to kill monsters, imbeciles, and hopelessly defective humans born from such research? This factor is why much research in embryo transplant has been done quietly. That apparently was the case with Bevis, who acknowledges that many attempts were required to produce three normal babies.

Mishaps certainly would occur. The risks in human implanting are only vaguely surmised. Edwards's associate, Patrick Steptoe, is reported to have said the decision to implant calls for a "brave decision." At this stage some would substitute the word *reckless* for *brave*.

In some *in vitro* experimenting we have had physicians and biologists working side by side. The biologist's commitment is primarily to seek the truth about living creatures. A physician's commitment is much more precise. He is first of all a healer, and has a strong obligation to do his best to do no harm. If he experiments he should have the informed consent of his patient. And the objective should be therapeutic, that is, to help the patient. Test-tubing is hardly therapeutic for the human-to-be. And consent is out of the question. For physicians the effort must be rationalized as therapeutic for the unhappy, consenting, childless couple.

Paul Ramsey, an amiable, pipe-puffing Princeton theologian with mutton chops, wrote two articles for the *Journal of the American Medical Association* (June 5 and 12, 1972), in which he coolly dissected for doctors the ethical problems with *in vitro* research. He based his analysis on the physicians' own professional standards and concluded that this kind of research "constitutes unethical medical experimentation on possible future human beings, and therefore is subject to absolute moral prohibition. . . . Either the accepted principles of medical ethics must give way, or fabricated babies should not be ventured."

I asked Ramsey if he had had any reactions from the researchers. He replied that at a seminar he and Edwards attended, Edwards called him "religious."

It is time, I believe, that research on *in vitro* fertilization and embryo transplant go public. I suggest this not only because of the attendant risks and ethical problems but also because the future of Man is involved. The issues are too large to permit often-ambitious individual scientists to make decisions, brave or otherwise, in the privacy of their labs on what research to attempt.

In each nation where such work is afoot, the public through its national legislature should establish a commission to regulate the whole area of human reproductive research as it unfolds. And it is unfolding rapidly. In the United States I am thinking of a regulatory agency. It should have powers comparable to those given the agencies that regulate the sale of food and drugs, control communicable disease, and regulate environmental pollutants. Some might suggest that such a commission is superfluous since the Department of Health, Education and Welfare has already established the National Commission for the Protection of Human Subjects of Biomedical and Behavioral Research. But this commission is merely an advisory, research-oriented group that has already recommended to the Secretary—in the form of a statement of the consensus of its "commentors"—that there is no need "to regulate research involving human *in vitro* fertilization." All that is needed is for proposed projects that would entail HEW funding to be cleared by the Ethical Advisory Board.

Something broader and tougher is needed, whether it is part of HEW or an independent agency. It should be a policing body, not an advisory one.

A good title for it would be the Human Reproduction Research Administration.

In every country, all research relating to human reproductive research should be registered with such a commission. In specified areas research should proceed only upon the commission's authorization.

In my view the commission should ban, at least for a period of years, all efforts to grow to natural birth a human embryo conceived in a laboratory. This would give the public a better chance to assess the hazards (as they emerge from primate research). And it would give the public time to grasp and assess the momentous social issues involved.

If eventually the ban is lifted, research projects should be handled as national projects, just as developing atomic power has been a national project. For humanity, the implications are as great as atomic power has been. The projects would be regularly monitored by the commission. It would have the authority to abort a fetus up to birth and even to destroy a grossly defective newborn baby produced by laboratory experimentation.

The various national commissions should seek as far as possible to arrive at an international consensus on whether to proceed, and if so, how. The geneticist James Watson—who predicted that "all hell will break loose"—has expressed grave concern about uncontrolled embryo research. He estimates that it is technically foreseeable that some kinds of *in vitro* manipulation of human eggs *could* become common medical practice in many major countries within a decade or so. He proposes international agreements to limit such research before "the cat is totally out of the bag." Presumably he was referring to the same cat that Shettles was going to "skin."

Such steps as national commissions and international agreements, it seems to me, are the least that humanity should expect of us.

Leased Wombs, Artificial Wombs, Nonhuman Wombs

14

If children can be had without being borne, working mothers need not be affected by childbirth. This is happy news for women.

> — *Jenmin Jih Pao,*
> newspaper of the Chinese Communist Party

The Chinese Communists were reacting to the news out of Italy that the test-tube growth of human embryos had proceeded to the heartbeat stage. The Russians were equally intrigued. Both societies have placed great emphasis on the idea that everybody — men and women alike — should engage in productive labor. To such societies any technology that offers to reduce the woman-years lost because of pregnancy is interesting news indeed. In Russia, for example, more than 80 percent of all women of working age have jobs.

Aldous Huxley's fantasy of growing babies in central hatcheries is technically almost halfway feasible already.

Meanwhile, another technique emerging from embryo research could soon make possible the use of substitute babybearers. This technique might prove to be more appealing to liberated Western women. The Communists may be less interested in it because it does not relieve the problem of lost woman-years. Still, there may be high cost factors of the mechanical incubators to consider.

If the techniques are permitted to proceed it would be interesting to speculate which system, by the year 2000, would offer the cheaper product (that is, baby). And which system would offer the higher-quality product. (Again, baby.) Using apes or cows to gestate humans-to-be might turn out to be the cheapest of all.

Let's look at the more technically feasible procedure first, the hiring of substitute mothers. This technique would push forward nine months the old profession of wet-nursing. That once-popular profession was made virtually obsolete around 1925. What made it obsolete? The development, after sixty years of frustrating experimentation, of an artificial breast, the baby bottle, and a tolerable substitute for mother's milk.

The obstetrician Landrum B. Shettles, who was singed in the controversy over laboratory-created babies, suggests that hatching babies in artificial wombs should be put on the back burner. The emphasis for the present, he told an investigator, should be to attempt straight woman-to-woman transplant.[1]

Worked in one direction — transplant from substitute mother to wife — this could give an infertile woman a chance to become pregnant and to bear a baby. Such an approach would bypass the explosive issue of manipulating human seed in a broth, and the increased risks of mishap. (Transplanting in itself, of course, can involve risk.) Shettles recently advised me that he hopes to make a transplant of an embryo from one sister to another.

Ethically, this procedure might raise as many questions for the married couple as test-tube fertilizing. In test-tubing, the couple's own sperm and egg can still be used. But transfer of the embryo necessitates Another Woman for early gestation, an aspect of the procedure that some find bothersome. She would be fertilized by artificial insemination from the husband. Once the embryo proved to be viable, it would be transferred to the wife's womb.

But the transfer could work in the opposite direction. And there might be more public interest in this alternative. The wife would be normally impregnated by her husband and a viable embryo would start to grow in her womb. There might be any number of medical reasons why it would be risky or impossible for the wife to try to carry the baby to term. She might have a discouraging history of miscarriages. She might be frail. And so a transfer would be made to a substitute mother.

Or the reasons might be nonmedical. Perhaps she has a career that is important to her and interrupting it to have a baby might create problems. Perhaps she is an actress or a TV personality or runs her own business, and doesn't want the bother of childbirth and early nurture. Perhaps she doesn't want to "lose" her figure or appear in public in a swollen condition. Perhaps she is not a woman who would enjoy pregnancy but rather would find morning sickness, anxiety, or frequent medical inspections a tribulation. Or perhaps she doesn't want pregnancy to interfere with skiing, tennis, or dancing.

While straight transplant in humans — womb to womb — is a less formidable challenge than implanting an embryo fertilized in a test tube, it is still a challenge. A near success was reported by two British doctors in the April 24, 1976, issue of the *Lancet*. Certainly it is not something that will be widely practiced with humans this year or next.

Transplant has been frequently achieved with livestock, but it was only in 1976 that a report appeared of a successful transplant in primates. This occurred at the Southwest Foundation for Research and Education at San Antonio, Texas. The transplant was between two female baboons. They had ovulated on the same day. The transplant occurred five days after conception. The embryo had descended to the womb but still had not implanted itself firmly in the lining of the womb of the natural mother. A normal male called Primero was born from the transplanted embryo. It came after ten tries.

The success rate with calves has been running about 50 percent when transplant is through the cervix. With humans, as with animals, synchronization of the two menstrual cycles is a major problem. Research with animals suggests that hormones can help bring a close synchronization. Temporary storage in a rabbit — or better still the temporary deep-freezing of the waiting embryo — might also help achieve perfect synchronization. As indicated, deep-freezing human embryos is still a challenge.

For the present, the best bet for the wife who wants a baby immediately without the bother of pregnancy (or adoption) would leave her completely out of the prenatal picture. A woman hired to bear the baby would be artificially inseminated by sperm supplied by the wife's husband.

THE COST OF LEASING WOMBS

But would any other female consent to be a human incubator? That is, spend three quarters of a year becoming blown up, and then go through labor and delivery, and maybe nurse the baby for a few weeks — all in order to provide some other woman with a baby?

Or, let's assume that the transplant of an embryo from wife to substitute does become feasible on a wide scale within the next few years. Would there be interested couples and interested substitute childbearers?

The answer to both questions seems to be "certainly," if the price is right.

As for women willing to lease their wombs, the gray market in white Caucasian babies offers a clue. A professor at a Houston medical school told me that some young women in Texas bear children with the deliberate intent of selling them to the adoption gray market. The mother's income per baby can be as high as $20,000. The CBS television network carried a report that a "baby farm" was operating northwest of Miami. The operators get up to $15,000 per baby. Sheila K. Johnson, an anthropologist, wrote in the *New York Times* that a number of lawyers were reported to be charging couples $10,000 to $25,000 for a healthy Caucasian baby. A former county crime commissioner was accused of running an international gray market in babies that promised to net him more than $3,000,000 in two years. The babybearers came for the most part from Yugoslavia and other Eastern European countries. They were given a round-trip holiday to a Caribbean island to have their babies, free medical service, token work, and about $3,000. Some of them have become pregnant accidentally, some deliberately.

As for interested clients, at least two couples unable to bear children because of the wife's infertility have advertised for "babymakers" in the *San Francisco Chronicle*. The woman to be chosen was expected to undergo artificial insemination with the husband's sperm and carry the baby to term. The price one couple offered was $10,000.[2]

In February 1977 several Michigan college newspapers carried an advertisement placed by "Al and Betty," a husband and wife. Betty was prone to miscarriage. They sought a "donor" who would carry and deliver a child after being artificially inseminated by Al's semen. A number of young women responded. Most asked fees in the neighborhood of $5,000 — just enough perhaps to cover all their college expenses for a year.

The newscaster Lowell Thomas told of a Los Angeles man who advertised in San Francisco for a woman who would undergo artificial insemination and bear his baby for $10,000. He was advertising in San Francisco, it was explained, because the going price in Los Angeles was around $20,000. Prices as high as $50,000 have reportedly been paid for custom-created babies in Los Angeles. One childless couple who paid $50,000 selected both the male inseminator and the female bearer from a collection of photographs of attractive young single men and women presented to them by their lawyer.

The traffic in babies in California has become so lucrative that law-enforcement officials suspect that members of the Mafia will enter the picture, if they have not already done so. The *New York Times* reports that one woman gave her baby to a baby broker for a used car.

In England the actor Richard Burton gave an English magazine a handwritten ad for a woman under thirty-eight who would bear him a child on a contingency-fee basis: $25,000 for a girl; $50,000 for a boy.[3]

In any leasing arrangement in which the woman would supply 50 percent of the genetic makeup she presumably would be carefully appraised. When she would simply be the bearer of an already-formed embryo, a reasonable price, according to the British embryologist Jack Cohen, might be around two thousand pounds.

The price of obtaining babies on the gray market has been pushed up in some U.S. states by laws requiring that all adoptions be handled through licensed agencies. There seem to be few laws as yet that would apply when a couple arranges specifically beforehand with another woman to help the wife have a child. The situation is still too new to inspire any law drafting (that I know of).

The high prices cited have also been influenced by the extreme shortage of adoptable illegitimate babies. In the United States that often means healthy babies of the same skin color as the adopters. The great increase in abortions, as a result of the drop in legal prohibitions, has shrunk the supply. Also, improved contraception has helped create scarcity. And the number of young unwed mothers who elect to keep their babies has recently and signally increased. The difficulties of adoption are further complicated by religion. In at least eleven states, adoption agencies are required to insist that the adopting couple be of the same religion as the baby's mother. And a childless couple seeking to adopt must often not only be investigated extensively but may have to wait from two to seven years for a baby to be offered to them. All this drives up the price childless couples are willing to pay on the independent market. And some lawyers specializing in baby sales push up the price still further by taking a large percentage of the fee for themselves.

An unwed girl already pregnant is not in as good a bargaining position as a prospective host mother, wed or unwed, would be. A professional mother-for-hire would regard carrying a child nine months for someone else as strictly a job. The someone would ordinarily be a married couple. But it could be a widow, a lesbian, a divorcee — or Richard Burton.

If the use of a substitute becomes a common, allowable procedure, a whole new occupational category may open up, one that would apply whether the woman becomes pregnant by transplant or by artificial insemination. What occupational labels will national departments of labor use for women performing this service? "Surrogate mother" offers the appeal of dignity and is a term that scientists tend to favor. "Contractual babybearer," "gestational mother," and "host mother" have also been suggested and have merit. "Mercenary mother" and "hired mother" have the merit of factualness. "Hired womb" and "walking incubator" might crop up among slang users.

The occupation would provide young women with a reasonably undemanding career. It might prove especially appealing to young widows or divorcees who have already borne a child. They would have proof of capability and would tend to be less anxious. It would help if the hired mother was of an easygoing nature and enjoyed pregnancy and TV-watching. There would presumably be some requirements imposed, such as robust health, no heavy drinking, no smoking, no drugs. The babies of heavy smokers tend to be small. The employer would probably bar vigorous athletics. If the hired mother were contributing 50 percent of the baby-to-be's heredity, there would probably be a lot of extra requirements, such as no previous history of the continuous use of LSD. (There have been assertions that this drug can affect genetic makeup.)

These hired mothers would be expected to get plenty of sleep and eat plenty of prescribed, nutritionally rich foods. They would probably be expected to curtail sexual intercourse, at least during the late phase of pregnancy. But otherwise they could lead a reasonably normal social life. There would be no time clocks to punch. They would no doubt be free to take on extra jobs for six months or more as long as the jobs were not physically demanding. They might, for example, make extra money selling tickets at movie theaters.

That is pretty much the job description for mercenary mothers. The pay? Competition, as with all careers, would tend to fix the price. As the career became recognized as a nonsinful service, any stigma attached to it would decrease. In the United States there probably would be plenty of applicants if the assignment offered $8,000, plus medical expenses and an attractive insurance policy. If the hired mother's egg — and thus genetic makeup — were involved in the conception extra requirements would be imposed. There would probably be an expectation of, say, an additional $2,500. Such an arrangement is already fully feasible.

But let us stay, for a moment, with the prospect of a transplant of an already-conceived embryo from mother to mercenary. There is no reason why that job should require much in the way of education, family background, good looks, or even skin color. If the woman is simply to be an incubator, the price would certainly be lower than if she contributed half the baby's heredity.

Undoubtedly, foreign competition would arise. I have met pleasant, conscientious Mexican-American girls in south Texas who might leap at such an assignment for $5,000. And I have observed girls south of the border who would leap at $2,500. The border might be a complication. The baby might be born in a Mexican hospital. There are some excellent Mexican hospitals for those who can pay the price. If lawyers can arrange Mexican divorces for Americans, they surely can arrange Mexican gestations.

The main advantage of using surrogate mothers over trying to adopt is that, however the child is registered at birth, it would be either 100 percent or 50 percent the married couple's own child genetically. Birth registration as it is now done could be a problem in many countries. The custom is to assume that the woman giving birth to the child is its mother. But doubtless pressure would build to issue birth certificates that show the name of the genetic mother as well as that of the gestational one. Or if there is documentary evidence that a transplant occurred — and there surely would be a contract — then it might become possible just to list the genetic mother.

CUTTING COSTS WITH NONHUMAN WOMBS

But what if the gestational mother was not human? That might still be awkward for the birth registration. We have noted the evidence that cow embryos have thrived for several days in a rabbit. Could human embryos thrive and grow inside a nonhuman animal? Impossible to contemplate? Maybe uncomfortable, but not impossible. The embryologist Robert T. Francoeur doesn't think it inconceivable. He has served as an advisor to the American Medical Association on reproductive technology, and has written:

"How is the fetus likely to respond to the mercenary mother, especially if his incubator is of a bovine or simian species? Would he recognize the substitution and subconsciously react, biologically, hormonally, or biochemically to the foreign womb? How is his human body and mind likely to accommodate itself to the hormonal differences of nine months in the womb of another species?"[4]

In practice the fetus would presumably be transferred to an incubator well ahead of term. Francoeur also speculates on the possible psychological complications if a cow-gestated child knew he had spent nine months in a pasture while his human genetic mother was busy giving parties. If a quick substitution were made after birth there should be no problem of imprinting to the genetic human mother, especially if she had been chemically processed to have nursing breasts.

Ideally — if the world is to turn to animal gestators — one of the primates might provide a better womb environment for a growing human fetus than a cow. Chimpanzees tend to be frisky and probably would be too small, unless the fetus were lifted at, say, five months and put in a mechanical incubator. Orangutans and gorillas might provide a better fit and disposition. Females of both species are quite gentle. The trouble with orangutans and gorillas is that they are quite rare — and thus expensive.

All in all, small cows would probably be the best bet if the chemistry could be worked out. The cost factor would be more favorable than with just about any nonhuman or human bearer you could use. The bovine breeders have the extra appeal that they could be slaughtered afterward to fill up the freezer. Cows don't smoke, eat foolishly, or carouse. The anatomy of four-footed animals makes giving birth a much more casual affair than it is for two-footed mammals. And cows have the added attraction of having a gestation period of about nine months, just as human females have.

Legal costs also might be somewhat less than with human breeders. But medical costs might, at least initially, run higher because of all the factors of chemical compatibility that would have to be watched.

Large-scale employment of human mercenary mothers might well open up a whole new legal field, one in which lawyers could exercise their ingenuity and reap their profits. The contracts could run for pages. And still they might have to be tested in the courts because brand-new kinds of human conflicts might have arisen. What if the genetic parents, on the advice of a genetic counselor, decided to abort the fetus? Would the gestational mother have the right of first refusal to continue the project on her own? Would the genetic parents have the legal right to require that the mercenary mother eat the kinds of food and live in the ways they prescribe? What if the mercenary mother became a loving one and refused to surrender the new-born baby? Or what if the genetic parents lost interest and refused to accept the baby? The Columbia University Law School, for one, has already staged seminars in which intriguing issues like the above have been kicked around.

CONTROVERSIAL ASPECTS OF HIRED WOMBS

The possibility of transplanting an embryo from one human to another does not strike quite as many sparks as the test-tube fertilizing of human sperm and eggs. It doesn't have the awesome potential for moving on to novel forms of reproduction that *in vitro* creation of potential humans has. (Transplant to nonhumans would be pretty controversial, however.)

The Shettles concept of using another woman's reproductive apparatus for a week or two in order to bypass a cause of infertility in the wife (such as a clogged Fallopian tube) might generally be viewed as preferable to adoption. The use of mercenary mothers to take over childbearing for working wives might well have substantial appeal, especially if the current trend toward careers for married women continues far into the future.

At first glance, the use of surrogates seems like a logical evolution. On closer inspection, though, it seems controversial at best. This is so whether the hired mother acquires the embryo by transfer from the wife or by artificial insemination in which the husband's semen is used. Mercenary mothering would threaten to weaken both marriage and the family. The wedding vow is based on an assumption of exclusivity in all matters relating to planting seed and bearing children. We would have to change our concepts of what a family is and broaden the definitions of such words as *father, mother, parent,* and *motherhood.*

How important is childbearing to the self-fulfillment of women? Have we overrated it because they have had to do it anyway? Or is it an important part of most women's concept of a complete life?

Can married women develop the love and sacrifice and dedication required to rear a child successfully if they deliberately abstain from bearing it? Would there be a diffusion of identity, an uncertainty about what it means to be a son or daughter or parent any more? Would mass educational programs be introduced to make the young less concerned about bloodline identity?

We don't really know. A few years ago the British Medical Association proposed that surrogate motherhood not be attempted until much more study had been given to its probable psychological demands on the wife, the child, and the hired mother.

Still another question is whether thinking of the fetus as a piece of property to be paid for in cash on delivery would contribute to the dehumanizing of Man.

And with mercenary motherhood, is it not probable that authoritarian regimes would soon start singling out physically and perhaps mentally superior girls and forcing them into careers as babybearers?

Certainly such questions make it all the more imperative that national commissions to regulate reproductive research be established.

THE HATCHERY APPROACH

In Huxley's novel *Brave New World* new humans were "decanted" at vast hatcheries. One of the characters chuckled about the quaint old days of the twentieth century, when humans were still born by "viviparous reproduction." According to Webster, viviparous means "producing the young alive from within the body, not as eggs: opposed to *oviparous*." Only by those two ways — womb and egg — have vertebrate creatures been recreated for millions of years. Huxley's third way, hatching human beings in an artificial womb, combined the two traditional modes of gestation with a lot of manipulation by Man thrown in.

Huxley was vague about the details of how his fictional humans were hatched. Today, hundreds of scientists are searching out clues, step by step, to successful procedures.

Earlier I mentioned the curious case of Daniele Petrucci, the Italian embryologist. Unlike Edwards and Shettles, he was *not* trying to keep an embryo growing long enough (four to five days) to attempt an implant in a natural womb. He apparently just let the embryo keep growing, in an artificial womb. This womb substitute, according to one account, was a silicone container filled with amniotic fluid withdrawn by syringe from the wombs of pregnant women. One embryo allegedly grew for twenty-nine days, at which point "a heartbeat was discernible."[5]

Petrucci said he then destroyed the embryo because "it became deformed and enlarged — a monstrosity." Later, he allegedy kept a female embryo alive for fifty-nine days. After he felt compelled to stop his work in Italy he went, as noted, to Moscow, where he received a Soviet medal and conferred with a number of Russian scientists, among them the brilliant reproductive biologist Petr Anokhin. According to Francoeur, Anokhin has been working on artificial wombs for many years.

In 1966 the Russians reported that they had succeeded in keeping more than 250 human embryos alive past Petrucci's alleged record of fifty-nine days. One fetus reportedly lived for six months and reached a weight of more than one pound, two ounces. Again according to Francoeur, the Russians expressed the hope that they would give birth to mankind's first totally laboratory-gestated human within a matter of years.[6] A decade has passed and they still have not, at this writing, held up to the world for inspection a laboratory-gestated baby.

Preposterous? Who knows? Aside from Petrucci's questioned claim, the most that the Western scientific world has been able to produce and document is a mouse embryo that survived to the first heartbeat stage. This was achieved by Yu-Chih Hsu at Johns Hopkins University with seven out of twenty-five mice. In America and Western Europe Hsu's feat was considered a major breakthrough because the period *immediately* after the new embryo would normally attach itself to the lining of the womb has been the major failing point. Most of Hsu's mice that survived to the heartbeat stage showed irregularity in one or more organs.[7]

D. A. T. New of Strangeways Laboratories in England has taken a *naturally conceived embryo* from a mouse's womb and kept it alive in an artificial womb for about one third the normal gestational period.

The problems of artificial gestation — whether with mouse or man — are monumental. Aside from deceiving the early embryo into believing it is embedded in the lining of a real womb, experimenters must produce in the artificial womb a fluid with continually changing ingredients that are similar to those found in a real womb during each day of gestation. The wrong amount of a single ingredient, if given only for a few days, could reduce a potential genius to a certain imbecile.

An even more formidable problem is to develop an artificial placenta. This amazing, permeable organ connects mother to embryo, or after three months to the embryo that has become a fetus. The placenta includes a membrane that permits such substances as amino acids, vitamins, proteins, oxygen, and sugars to pass to the fetus in the right amounts. But it also rejects many substances in the mother's blood — blood cells and certain proteins, for instance — that would harm the growing embryo. At the same time it permits the passage back to the mother of waste materials such as carbon dioxide.

It took scientists at Cambridge University five years just to learn how two substances, oxygen and carbon dioxide, cross the placental barrier of a mother pig. Artificial placentas have been developed at King's College Hospital, London, for experiment on aborted human fetuses. The fetuses were kept alive for a few hours.

LESSONS FROM POSSUMS?

Research on ways to sustain life outside the womb has taken an interesting turn to the study of marsupials. The animals, which lack a placenta, include the opposum, the kangaroo, and the wombat.

At a very early stage in gestation the tiny immature marsupial is discharged from the mother's womb. Somehow it manages to climb into the mother's snug pouch, where it clamps its mouth onto a substitute for the placenta, a nipple. An opposum is the size of a cat, yet a dozen of its newborn could fit into a Ping-Pong ball.

Scientists in Van Nuys, California, have been deceiving newborn possums into attaching themselves to nipples inside a plastic, transparent pouch.[8] Various nutrients and drugs are fed through the nipples and the reactions of the baby opposums are studied. (In Huxley's imaginary laboratories, the babymakers varied nutrients depending on what brand of humans the regime wanted to produce at that particular time.)

The French biologist Jean Rostand saw the kangaroo's reproductive pattern as a possible model for humans. Since in humans the early weeks of gestation are proving to be by far the most difficult to manage outside the womb, and are the most error-prone, he suggested that we adopt the marsupial approach. The human mother would undergo a normal pregnancy for a few months. At that point she could be chemically stimulated to go into labor and discharge the fetus. This fetus would go into an artificial womb or incubator. (An alternative would be to surgically lift out the fetus.)

At present, scientists are working from both ends of the human gestation period to reduce the time the fetus needs to be in the womb. They assume that by some point in the future the original nine-month period will have shrunk to zero days. They are making spectacular progress in being able to save prematurely born infants at ever earlier ages. Vincent Collins, chief of anesthesiology at Cook County Hospital in Illinois, told me:

"We can now take over a fetus at five months, and soon I believe we can do it at four and a half months. We can keep a one-pound baby alive outside the mother."

Four and a half months is half the gestation period. However, at the other end, the early embryonic phase of gestation, progress is measured in hours, at least in the Western world. And the hazards of defects and deformity in the first days and weeks are vastly greater.

All this suggests to me that the drive to create a completely artificial womb is folly, and should be brought under social control. It is folly just as the dream of colonizing the moon is folly. It could eventually be done, but at considerable moral, social, psychological, and economic cost. Thousands of near-term fetuses and perhaps live infants with grave physical or mental defects would have to be slaughtered. The family would be further under-mined. The first generation of children so born would consider themselves freaks. As for the monetary cost to create a human being in a laboratory and provide the necessary equipment, nutrients, and hourly monitoring by highly paid technicians would certainly run to thousands of dollars.

And it is all so unnecessary! The only social gain of any significance would be to free the workingwoman from having to interrupt her job in order to produce her baby. But for most workingwomen pregnancy is no problem at all for the first seven months. And the womb gives a baby a vastly safer haven than any artificial womb could conceivably provide. I have known successful career women who stayed right on the job until labor started, with no ill effects.

But let's consider lifting the fetus from the mother at six months. If a workingwoman would prefer to have her baby removed then, the incubator people are reaching the point where they can handle it. It would be expen-sive, but they could do it. The mother would be wiser to wait till the seventh or eighth month. The baby would be less at the mercy of Man's ingenuity in manipulating knobs and nutrients.

The earlier the infant goes into the incubator, safe as it is becoming, the greater the statistical chance of death, defects, and mental impairment.

A mother's womb will, for the great majority of people, always be the best, the healthiest, place in which a baby can grow.

15

Males or
Females to Order

I have always thought that this is the type of discovery that should not be made.
— Hans Zeisel, professor of law and sociology, University of Chicago, in speaking of the new science of controlling the sex of babies[1]

Only in the fourth month of growth can you look at an aborted fetus and tell what sex it would have been. It is then that you get the differentiation between the penis and clitoris. But if the fertilizing occurs in a laboratory under glass you can tell within a matter of hours after conception what sex the infant will be.

It is the husband's sperm that determines the sex of an offspring. The wife's egg plays no part in sex determination. Husbands, royal and otherwise, who have been furious with their wives for bearing female children should be ashamed of themselves.

All eggs are similar. They have an X sex chromosome. Sperm, on the other hand, are of two kinds: those with an X chromosome (female-producing) and those with a Y chromosome (male-producing). In normal intercourse the sex of an offspring is determined by which type of sperm wins the long, seven-inch race up the woman's reproductive tract. It fertilizes her egg somewhere near the top of her Fallopian tube.

If an X (female-producing) sperm wins, you get an XX chromosome sex pairing in the fertilized egg, or a female. If a Y (male-producing) sperm wins you get an XY pairing, or a male. Every cell of a normal male body has an XY as one of its twenty-three pairs of chromosomes. And every cell of a woman has an XX pair of chromosomes.

Scientists have learned that most Y-bearing or "male" sperms have a somewhat different shape as well as other differences that distinguish them from X-bearing or "female" sperms.* They are trying to use this knowledge in order to rig the race in favor of the sex of choice. According to some observers, this capacity to rig the sex could have far-reaching consequences for society; in fact, it could shake humanity to its roots.

Landrum Shettles has been studying human sperms and eggs for nearly forty years. Another of his main interests at present is developing techniques for predetermining sex.

Even under a powerful microscope male and female sperms look pretty much the same. One day Shettles tried studying immobilized live sperms under a phase-contrast microscope. It tends to throw an illumination around dark objects. He was delighted to find that there did seem to be clear differences in size and shape between the male and female sperms. Other scientists have advanced his knowledge of the differences in shape and behavior of the two.

On the average, male sperms are smaller than female sperms. They have longer tails, which seem to help them swim somewhat faster. Male sperms have round heads, whereas the females have oval heads.

Now, wonder of wonders, scientists have learned how to put tags on the male sperms. If you take a drop of semen, put it in a hollowed-out circle on a microscopic slide and add a fluorescent chemical called quinacrine, something amazing happens. The Y chromosome in each male-determining sperm lights up; the X chromosome of the female-determining sperm does not light up.

Landrum Shettles showed me one of the ways he studies the behavior of sperms as they swim toward the egg. He picked up a glass pipette (tube) about eight inches long and held it over a Bunsen burner flame. As he held its two ends in his hands it began to melt. When it was just right he made a graceful swoop, like a symphony conductor making the opening beat at a concert. The melted glass now looked like a yard-long transparent string. But it was still hollow its entire length! The tunnel had narrowed until it would barely admit the end of a paper clip.

This, he said, beaming, was his racetrack for sperms.

He varies the liquid conditions inside the tiny tube. Then he introduces male and female sperms and watches with fascination through his microscope to see which sperms win the race. Sometimes the "track" will be filled with secretion from a woman's vagina. Sometimes it will be filled with secretion from her cervix or womb. The sperms react like trout going upstream, he said. He adds other ingredients experimentally. Sometimes it is a bit of white vinegar to make the secretions more acid. Sometimes it is dissolved baking soda to make them more alkaline.

* Technically, the Y-bearing or "male" sperms are known as androsperms, and the X-bearing or "female" sperms are known as gynosperms. They aren't sexually different, but in laboratory talk they are commonly called male and female sperms. I will use the latter terms in this book.

Male sperms tend to be flashy performers for the short run, but the female sperms have more staying power. The female sperms are more robust.

In sexual intercourse, he said, "everything you do to make it easier to conceive, the more likely you are to have a male baby. When you make conditions most difficult, you are much more likely to have a female baby."

How do you make it more difficult to conceive, and thus favor the female sperms? Shettles offers these suggestions:

- Engage in intercourse as much as you want until two days before ovulation but then stop for a week. The female sperms usually can survive until the egg emerges from the ovary. The males mostly die off.
- Immediately before intercourse the wife should use an acid douche. Acidity bothers and inhibits male sperms more than females. The douche can be made by putting two tablespoons of *white* vinegar in a quart of water.
- During intercourse the husband should seek to ejaculate while his penis is just an inch or two inside the vagina rather than at deep penetration. This forces a longer race. Also, most of the male sperms will be killed off or discouraged by the naturally hostile acid condition of the vagina. Conditions become more alkaline (and favorable to fast swimming) once the sperms reach the cervix and womb.

Shettles also suggests that the wife try to refrain from orgasm because the fluids emitted by her orgasm are mostly alkaline and thus make a faster racetrack.

On the other hand, if you badly want a boy baby, do things pretty much in reverse. Hold off on intercourse until the very day ovulation is scheduled to begin. (A doctor can advise you how to get precise timing on this. It may involve applying cervical mucus to get a color chart, or using a thermometer.) On that day Nature tends to make everything ideal for fast swimming and longer survival. You can further help make for a fast track all the way by employing an alkaline douche before intercourse. (Two tablespoons of baking soda fully dissolved in a quart of water.) And during intercourse attempt deep penetration and female orgasm.

Shettles told me that the procedures he outlined produce the desired results between 80 and 85 percent of the time. Some of the points are still being argued by medical scientists.

Of the procedures mentioned, timing is apparently the most critical. Sophia J. Kleegman, a New York authority on reproduction, has reported up to 80 percent success in predetermining sex just by using the same time scheduling that Shettles advocates.[2]

Shettles has another idea, quite ingenious, for further shaping the odds.

The general idea has been around for some years, but he thinks he now knows how to make it work provided the desire is to conceive a male. He finally described it in the October 1976 issue of the *Journal of Urology*. The idea is to set up a screening device that will permit male sperms to wiggle through but will stymie female sperms. It is a spaceship-shaped filtering device that can be placed, like a diaphragm, at the entrance of the cervix. In it is a maze of passages. The smaller-headed male sperms can get through but the females have difficulty.

Some people might assume that sperms — male or female — after bumping their heads against a screen, would turn around and go elsewhere. That apparently is not so. "Sperms go wild to get into the mucus of the cervix," Shettles said. "To them, it's like smelling ham and gravy cooking."

Shettles smears cervical mucus on his filter. In twenty-eight tests the sperms that got through were 89 to 97 percent male. Whether they still have the capacity to reach and fertilize an egg is being tested.

SEPARATING SPERMS IN THE LABORATORY

Another approach is to separate the sperms by sex and then insert the desired sample into the wife by artificial insemination. Here are five of the ways this is being attempted:

Centrifugal separation. The idea here is that if you put sperms in a centrifuge and spin them, the heavier ones (they are usually female) will end up at the bottom. And the lighter, male ones will be on top. One might assume that a sperm which survived sufficient spinning to achieve separation by size would be one dizzy sperm. Could it find its way to an egg? Artificial insemination would, of course, give it a big head start.

Swedish scientists claim that in experiments with livestock the centrifugal separation of semen has produced ten or more bulls in a row. Some scientists are concerned about possible chromosomal damage that might be inflicted on human chromosomes by the spinning. Others note that individual variations among both male sperms and female sperms complicate any effort to separate them purely on the basis of mass. Tail lengths among male sperms, for example, vary considerably. Centrifugal separation is probably one of the poorer ideas for presexing humans.

Electrical separation. Here the idea is to exploit the finding that XY (male) and XX (female) chromosomes have different electrical charges. This approach has been tried for the most part in Russia and the United States.

When an electrical charge is passed through semen the male sperms tend to drift to the negative pole and the female sperms to the positive pole. But usually a bunch stays in the middle. A Michigan State University researcher, Manuel Gordon, was able to produce female rabbits 71 percent of the time by using sperms taken from the vicinity of the positive pole.

Then in 1973 a Pennsylvania biochemist, John L. Lang, created a most impressive breakthrough by altering the technique.[3] He passed semen through a resin, a ground-up plastic material carrying an electric charge. The male sperms stuck to the resin, the female sperms swam through it. In tests with rabbits and other small animals he was able to produce progeny that were up to 95 percent female by this method! If you want males instead of females apparently you can do almost as well by simply washing free the males that are left stuck on the resin. Cattle breeders in Germany and France, among other countries, are now conducting large-scale tests of this technique.

Separation by swimming contests. In 1972, A. M. Roberts at Guy's Hospital Medical School in London noted a difference in the speed at which sperms swim downward. He filtered the sperms out of a sample of seminal fluid and then introduced them back into the fluid at the top. He found that the male sperms moved down through the fluid faster than the female sperms. He observed that an enriched concentration of males was soon at the bottom. Presumably they were also the more purposeful males.

A team of scientists headed by R. J. Ericsson at the A. G. Schering Pharmaceutical Company in Berlin followed up on Roberts's clue. They decide to make the swimming harder by using a dense, viscous fluid, a bovine albumin. Only about a third of the human sperms managed to get down through it. These tended to be predominantly males. Those that got to the bottom were removed and introduced again at the top of a fresh fluid. Again those that got to the bottom were reintroduced at the top of fresh fluid. The investigators reported: "We can isolate repeatedly up to 85 percent Y (male) sperm."[4]

They made the qualification that final proof would have to await the actual sexing of offspring. Survivors might be tuckered out. In 1975 Scottish investigators using "essentially the same technique" reported some difficulty in replicating the experiments.

A few U.S. fertility clinics are already using the technique with couples desiring a male offspring, according to the Population Reference Bureau. The Ericsson concept has now been patented in many countries. Kits containing the necessary materials for the separation are, at this writing, about to be marketed to the medical profession.

Separation by sinking rate. Simple sedimentation of sperms seems to produce a separation pattern the direct reverse of that of swimming, and for good reason. B. C. Bhattacharya, an Indian zoologist, apparently noticed it first. He found it curious that peasants in the country preferred to bring their cattle in at dusk for artificial insemination. The can of inseminate had been sitting around all day. The peasants told him that late-in-the-day inseminate produced more male cattle. He found that this seemed to be true.[5] The reason, he concluded, was that the heavier female sperms had tended during the day to sink to the lower part of the can. The inseminators who were serving the peasants drew the fluid from the male-dominated, upper part.

Later, Bhattacharya put his discovery to the test with thousands of rabbits at the Max-Planck Institute for Animal Breeding at Hagen, Germany. He inseminated the rabbits from semen drawn from various levels in the can. To prevent aimless swimming, which would affect the sinking rate, he chilled the sperms into immobility by refrigeration. He then impregnated 176 rabbits. Sperm from the top of the container produced males 78 percent of the time; sperms from the bottom produced females 72 percent of the time.

Separation by professional killers. Cancer researchers in New York noticed that male sperms exposed under glass to H-Y antibodies could be killed off in substantial numbers. These antibodies had been developed specifically to lock onto the H-Y antigens on the cells, including sperm cells, of male mice. After a long series of tests, the researchers reported an 8 percent decrease from normal in the male mice born.[6] They feel that with refinements they could achieve a pronounced increase in the percentage.

Of the five types of separation cited, two seem superior: the use of electrically charged resin filters and the exploitation of differences in sedimentation rates of chilled sperms. They seem to produce the surest and safest desired results. If you combined the two methods you could have ejaculates that would be quite close to being either all-boy or all-girl, as you chose.

INTERVENTION AFTER CONCEPTION

In the first days after conception it is often possible to upset and reverse the indicated sex of a new embryo with hormones. Another approach, one accomplished under a microscope, is to nip off a few cells of a newly formed embryo and apply a dye called chromatin to them. The chromatin will show up only in female cells. Or if you have the time and patience you can, under magnification, locate and identify the actual chromosome pair that determines sex.

Is the embryo male or female? The anxious couple may be hovering nearby. If they decide that the speck that constitutes the future human being is of the "wrong" sex, it could be discarded. Robert Edwards and his associates at Cambridge University would not force a couple to give a thumbs-down sign. Instead, they propose that several embryos be grown for each couple. Theoretically, doing so would not be much more bother. The couple could be given a briefing of the known characteristics of each embryo, including its sex, and could then be invited to indicate which one they prefer. The decision need not be made just on whim. Perhaps the couple has been warned that some sex-linked genetic disease runs in the family.

If the wife is already pregnant and the couple develops a deep curiosity to know the sex of their child-to-be, there are four possible ways to find out. Some doctors are reluctant to cooperate, especially if risk is involved. Presumably, if the fetus is of an undesired sex the parents might consider aborting it. The four ways:

Analyzing the amniotic fluid. The technique involves puncturing the abdomen of the mother with a four-inch-long hypodermic needle and withdrawing some fluid from the sac surrounding the fetus. The process is called amniocentesis (*am*-nee-o-cen-*tée*-sis). As a fetus grows, it sloughs off some of its body cells. It is these cells that are of diagnostic interest.

The technique was first developed to study whether the cells might reveal serious genetic disorders, but the same cells can, by sex chromatin analysis, reveal the sex of the child-to-be. A highly accurate assessment can be made by the fourth month of pregnancy. Some couples eager to learn the sex of their fetus have invented fake medical histories about sex-linked genetic disorders in parents or grandparents in order to have the cell tests taken that will reveal the sex of the fetus.

There are two major problems. One is risk. There is always a chance, though relatively small, that the probing needle may accidentally jab the fetus or the placenta and cause damage. In the hands of a skilled obstetrician this is highly unlikely unless the fetus is old enough to move suddenly.

The other problem is that there is not a great deal of leeway in terms of time if abortion is considered as an option. By the time the fetus is sloughing off enough cells to make amniocentesis feasible, it may be within a few weeks of being so old that abortion would be considered only reluctantly.

Taking a cervical smear. In 1971 the ever-curious Shettles reported in *Nature* that he had been having some success in picking up fetal cells for sex analysis by taking smears from inside the cervical plug. The technique was simple and totally without risk to the fetus. In 1973 a Japanese team from Hokkaido University, who tested a number of women in their first three months of pregnancy, had no luck with this method. Then, on

December 1, 1974, the *Ob. Gyn. News* carried the headline SHED FETAL CELLS RELIABLE INDICATOR OF SEX. The method had been to take smears from within the cervix, just as Shettles had done.

Checking the mother's blood. A research team at the University of California Medical Center in San Francisco came up with the surprising discovery that the white cells in the blood of some pregnant women contained male chromosomes. Where on earth had they come from? It had been assumed that the placenta prevented the white cells of the fetus from escaping back to the mother. But apparently not completely. The team found that if the fetus was a male, then after three months of age there was a 95 percent chance they could find at least a couple of male white cells in the mother's blood, provided they had available at least twelve hundred cells from the mother's blood for analysis.

This kind of analysis is time-consuming. But the embryologist Robert Francoeur believes that computers can be programmed to scan the cells swiftly. In fact, recent reports indicate that the desired speed has already been reached.

Analyzing frogs injected with a pregnant woman's urine. This method is perhaps the most improbable of all. Just as rabbit and frog tests can tell whether a woman is pregnant, a more refined frog test can tell the pregnant woman what sex her child will be.

According to a report from the Ministry of Health in Tbilisi, Georgian SSR, if frogs are injected with the urine of a woman who is pregnant with a male fetus, some of the sperms that the frog produces will have positive electric charges and others will have negative charges. If the mother's fetus is female, all the sperms will carry the same electric charge. Predictions of sex based on this difference proved to be about 95 percent correct in 370 cases. And, significantly, the percentage held true as early as the eighth week of pregnancy.[7]

We see, then, that scientists have now developed at least a dozen ways of predetermining sex. Some procedures are more accurate than others. Some are more complicated or risky. Some can be employed before conception. Some must await conception or pregnancy. But whatever the method after conception, the predetermining can be achieved only by discarding the undesired embryos or fetuses.

Although Man has been concerned about the sex of offspring for many centuries, virtually all the techniques cited have emerged within the past several years. Surely within a few additional years there will be more discoveries and refinements. Perhaps there will be capsules — pink or blue — to be taken by husband or wife within a few hours before procreational intercourse. At any rate, it is virtually certain that a safe, simple, inexpensive means of choosing the sex of one's offspring will soon become generally available.

In 1976 Park S. Gerald, a pediatrician at the Harvard Medical School, expressed concern that there was still no consensus among obstetricians or the public about the ethical problems of aborting fetuses of an undesired sex. The problem was urgent, he said, because several quick and inexpensive ways to determine sex early in pregnancy are about to become available. He indicated that within a year or two a simple blood test might be perfected for use soon after conception.

He said, "I am terribly concerned about the propriety of obliterating a fetus because it is not the sex you want." As quoted in the *New York Daily News,* he said further that while most obstetricians object to abortion for sex selection, "a significant fraction" believe it is justifiable.

WHO WANTS IT?

Two questions arise. Who would want to utilize such techniques? And what would be the impact on society if they became widely utilized?

Some people feel that the excitement of having a baby is enhanced in *not* knowing — and in being able to wonder — what the sex of their forthcoming child will be. They appear to be in a minority. Most parents would be very curious about the sex of a forthcoming child.[8]

But actually choosing the sex of one's baby produces differing views. The Princeton University Office of Population Research surveyed nearly six thousand women of childbearing age in 1970. One question asked was whether the responder would want to choose her child's sex. Only 39 percent of the women acknowledged that they would.[9] Perhaps many of them were just saying what they thought they should say. And bear in mind that the survey was confined to women. (In guessing at the possible mass market for a sex-selecting procedure, we must take into account the fact that four out of ten pregnancies are unplanned.)

A survey of as many men would probably show a higher interest in choosing the baby's sex than the survey of women showed. A large majority of both young men and young women with an opinion want their first child to be a boy, according to a 1974 survey. The results differed little from those of a similar survey made twenty years earlier.

Apparently the nature of the sex-controlling strategy would affect willingness to try it. Among the women who say they would like to try choosing the baby's sex, interest falls off sharply if the strategy would involve artificial insemination. As for abortion, some evidence suggests that only people with a real concern about the sex of their next child would accept abortion as a strategy. A married couple with three daughters and no sons might be willing to abort a fetus if there was evidence that it was female.

Some would want to choose the baby's sex because there is a family history of one of the twenty-odd, sex-linked genetic diseases. Hemophilia, for example, is "carried" by females but usually afflicts only males.

Entrepreneurs in the livestock breeding business are, of course, enormously interested in any workable way to control the sex of offspring. In the dairy industry, profits would soar if almost every offspring were a milk-producing female. Most male calves have to be sold off as meat.

THE SOCIAL IMPACT OF
A SEX-CONTROL TECHNOLOGY

In underdeveloped countries like India and Venezuela an inexpensive sex-selection technology would attract intense interest. The average citizen would be interested because of his (or her) enormous concern for having sons. This concern springs from economic and cultural sources. Government planners would be deeply interested because sex control would be a breakthrough in their long-frustrated efforts to interest the people in birth control. Much of the overpopulation in these countries is caused by the fact that citizens keep having children until they obtain the number of sons they desire.

Over millions of years Nature has worked out a precise formula for keeping the two sexes equal in number. They are not equal at birth. Of all newborn babies, 51.5 percent are male. But males do not survive as well as females, so that by reproductive age the two are almost precisely in balance.

What effect would the wide-scale availability of a sex-control technology have on this balance? A number of sociological surveys of parental attitudes have been made. But these could be misleading because, as noted, 40 percent of pregnancies are unplanned. Perhaps the unplanned figure would drop as both birth control and sex control became widespread.

For those taking advantage of a sex-control technology the main impact would be on the sex of the firstborn. About a third of the young men and women surveyed in America seemed to have no preference. But of the two thirds who did have a preference, about 90 percent wanted the firstborn to be male.[10] If they could have only one child, 72 percent of those with a preference wanted a male, the men being more emphatic. If all these figures are combined, 75 percent of firstborns and about 64 percent of onlies-by-choice would be male.

After the first child is born most middle-class people desiring a larger family would seek balance in the second child. Beyond the second child there would be a tendency to favor males.

There are no indications of how — with sex control — a six-child family would balance out on the average, at least among families in the Western world. With more women working, girl babies are no longer seen as liabilities (they may be regretted in underdeveloped societies). Probably the average six-child family would have four boys and two girls, because now most large families are the poorer families who are more likely to have the traditional pro-male values.

In the Western world the family of "average" size has just over two children, or close to zero population growth. That fact suggests that a sex-control technology might have very little effect on the sex balance of a population. There would be one boy to one girl.

But this assumption ignores the predictably heavy proportion of males among families with only children and among large families. It also ignores the fact that the two-child average includes millions of families in the Western world that have *no* children. There are still more four-children families around than the two-child average would indicate.

My estimate is that if all people could start choosing the sex of each of their children, the preference for males would run close to 60 percent. The proportion of males actually born, however, would drop to 56 or 57 percent because some parents would conceive children accidentally or wouldn't bother to exercise sex control.

And don't forget the male baby's greater proneness to die off. By reproductive age the male-female ratio would settle down to around fifty-five males to forty-five females. Of every one hundred people there would be ten more males than females.

That difference would change the appearance of the landscape quite a bit. Certainly the girls would notice the change: they would be in a chooser's market for mates. It would be a harder-edged world. Males tend to be harder-edged people. Alaska has long had a predominantly male population but has somehow survived.

Some gloomy forecasters see a male-ward shift as bringing with it sharp rises in homosexuality and prostitution. The overt homosexual scene has been changing dramatically in just one decade without sex-control technology. Any technology-induced changes would not be felt for at least two decades. And quite possibly the changes would be no more dramatic than the changes in sex mores we have just witnessed in the past decade.

The Columbia sociologist Amitai Etzioni anticipates that many significant changes would occur. For one, criminality would spread. Lately there has been some increase in criminality among females; but still, 85 percent of Americans who are convicted of serious crimes are male. They come mostly from lower-class families, in which the big swing to maleness would occur if a sex-control technology were to become available.

Etzioni credits women with exerting a civilizing effect on societies by their greater interest in such cultural activities as reading books and going to plays. (He might have added that women are more skillful peacemakers and are more likely to be spiritually oriented.) Etzioni even suggests that the two-party political system might well be disrupted in the United States with a shift to male dominance. Males, far more than females, tend to be Democrats. With a male-ward population shift, the Republican Party — which has been losing ground among registered voters for decades — might end up as a permanent fringe party. Etzioni thinks it could suffer a five-point loss in total registration as the result of the shift.

One clear, and saddening, effect of a sex-control technology could be a renewed emphasis upon male-female differences. We might not see again the extremes that existed on the predominantly male U.S. frontier, but the current trend toward fluidity of sex roles and minimization of sex differences would certainly be undercut. Sex-consciousness would grow. The predetermination of sex, so widely practiced, might well make millions of women feel that they were somehow less desirable than males, a feeling that has been evaporating in modern societies, for sound reasons.

On the positive side, parents would have the satisfaction of determining the shape their family would take. Millions of children would not have to grow up being less loved (and usually feeling less loved) because their sex was a disappointment to their parents. All this could have a sizable impact on the mental health of an entire society. The females who were born would know they were wanted.

But the momentous change for the better would occur outside Western societies. The continued sharp population growth in most of the world's underdeveloped countries certainly looms as Man's most urgent and frightening social problem. Efforts to persuade poorer families to voluntarily stop having large families have met with only modest success. The main reason is that in these countries, which usually have inadequate social security, poor parents looking toward the terrors of old age see security in having sons.

China may no longer be considered underdeveloped, but in a hospital in Tietung during the mid-1970's, the fetuses of one hundred women who had recently conceived were checked to see how many were male and how many female. After hearing the results of the tests, thirty mothers asked for abortions. Of the thirty fetuses aborted, twenty-nine were female.[11]

In underdeveloped countries a simple sex-control technology might be of enormous benefit. Wives would not have to keep having babies until they had the desired number of sons. They could have a son or two immediately, and be more relaxed about having other children. This fact would heighten their interest in birth-control techniques. Within a single generation we might well see a worldwide reduction in population growth.

The probability that a higher ratio of males would also be born in modern countries would, in less than a century, ease the current great surplus of females over males among the elderly. This surplus appears in many countries. In the United States there are now in the over-sixty-five population only sixty-nine males for every hundred females. And the gap in widening.[12]

Well, those are some of the possible benefits and hazards. For some societies sex control will probably be more of a boon than for others. Ethical problems loom more clearly if sex control is attempted after conception than before conception.

The varying potentials are one more reason why every nation should establish a commision for regulating human reproductive research. The commission in each country should monitor and regulate the introduction of sex control when one is authorized. If serious social dislocations threaten, the commission might need to regulate the availability of any devices or chemicals involved. They might be made available, for example, only to families with more than two children.

At any rate, this is a technology of such awesome potential that we should not allow ourselves to drift into its employment haphazardly.

Modifying Our Genetic Blueprints 16

In Homo sapiens . . . *something new appeared on this small globe. The next step for evolution is ours. We must devise that once again on this sweet planet a fairer species will arise.*
— Robert L. Sinsheimer, biologist

This "fairer species" of malleable man is to be created primarily by the strategies of geneticists and their scientific allies. It won't be created next year or in a decade. But a mere quarter century ago very little was known about the mysteries of heredity. In that light the recent breakthroughs toward manipulating our hereditary equipment have been astounding.

We literally have been thrust into a new biological epoch. The American genetic philosopher Theodosius Dobzhansky wrote in *Science* that the emerging genetic technology "would represent an instrument of scarcely imaginable power for guidance of the evolution of the human species."

But all is not celebration on the frontiers of genetics. At the International Congress of Genetics in 1973 the recitals of progress were interrupted by anxious questions and accusations of irresponsibility. Would a new eugenics lead to genetic discrimination against certain types of people? Would it provide subtle tools for social control? Soon thereafter the scientific world was stunned when a group of the world's leading geneticists called for a moratorium on certain types of experiments in which genetic materials were being put into new combinations. Brand-new creatures about which little was known were being assembled in laboratories. They were microscopically small but some of them could possibly be deadly if they got loose in the world. There was a long, restless pause throughout the world before such work was resumed under a cautionary code. But the dangers are still being argued.

The geneticist James J. Nagle of Drew University points out that Man is now on "the threshold of consciously controlling and engineering his future in terms of the very composition of the human population." That fact, he contends, raises dilemmas requiring decision.

Some scientists are philosophically disturbed by any action that could put restraint on their innovative efforts. Robert Baumiller, a geneticist at the Center for Bioethics of the Kennedy Institute in Washington contends: "If you're dedicated to the truth, you have to say that there are no truths not worth seeking."

Is creating a dangerous new life-form a "truth"?

For the immediate future any effort to manipulate human genes by working on them *directly* will of necessity be on quite a small scale. The technical problems are fantastic. But there are strategies for modifying genetic endowment *indirectly*. These involve manipulating cells or modifying the cells containing the genes you have in mind . . . developing new kinds of human seed . . , selective breeding . . . eliminating or treating genetic defects at the prenatal, natal and postnatal stages, and so on.

We will get to them. But as a start let's look at our incredible genes, and the ingenious ways that scientists are learning to modify them.

HOW THE GENES SHAPE OUR BODIES

No one had ever seen a gene until very recently. Even under enormous magnification it shows up simply as a tiny, almost meaningless thread.

Late in the nineteenth century the Austrian monk and botanist Gregor Mendel worked out his universal laws of heredity for living matter. He deduced that there must be *some* physical substance that transmitted ancestral characteristics over generations. But he had no idea what it was.

Early in this century it was widely assumed that man's genetic endowment, however transmitted, was immutable. This was a comforting idea. It fell in 1927, when the geneticist Herman J. Muller reported that he had changed heredity patterns by the use of X rays. Mutations showed up in the next generation. This evidence led him to advocate exploiting the apparent malleability of Man by changing Man for the better through genetic manipulation.

What the heredity-transmitting material was remained a mystery. As microscopes became more powerful, scientists began assuming that thin strands called chromosomes, which they found in the nucleus of every cell, probably had something to do with heredity.

Then in the mid-1940's experiments by Oswald T. Avery at the Rockefeller Institute produced a stronger clue. He concluded that the molecules of an obscure acid known as DNA (deoxyribonucleic acid), which was found in the strands of chromosomes, seemed to have something to do with heredity. Just how DNA was involved — and how it performed its miraculous chores — remained wholly unclear until the 1950's.

Two flamboyant young biologists at Cambridge University—an Englishman, Francis Crick, and an American, James Watson — put many clues together and constructed a giant architectural model of a DNA molecule. This was the master molecule of life, invisible except under strong magnification. But in their laboratory model it reached from floor to ceiling and looked literally like a spiraling staircase. It was made of heavy wire, metal strips, and knobs. If they had attempted to construct an entire DNA molecule it would have reached up through the ceiling and gone higher than a skyscraper. But their sample was enough to depict their well-informed guess. They called it a double helix.

The two twisting banisters in a real DNA molecule, they explained, are made of phosphates and sugars. And each stair tread is made of two chemicals separated at the center of the tread and lightly held together by a hydrogen bond. Altogether, only four organic compounds (bases) make up the treads of this spiraling staircase. They are adenine, guanine, thymine and cytosine, in various combinations. Called, for short, AGTC, they constitute the genetic code. As Jean Rostand, the French biologist, eloquently put it:

"The properties of an individual's heredity depend on the way in which those four bases are arranged and ordered in their molecules; all the genetic diversity of the species stems from them, just as all our literature is written with twenty-six letters and all our music with seven notes."

Within a DNA molecule a particular gene is the stretch of genetic code that is required to call up all the building blocks (amino acids) needed to produce a protein — often an enzyme — for a particular purpose in bodily structure or function.

These master DNA molecules containing the genes are found in every cell of a human embryo. And how does all this information get passed on as the future human body grows? Guess! The answer is one of the great wonders of Nature. When a cell divides, this spiral staircase splits right down the middle of the treads along the line of the weak hydrogen bond. In the newly formed DNA molecule each half of the old molecule picks up from free-floating materials the ingredients it needs to construct its exact complement. Soon there are two spiral staircases where there was one. And the dividing continues. These extraordinarily long, thin DNA molecules carry masterplans for building a unique human being. And the plans get passed on to the next generation by way of germ cells (eggs and sperms).

A thimbleful of male semen contains tens of millions of sperm cells. And each sperm carries half of the male's DNA in its chromosomes. Likewise, each female's egg carries half of her DNA. In a fertilized egg the combined DNA provides the master plan for a new individual, complete with specifications for freckles, a weak chin, premature baldness, and a brain of a certain physical magnitude. Nature's achievement is comparable to printing the Bible on one's fingernail. The task of geneticists who aspire to delete or replace genes directly is at least as difficult as making editorial corrections on that fingernail.

Crick and Watson won Nobel Prizes for their conception of the structure of the DNA molecule. Their concept was soon substantiated as essentially correct by laboratory experiments. It remained for another little-known American biologist, Marshall Nirenberg of the National Institutes of Health, to crack the genetic code. In 1961 he matched the components of DNA with their functions and demonstrated how the building blocks get called up as needed.

By 1967 a strand of DNA was actually synthesized in a test tube. This was the feat of Arthur Kornberg of Stanford University.

And ever since then there have been fast-breaking achievements in getting to know and modify the gene.

In 1970 the first complete gene was synthesized. This was a yeast gene involving a stretch of DNA seventy-seven base pairs long. It was put together at the Massachusetts Institute of Technology by a team headed by the Indian-born biologist Har Gobind Khorana.

By 1975 the first gene of a mammal (rabbit) was totally synthesized, this by Harvard biologists.

And in 1976 Har Gobind Khorana was back in the news again. His group had put together a gene out of shelf chemicals that actually did its work as a gene when implanted in a bacterial cell. This new gene had the crucial "start" and "stop" signals that, in a living cell, make a gene function. The gene was a linear sequence of 199 components—all involving a combination of the four ingredients of the genetic code (AGTC). The genetic mechanism was just as Crick and Watson had surmised twenty years earlier.

MAPPING THE GENES

Meanwhile, in dozens of laboratories geneticists began trying to pinpoint the location of individual human genes on Man's twenty-three pairs of chromosomes. This knowledge is crucial if change or repair is to be attempted on a particular functioning gene. The main achievements in mapping have been performed by a team of thirty geneticists, biochemists, and cell biologists at Yale University, headed by an amiable, lean giant named Frank Ruddle. In 1970 his group found their first human gene. Its function was to carry information for the production of a certain enzyme. By 1977 they had located more than two hundred. They had not only found which chromosome the genes were on, but in many instances approximately where on the chromosome they were located. Ruddle and his colleagues do this mapping, not by studying specific genes, which is still impractical, but by studying the enzyme product created by each gene. Then they diagnose the area on the chromosome where the action that created the product was triggered.

Ruddle hopes that by 1980 his group will have mapped at least one thousand human genes. That still is only a small fraction of the 50,000 or so genes now believed to exist on human chromosomes. But some of those already mapped are known to be implicated in serious genetic disorders, such as Tay-Sachs disease and the Lesch-Nyhan syndrome.[1]

The kind of analysis required to map the genes would not have been possible except for another major development. This permits genes to be manipulated by fusing two different cells. The fusion technique was developed after it was discovered that a virus called the Sendai virus will fuse cells.

Cell fusion was pioneered by a team headed by George Barski at the Institut Gustave Roussy in Paris. Mouse cells were fused. This was in 1960. By 1967 Mary Weiss and Howard Green at New York University were fusing the cells of mice and men!

Fusing the cells of mice and men appealed to the Yale gene-mapping team. When these newly created hybrid cells were placed in a nutritive broth, they divided in the usual way cells do. After many divisions all the mouse chromosomes remained but only certain of the human chromosomes remained. This was apparently not due to any dominance of mouse over man but rather to differences in the rates at which cells of the two species mature and divide. The systematic isolation of different human chromosomes — and the order in which they were shed — made it easier for the mapping team to spot where actions of specific genes originate.

ADDING TO CELLS AND MASS-PRODUCING GENES

Meanwhile, more compelling reasons have emerged to fascinate scientists with the potential that cell fusion offers for actual genetic engineering.

In 1971 Henry Harris and his group at Oxford University startled the scientific world by correcting a genetic defect in a mouse cell. The defect was that the cell was unable to manufacture a certain enzyme. In humans this same enzyme is involved in the usually fatal Lesch-Nyhan syndrome.

The Harris group fused a mouse cell carrying this genetic deficiency with a normal chicken cell. The hybrid cell with two nuclei thrived and began dividing. As hoped, the nucleus from the mouse, by maturing faster, took charge. Soon the chromosomes from the chicken began to pulverize. They were smashed into bits of genetic material. And from these bits the mouse nucleus picked up the genetic material it needed to overcome its deficiency!

In the first attempts the chick gene seemed only loosely integrated in the mouse cell. But the journal *Nature New Biology* concluded that in principle at least, the achievement opened up the possibility of treating genetically defective human cells.[2]

Still another spectacular feat was reported in 1974.[3] It involved a bold strategy that may ease the formidable problem of having to deal with one cell at a time in making genetic changes. A group at the University of Colorado took advantage of the discovery that when a cell is exposed to a certain fungus its nucleus drifts toward the edge of the cell body and protrudes. Imagine the yolk of an egg that protrudes from the edge of an egg ready to be fried.

Such a protruding nucleus is easy to dislodge from the cell body. The Colorado group did just that on a large scale. They put these lopsided cells in a low-speed centrifuge by the thousands. The nuclei and cell bodies were separated. Then the investigators took large batches of the nuclei and large batches of the cell bodies and put them into a bath together. In this bath the cell components were exposed to the Sendai virus, which causes cells to fuse. The nuclei and cell bodies fused by the thousands into new combinations.

The success of the Colorado group immediately brought suggestions to put young nuclei in old cell bodies, healthy nuclei in diseased cell bodies, and so on. By working with tens of thousands of cells simultaneously, the scientists hope to effect enough change to alter small ailing areas of the body for the better. Some experimental work is also being done to try to grow a desired type of replacement cell on a mass basis in a tissue culture of one person's cells.

In addition, scientists are working on an infiltrating strategy called transduction, which takes advantage of the fact that certain harmless viruses can carry genes into cells. It is hoped that if genes of the right sort can be introduced, they can take over functions of genes that are malfunctioning and cause a genetic abnormality. For instance, the cells of diabetics are defective in that they do not carry the correct form of the gene that manufactures insulin. Transduction could conceivably end this state of affairs and "cure" diabetes.[4] At this writing, the cells of a number of small organisms have been successfully infiltrated, and apparently the method has worked in a few instances with human cells.

CREATING NEW FORMS OF LIFE

The strategy for gene manipulation currently causing the most commotion arises from the growing sophistication of gene scientists. They can take the long, thin DNA molecules and wrap them in circles. Or they can take them apart and put them back together again in new combinations, and with bits of new organisms added. These techniques have come to be known as gene-splicing.

In this way they transplant genetic information from one unrelated creature to another and brand-new organisms emerge. Scientists from Stanford and the University of California managed to transplant genes from a toad into the DNA molecules of the common laboratory bacterium *Escherichia coli*. These bacteria are found widely in the human gut.

One of the controversial strategies goes by the name of plasmid engineering. Plasmids are something quite new in scientific shoptalk. It seems that all genes are not on chromosomes, as had been assumed. Some float around in small DNA rings called plasmids. The scientists have found an enzyme that splits open the ring long enough for foreign genes to be introduced. The altered plasmid ring is then inserted into a life-form that has attracted the scientists' curiosity. The new life-forms created begin carrying the hereditary instructions contained in the altered plasmids.[5] At this writing, insertions have been done mostly with bacteria, such as *E. coli*.

The stunning versatility of the gene scientists should not persuade us that we are on the verge of genetically reshaping man. Not at least by direct manipulation of genes, such as has been described. We are still some decades removed from the day parents will go to the genetic supermarket for gene seed mixed to their specifications.

Most of the traits the parents would want to specify — intelligence, disposition, type of hair, physique, bust size, personality, nose shape, proneness to long life — are not found in a single gene. They are the result of the interaction of many genes. (And some involve also the individual's interaction with the environment after birth.) Scientists still have only the foggiest ideas about how even to start identifying all the genes that would be involved in prescribing shape of nose, and how they interact.

Still, the recent successes in manipulating genes in small organisms — and discovering the products that some individual genes create — are of considerable interest to Man. Scientists hope that with new kinds of bacteria they can create biochemical factories for producing such things as antibiotics and antibodies. In fact, a number of scientists have already been applying for commercial patents.

Specifically, they are intrigued by the possibility of creating bacteria that will stimulate the growth of grain foods. This would reduce the need for expensive and increasingly scarce fertilizers. They also hope to introduce a cow gene into bacteria to create a curdling substance that will cut the cost of making cheese. It has been reported that General Electric is trying to patent a process to create bacteria that will eat oil, such as in oil slicks. And one of the GE scientists hopes to develop an organism that will extract precious metals from waste substances.[6]

That is the bright side. But the creation of new life-forms in bacteria by such techniques as plasmid engineering gives many people — including prominent scientists — the shivers. Some wonder, for example, what would happen if oil-eating bacteria got loose in the lubrication of our automobiles and industrial machinery.

Eleven distinguished gene scientists pleaded in 1974 for a moratorium on certain dangerous manipulations that were going on or were being planned. They had the encouragement of the National Research Council. In their plea they warned, for example, that "new DNA elements introduced into *E. coli* might possibly become widely disseminated among human, bacterial, plant or animal populations with unpredictable effects." A moratorium was widely recognized.

The next year 140 gene scientists from many countries, including the USSR, assembled in the Asilomar State Park in California to hash things over. It was an unprecedented conference. Many were wary of any serious self-regulation. In reporting on the conference, *Science News* summed up the hazard they were wrestling with in these words:

"Both by accident and by purposeful manipulation, genes for drug resistance or cancer or lethal toxin formation could be inserted into common organisms. Biological warfare agents and massive epidemics could be created too effectively."

After three days the conferees came close to unanimous agreement on a set of guidelines. Many went along with the agreement just to get rid of the moratorium. Meanwhile, in America the Department of Health, Education and Welfare, which funds much of the academic research, made efforts to form a public policy for issuing its grants in gene research. Any policy HEW announced would, of course, not affect what researchers for the Department of Defense or for industrial firms or for foreign countries are doing.

The need for public control through laws or a regulatory body, not by guidelines, was indicated by one sentence in the *Science News* report on the Asilomar Conference:

"Ten years from now, genetic engineering kits might be standard equipment in high school classes."

That would be something. I say that as one who accidentally blew up a part of his high school chemistry laboratory.

The Quality Control of New Humans

What may start as the biological control of illnesses could become an attempt to breed supermen.
— Amitai Etzioni, sociologist

In the opinion of H. Bentley Glass, the geneticist, leaders of the future will decree that parents "have no right to burden society with a malformed or a mentally incompetent child." Future societies, he states, will regulate who gets born and who does not. If nothing else, population pressures will make regulation inevitable. And he adds: "The once sacred rights of man must change in many ways."[1]

Changes will come mainly through eugenic engineering. Eugenics is the name of the long-discussed movement to improve the human species through the control and manipulation of hereditary factors. This can be done without working directly on individual genes. As we have seen, the direct manipulation of genes is still quite a problem.

Eugenic engineering has several aspects, some more controversial than others. In its simplest form, sometimes called euphenics, it helps modify or halt the biological development of an individual. Parents are helped to decide whether to go ahead with a baby. A seriously defective fetus is screened out. A baby born with a defect is given help.

A broader form of eugenic engineering is negative eugenics, the *mass* screening of babies, or of aspiring parents. The aim here is to reduce the number of defective genes at large in a human society.

Finally, there is positive eugenics. This involves purposeful efforts to produce on a large scale improved or novel kinds of human forms. We will discuss these efforts in subsequent chapters.

First, a word about the scary phrase "genetic defects." Before we, the people, plunge too energetically into purging all offspring who have genetic defects we should bear in mind a couple of facts: Every one of us has at least a few genetic defects. Eight defects per person might be a fair average.

In any efforts we make we should focus only on the real cripplers, not the kinds of defects that cause a stiff big toe or premature baldness. And we should remember that some of the defective genes we carry are not evident in our lives. They are recessive defects, which must be carried by both the mother and the father to show up in offspring. If the defective gene is "dominant," one parent can transmit it.

About one newborn in fifty is sufficiently defective genetically to be a burden on its parents and perhaps on society. Some of these babies have gross chromosomal defects that are clearly visible under a microscope.

Some serious genetic defects occur because of the failure of a single gene to order an enzyme. When a single gene is responsible, there now is a good possibility that the gene manipulators may solve the problem before too many years pass.

Of the two thousand or so disorders that can be inherited, some of the more familiar are

Epilepsy	Feeblemindedness
Cleft palate	Dwarfism
Male baldness	Harelip
Color blindness	Muscular dystrophy
Cataracts	Parkinsonism
Albinism (lack of pigmen-	Diabetes mellitus
tation)	Clubfoot
Gout	Extra fingers or toes
Hemophilia	Pernicious anemia
Sickle-cell anemia	

A few genetic disorders cannot be predicted on the basis of hereditary laws. Consider the genetic botch-up that creates mongolism (Down's syndrome). The cause seems to be not inheritance but a copying error involving the chromosomes, either in the formation of sex cells or in the early stages of cell division. An extra chromosome ends up on the twenty-first set. The result is an unfortunate human with a broad head, slant eyes, tiny hands, and minimal potential for developing human intelligence. About one newborn in six hundred is so afflicted. In the United States mongoloids cost society more than $1.5 billion a year.

The best indicator that such a gross copying error may occur seems to be the age of the mother. If she is under thirty the chances are 1 in 3,000 that she will give birth to a mongol. If she is thirty-five the risk rises to 1 in 280. By the time she is forty-five, it is 1 in 40. In short, the chance of giving birth to a mongol rises about seventy-five-fold. Even so, the risk is still only 1 in 35. It can be almost totally eliminated by fetal screening, if the mother is willing to have an abortion should the fetus prove defective.

There are a number of strategies for coping with genetic disorders. They can be compared to the checks on quality that a manufacturer makes before, during, and after the manufacturing process. In checking for genetic defects, yet another control has been proposed: inspection by marriage license bureaus. In the sections following we will describe these four procedures.

SCREENING BY GENETIC COUNSELORS

In 1951 there were only ten known genetic counselors in the United States. Today there are at least a thousand. They can be found in hundreds of genetic counseling centers. Most of them are pediatricians who have boned up on genetics. Some are psychologists or internists. A few are obstetricians or gynecologists. You will find them mostly in medical centers, medical schools, hospitals, public health agencies. Your doctor will know the nearest qualified center.

The people who seek counseling are usually scared. A girl considering marriage is worried because her boy friend's father has a harelip. A married couple is concerned because their first child has a clubfoot, and they want to know what the chances are that any other children they conceive might have the same defect. (Answer, in this case, slight — about 5 percent. A congenital clubfoot is what they call a "multifactorial" disorder, which is caused by the interaction of numerous genes.)

It helps if a husband and wife, before going to a counselor, work up a three-generation medical history of their two families. The National Foundation–March of Dimes provides without charge a form for couples to fill out that will give the counselor the necessary information. Requests for copies should be accompanied by a stamped, self-addressed envelope (large size) and should be mailed to the foundation at the following address: P.O. Box 2000, White Plains, New York, 10602.

In stable communities it is fairly easy for young couples to work up the histories. But in our nomadic society millions of young people know little about their grandparents or uncles. It may be necessary to get on the telephone and call an elderly aunt for information.

The genetic counselor will in any case get as much background as possible from a couple, and will then draw up diagrammed pedigrees. These indicate any known genetic disorders and note whether they were dominant, recessive, or sex-linked. Then the counselor figures the odds that the defect will be inherited. To take the simplest situation: If either partner is afflicted by a dominant genetic disorder (usually visible), the odds are 50–50 that each child will inherit it. If only one partner is the carrier of a recessive disorder (it may not be visible), none of the children will exhibit the disorder (though one or another may be a carrier). If both husband and wife are carriers, then the chances are one in four that each child will have the defect. To put it another way, if they have four children, the odds are that one will become afflicted, two will be normal but will be carriers, and the fourth will be completely normal.

Most people don't know if they are carriers of a genetic defect or not. If some genetic disease has shown up often in the family tree of one of the partners, the counselor may recommend that a test be taken. A considerable number of genetic disorders that people carry — but are not afflicted by — can now be identified by tests.

Advice from a counselor may also be sought after the wife has become pregnant and there is a chance that the fetus may have inherited a serious disorder.

SCREENING THE FETUS IN THE WOMB

One of the more interesting if little-known aspects of the biological revolution is monitoring the growing fetus. One method is testing with ultrasonic beams to locate the placenta and provide an image of what is developing. Another is the use of the fetuscope (still being perfected), which permits the doctor to peer into the womb during the early months to make a visual check on progress. A third is testing the mother's blood for fetal abnormalities. For example, a chemical called Alpha feto protein leaks into the mother's blood if the fetus has a damaged nervous system.[2]

At present, the most promising monitoring tool is amniocentesis, already described. As recently as 1973 many doctors were totally unfamiliar with it. By now, thousands of women have undergone the test. Average price: around $250. More than one hundred genetic disorders can be spotted by analyzing the amniotic fluid. Some checks on the cells can be made fairly quickly — those on chromosomal irregularities, for example, and those on sex. If the fetus is female, the parents can stop worrying about a sex-linked genetic disease like hemophilia, because it afflicts only male children. Other biochemical tests, however, require growing the fetal cells in culture for some weeks in order to make a diagnosis.

A study was made by the National Foundation–March of Dimes in 1974 of 2,187 women who had been tested by amniocentesis in their second trimester (fourth to sixth months of pregnancy). Sixty-two of the fetuses (3 percent) were found to have a significant genetic disorder. All but two of the mothers elected to have the fetus aborted.

The National Institute of Child Health and Human Development reported in 1975 on a four-year study of 2,000 pregnant women, of whom 1,040 underwent amniocentesis. The diagnostic results: 995 fetuses were judged normal and 45 (3 percent) genetically defective. Only six diagnoses proved to be wrong. Two of them involved mongoloids who had apparently been missed because of human error. To put it another way, the diagnoses were more than 99 percent correct.[3]

The results emboldened the U.S. Department of Health, Education and Welfare to declare that amniocentesis is a safe and accurate strategy for detecting genetic diseases in the unborn. Accordingly, an official of HEW said in 1975 that the government was planning to help make amniocentesis available to anyone who might want it. At this writing little has been done at HEW to implement this promise. In 1976 Congress allocated $90 million for a broad three-year program to detect genetic disorders. Cited for support were counseling, various types of genetic testing, and public and progressional education.

The government's approval of amniocentesis inflamed the right-to-life enthusiasts. They assume, I think correctly, that widespread use of amniocentesis would lead to more abortions. The movement considers all abortions evil except those performed to save the life of the mother.

A medical geneticist at the Mount Sinai School of Medicine in New York estimated that four thousand pregnant American women underwent amniocentesis in 1975. Five percent were found to have fetuses with serious defects. Most of these were aborted. Many of the mothers were reported to have suffered afterward from deep remorse. Some of them terminated sexual relations. Others were divorced. By 1976 the number undergoing amniocentesis had reached ten thousand.

The positive finding of a serious genetic disorder in a fetus is not necessarily a prescription for abortion. For some disorders there are treatments which, if begun early in life, can ameliorate the damage. The bleeding of hemophiliacs can now be controlled by a new blood-plasma product. Surgery can overcome the effects of cleft palate. But treatment does not, as yet, affect the damaged genes in any way and they are passed on by reproduction.

THE LEGAL UPROAR OVER FETUSES

Abortion is always an option for genetic screening, and abortion has become an explosive issue in many countries. In Great Britain there was a considerable commotion when it was discovered that aborted living fetuses had been sold for scientific research. Rules on treating fetuses were tightened.

In the United States, as recently as 1969 and in all but a few states, it was illegal to perform an abortion to prevent the birth of a defective child. In a wave of legislative action abortion laws were loosened. And in 1973 the U.S. Supreme Court ruled (in *Roe* v. *Wade*) that the abortion of a fetus is not a proper concern of the state unless the fetus has reached the stage where it can become "viable outside the mother's womb, albeit with artificial aid." The Court held that the "viable" stage is "usually placed at about seven months (28 weeks) or may occur earlier, even at 24 weeks."

The Court's inclusion of "artificial aid" in the definition of viability was, in my view, unfortunate. Each year science is increasing the length of time that a fetus can be sustained in an artificial womb environment filled with pumps, fluids, and hoses. Later, in a 1976 case the Court decided that the question of when "viability" begins should be left to the responsible physician.

At any rate, many of the right-to-life forces assailed the Court's 1973 decision as a "day of infamy" and sought ways to slow down fetal research and abortions. In America the center of the battle focused in Boston, which is both strongly Catholic and medically oriented.

The most controversial case centered on the chief resident in obstetrics and gynecology at the Boston City Hospital. He had aborted a fetus which he understood to be between twenty and twenty-two weeks old. The prosecution claimed that the fetus's age was between twenty-four and twenty-eight weeks. Pathologists estimated the age at twenty-two to twenty-four weeks. At any rate, the fetus weighed about one and a quarter pounds. The doctor had been unable to induce abortion by the usual injection of a saline solution, so he performed what amounted to a Caesarean section. The charge against him was that he was guilty of manslaughter in not trying to save the fetus's life when it was so close to "viability." Of the twelve jurors, ten were Catholics.[4] They found him guilty. He could have been sentenced to twenty years in jail. Instead, he got a one-year suspended sentence, went back to work, and appealed the conviction. He was upheld twenty months later, in 1976.

Meanwhile the Congress, under pressure from the right-to-life forces (which include many non-Catholics), imposed a temporary ban on research on live human fetuses. It established a commission to study the problem and make recommendations to the Secretary of HEW. In 1975 the commission recommended that the ban be lifted but that research be conducted within certain prescribed limits. Among other stipulations proposed was one that would lower the age of "possible viability" to twenty weeks (about midpoint in pregnancy). It should be stressed that the Supreme Court had given no support to the claim that a viable fetus, no matter how mature, was entitled to be considered a "person" and consequently eligible for legal protection under the criminal laws against homicide.[5]

Whatever the legal situation, the right-to-life advocates believe it is immoral to end the life of a fetus. And this brings us to the question of when does human life begin, which is at the heart of the uproar over curbing fetal experimentation and abortion.

WHEN DOES HUMAN LIFE BEGIN?

Some purists, particularly some theologians, say at the hour a human sperm manages to fertilize a human egg. This view certainly has validity. During that hour the blueprint is fixed that will largely shape a future human being. The fact that the blueprint is on a speck invisible to the naked eye is perhaps not relevant. Other theological purists relent by a few days and assert that life begins when the embryo implants itself in the womb. By this time it is slightly larger, but barely visible.

Whether either speck deserves to be called human *life* is in dispute. Some say you don't have a human life until you have a functioning human being. A Princeton theologian, Paul Ramsey, perhaps for the sake of controversy, has offered the suggestion that the beginning of human life be set at when the baby is about one year old. Then it has the power of speech and self-awareness. He reasoned that a baby should not be regarded as human, at least in the sense of having legal rights, until it is conscious of itself.

Perhaps the ultimate consensus will end up somewhere between these two extremes. At any rate here are a few of the criteria, past and present, for judging when human life has begun. I should add that I talked with a good many experts on this subject and was surprised to find how far apart some of the opinions were.

Movement of the fetus. A common Protestant view is that life begins with "quickening," that is, when the pregnant woman feels or notices fetal movement. The Catholic Church during the Middle Ages permitted abortion prior to "fetal quickening."[6]

The British common law has at times held that quickening occurs after four months. That pretty much coincides with when a pregnant woman notices movement, usually between the sixteenth and twentieth weeks. In observing fetuses through a fetuscope, however, doctors have noted stirrings by the tenth week. And there is in existence a film showing mouth movement by the eighth week.

So quickening, once regarded as a criterion, isn't much help.

Fetal heartbeat. Historically, our concept of human life has been tied up with the heartbeat. It was always assumed that when the heart stopped beating, life ended. So, it has been argued, life starts when the first heartbeat occurs. A doctor's stethoscope can pick up a heartbeat at some time between the twentieth and twenty-fourth weeks. But an ultra-sensitive stethoscope or an electrocardiogram machine can pick up a heartbeat by the fourteenth week. Dissection shows a very primitive circulatory system by six weeks.

Human appearance. We talk about the image of Man. If you came across a mixture of small animal and human fetuses you could surely pick out the human ones if they were over two months old. Photographic evidence shows that the arms, feet, head, and ears are all clearly discernible even though the fetus is only one and a half inches long.

Brain function. Increasingly it is being recognized that the essence of human life is in the brain, not the heart. For scientists today, the surest sign that life has ended is when the line on the brain-wave machine (EEG) stops bouncing and remains flat. They call it a "flat EEG." In aborted fetuses "brain" waves can be detected by the seventh week. If the fetus is in the womb, the first time that brain waves can be recorded is at about the thirteenth week. And they come from the primitive brain system. Waves from the cortex can't be recorded until three weeks later.

But all this brain-wave activity is much like that which might come from a mouse. The embryonic brain can hardly be viewed as human in terms of function. The exquisite circuitry required for mental activity has not developed in the uniquely human part of the brain, the cerebral cortex.

Studies at the Albert Einstein Medical School of Yeshiva University indicate that the brain is not ready to function in a human way until the twenty-eighth week at the earliest. And in the following four weeks it undergoes a great leap in sophistication. In this period the thinking portion of the brain develops the latent capacity for functioning in ways that we regard as signs of human intelligence.

The science writer Maggie Scarf found that in the opinion of every medical expert she consulted there was virtually no chance a fetus could suffer pain before the age of twenty-eight weeks. This is because the capacity to feel pain is linked to the development of the nervous system and the brain.

Leaving the mother's body. Perhaps the best judge of all when human life begins is the mother. Most mothers think of a baby as human only after it stops being a part of her body. This definition, of course, would not please the right-to-life people, and perhaps it should not. There have been abuses in fetal experimentation and it is probably not good public policy to permit abortions after the fetus attains consciousness or the capacity to suffer pain at about twenty-eight weeks. That might be a reasonable time to think of a fetus as a person with legal rights.

NEGATIVE EUGENICS

Thus far we have been talking about *individual* actions to reduce the chance of having a seriously defective child. Genetic counseling certainly is laudable. And individual decisions to abort fetuses that fetal monitoring has proved to be seriously defective can be argued as morally defensible, I think, assuming that the abortion would be performed before the twenty-eighth week of pregnancy.

But what about compulsory, broad-scale programs to prevent the conception of defective babies or to identify fetuses with poor genetic endowment?

I am not talking about Hitlerian programs to kill off the defective and weak at birth. I am talking about compulsory counseling, compulsory sterilization, denial of the right to reproduce without a license, and mass screening of newborns to spot those that are defective. These add up to negative eugenics.

Some persons are of the opinion that given our new knowledge of genetics the State has the right — in fact, the obligation — to reduce the number of genetically unhealthy citizens. One argument for this, often advanced, is that in recent decades Man has undermined natural selection. He has permitted the gene pool to deteriorate, and seriously. What are the facts?

IS THE HUMAN GENE POOL DETERIORATING?

A number of well-known biologists have answered yes. They include Sir Julian Huxley, Bentley Glass, Joshua Lederberg, H. J. Muller, Leon Kass, and Theodosius Dobzhansky.

Among the causes, some scientists cite the advance of medicine first. We can now keep alive many people who suffer from mental or physical genetic ailments, or who are genetic carriers, until they are able to — and do — have children.

The number of people suffering genetic mutations, the scientists contend, is mounting. Marc Lappé, a biologist at the University of Pennsylvania, cites the fact that the incidence of a once-fatal and dreaded cancer of the eye, retinoblastoma, doubled in the population of the Netherlands in just thirty years. It is transmitted by a dominant mutant gene and is now treatable, in many but not all cases, by removing part of the eye.

Lappé, incidentally, is less certain than Huxley and the other scientists cited above that there is any significant gene-pool deterioration. He points to changes in Western family patterns—less intermarriage among relatives is one—that tend to improve genetic quality.

The number of genetic mutations, however, is certainly growing because of the increased amount of radiation in our environment. Radiation triggers gene mutations. First, there was fallout from nuclear explosions. Now, there is the small but constant amount of radiation from atomic energy plants and television sets. And because X rays have become such a necessary diagnostic tool, each of us is absorbing more radiation from this source than ever before.

Some point as well to our easier style of life. The automobile, central heating, and packaged foods make it easier for the physically weak and defective to survive. Natural selection today is not as rigorous in its people-culling effect as it was a mere half century ago.

Dobzhanksy has predicted a "genetic twilight" if present trends continue. Some put the pollution of our genetic pool more specifically. They contend that in each future century the number of defective people dependent on medical technology to survive will increase by 8 percent. If they are right, in six hundred years — the same amount of time that has elapsed since A.D. 1400 — the bulk of the population will have become seriously defective genetically.

Others make the more controversial assertion that we are also trending genetically toward duller-wittedness. Through the ages, those endowed with genes that contributed to higher mental ability had a better chance of surviving. Today, these brighter people, it is argued, are more often involved in family planning and having small families than the less-endowed mentally, who are more likely to be among those with large families.[7]

Some suggest that welfare policies, in the United States at least, have the effect of encouraging many of the poor to have children they might not otherwise have. In a much-argued-about article on IQ differentials published in 1969, the educational psychologist Arthur Jensen asked: "Is there a danger that current welfare policies, unaided by eugenic foresight, could lead to the genetic enslavement of a substantial segment of our population? The new U.S. government program to encourage and support fetal monitoring might be seen as 'eugenic foresight.' " But not to the right-to-lifers, of course.

MASS SCREENING AT BIRTH

In the long run the mass screening of infants at birth might also make some contribution to "genetic foresight." The primary aim is to spot newborn infants with genetic defects immediately. In many cases there may be some alleviating treatment. But more important for the long run, healthy babies that are carriers of one or another genetic defect could be identified. It is now possible to identify the carriers of nearly fifty genetic defects. If by childbearing time carriers were alerted to their condition it could affect their marital and procreational plans. It might be more humane, however, to postpone the mass screening until people are of reproductive age. Then they wouldn't need to worry about the matter until they are capable of reproducing.

Although most parents of infants are not aware of it, mass screening of some sort is already being done in most states, and in many foreign countries. Forty states have laws requiring hospitals to test all newborn babies for phenylketonuria (PKU). This is a genetic disorder that occurs in one in ten thousand births. A couple of drops of blood are taken for analysis from the baby's heel. PKU involves a metabolic failure that causes serious mental retardation if the child is not put on a special protein-free diet for several years.

About a dozen states have compulsory screening of the newborn for sickle-cell anemia, which afflicts many people, including some of Mediterranean ancestry, but especially blacks. New York State has recently instituted mandatory screening for seven different hereditary disorders that could retard brain development.[8] With parental consent, Massachusetts hospitals routinely screen babies for several dozen genetic defects. Most states that have compulsory screening of some sort have no statutes to protect the confidentiality of the information. In later years the carrier might be identified and stigmatized in some way.

The philosophy of many lawmakers is "if you can screen for it cheaply, then do it." Machines are now available that can make dozens of tests in a minute.

Still, reservations arise about this sudden legislative fascination with passing laws for every new disorder that can be easily diagnosed.

Why require tests at birth for genetic defects that are incurable and for which no promising treatment is available? This occurs.

Why require tests for disorders that are rare? A Texas legal researcher, Philip Reilly, points out that the small state of Rhode Island screens for something called "maple syrup urine disease." It leads to mental retardation. He estimates that it might show up once in a decade in Rhode Island.

Why require tests at all? Lawmaking bodies are moving into areas where their mandate is fuzzy when they pass laws requiring tests. Their only clear mandate is to protect the public health. But genetic defects are not contagious or poisonous. So what is the legal justification? The lawmakers' excuse for intruding into private lives is that they are trying to reduce the cost to the taxpayers of the custodial care of seriously defective persons. That is a reasonable concern, but hardly one that requires compulsory screening. It could be done effectively on a consent basis.

If testing is mandatory, then the record of one's inherited defects will be on tap somewhere for life. Undoubtedly the record will end up in some central electronic data bank, perhaps even a federal data bank. Some U.S. congressmen have been pressing for a uniform testing law.

The future would look more cheering if testing were done only by parental consent and if the only record available is turned over to the parents. Government units need only *statistics* on the prevalence of genetic disorders, not the names of the persons who carry the disorders.

And why test if it does more harm than good? Certain of the ailments tested for are dimly understood and hard to identify. Serious questions have, for example, arisen about the compulsory testing for PKU. Of the millions of babies tested each year, only about three thousand show blood symptoms *suggesting* that PKU is present. But only two hundred babies actually prove to have classic PKU.[9] If all three thousand, most of them capable of developing normal or near-normal minds, are put on a protein-free diet for five years, what happens? They develop a lowered mental capacity, which they are purportedly being protected against by the testing! Proteins, as noted earlier, are crucial to sound mind-building.

Finally, why make compulsory tests if the results unnecessarily frighten or stigmatize people? Sickle-cell disorders are an excellent case in point.

The problem has arisen because of confusion between the usually lethal sickle-cell anemia (which shows up in one out of five hundred newborn black babies) and the sickle-cell *trait,* which is harmless. The person with the trait is simply a carrier. And the trait will only be a problem if the carrier has children by another carrier. Many blacks carrying the trait have been distressed to learn that they were carrying "bad genes." They were not the only ones confused about the nature of their ailment. Some insurance companies assumed that carriers were poor health risks and refused to insure them at normal rates. Many airlines refused to hire carriers as flight attendants for the same reason. Even the U.S. armed forces worried about accepting carriers until a committee of the National Research Council advised the military people that their worries were groundless.

Another stigmatizing genetic disorder involves males with the now-notorious extra Y chromosome. A noted sociologist made this widely read statement: "There is accumulating evidence that people born with XYY genes . . . may have a predisposition toward criminal insanity." A specialist envisioned the possibility that in the future there might well be state-controlled monitoring and abortion to "rid us" of the XYY type.

On a nationally televised show the host referred to the "criminal gene." Some public officials have advocated that all XYY males born be promptly registered at birth and be kept under scrutiny over the years until they have passed through young adulthood.

THE XYY PERSON AS A SPECIAL TYPE

Who is the XYY person who has so stirred up the scientific world? Until twelve years ago he was just a mild scientific curiosity. As noted, boy babies result when the father's sperm adds a Y chromosome to the mother's X chromosome. Girl babies are born if the sperm doing the fertilizing is a female or X sperm. But in some cases an extra X or an extra Y get into the fertilized egg. In about one male birth out of nine-hundred an extra Y ends up in the sex chromosome set. The odds are about the same for boys to have an extra X.

Some people have speculated that the double Y might make the growing male unusually masculine, and that the extra X (XXY) might have some feminizing influence on him. There is little evidence of the latter occurring. Some expert observers have suggested that an XYY male may be a bit more likely to be rambunctious or impulsive in the nursery than average, but many are "perfectly fine." Few differences show up on personality tests. As young men they are likely to be quite tall and to have acne.

The commotion over the XYY males began in 1965, when *Nature* carried a report by Patricia A. Jacobs and her associates on prisoners in a Scottish special-security prison. The prisoners were considered to be dangerous, violent and to have "criminal propensities." Of the total, 3½ percent possessed the XYY chromosome makeup, a far larger proportion than is found in the general population. An early study in Denmark came up with roughly comparable findings. A study carried on at several general prisons in Pennsylvania was confined to tall prisoners (over 5'11"). Four percent were XYY. (Interestingly enough, 5 percent had the supposedly unimplicated XXY imbalance.) In all, 9 percent of these *tall* prisoners had one or the other unbalanced set of sex chromosomes.[10]

The great majority of XYY males lead relatively normal lives. Only a small fraction ever show up in prisons or mental institutions. But XYY males did seem to show up considerably more often as problem people than their prevalance in the general population would seem to warrant.

In many parts of the world criminal lawyers have been checking to see if their clients have XYY chromosomes. If they do, the lawyers try to get them freed or have their sentences reduced. Their clients, they argue, are prisoners of their genes. In Paris, a lawyer got a reduced sentence for a client. And in Australia, the XYY argument was one factor in having a client sent to a mental hospital instead of to a prison.

A number of scientists have shown a fascination with mass-screening programs that would enable them to identify and study XYY males.

In Maryland, more than fifteen thousand juveniles—some delinquent, some not — were screened. The project was federally funded and handled by Johns Hopkins University scientists. In most cases parents were not advised that their children were being tested for this stigmatizing disorder. The same blood taken routinely to screen for anemia was used to test for XYY chromosomes! Protests by civil liberties groups brought an end to this project.

Meanwhile, a psychiatrist and a geneticist at the Harvard Medical School had embarked on a long-range project to identify XYY male children by mass screening. For seven years they tried to spot all such children born at the Boston Hospital for Women. More than ten thousand babies were screened. At first, at least, informed consent from the parents was either not sought or sought in a vague way. The study was largely funded by the crime and delinquency division of the National Institute of Mental Health. A dozen or so XYY male babies were spotted.

As the investigators began identifying XYY infants they faced a dilemma. Should they advise the parents or not? They concluded that they would be morally delinquent if they did not do so. They went ahead and offered to counsel the parents on child-rearing strategies in order to offset any tendencies the children might show toward antisocial behavior.

Critics at Harvard, and elsewhere, argued that this intervention into family life would hopelessly warp any findings the two scientists would make about the way XYY boys develop. They argued further that notifying parents that their newborn babies were somehow abnormal put a strain on the parents and worsened what might have been normal parent-child relations. And the children, as they grew up, might have reason to feel stigmatized.

For months Harvard was an XYY battleground. The faculty officially voted to support the project. But the psychiatrist in charge finally abandoned it on the grounds that the controversy was exhausting him. He said: "My family has been threatened. I've been made to feel like a dirty person."

In August 1976, *Science* published a report by a team of thirteen Danes and Americans that may well put to rest the whole argument linking XYY to aggressiveness. They made a study of 4,139 male Danish citizens in their early thirties who were over 6'1". Twelve of them, or about 1 in 333, had the XYY chromosome. These twelve — not a very big sample — did have a significantly higher arrest record than the normal

XY's. But almost all of them had been arrested for stealing. Only one had been involved in a crime of violence, and that was relatively mild.

The XYY's in the sample scored well below average on the Danish army's intelligence tests. This led the investigators to wonder if the elevated arrest record of the XYY's might be due to the fact that "people of lower intelligence may be less adept at escaping detection." The sixteen XXY's discovered in the same sample were also of below-average intelligence and had a slightly elevated arrest record.

Certainly the criminal-justice systems of the Western world have more pressing matters to deal with than to hunt down and watch the rare and usually mild XYY male in our population. His main problem, if he has one, is likely to be that of coping with an intellectual apparatus that functions below normal. And any good school system should spot that without putting a stigmatizing XYY label on him.

As for the general concept of mass screening at birth for genetic defects I agree that it can have a useful place in health care. But the screening should only be for severe ailments that are fairly widespread, treatable, and clearly understood. The screening must be with informed parental consent. There must be no records kept that are identifiable by name. Government agencies should be kept completely out of any screening procedures.

OTHER APPROACHES TO WEEDING OUT DEFECTIVE HUMANS

There have been a number of proposals to require or encourage the sterilization of persons with serious genetic defects. It would certainly be wise policy — both for the family and for society — if persons with serious genetic defects were systematically cautioned by their doctors about reproducing. Their attitude toward abortion, if fetal monitoring turns up evidence of a defect, might also be explored. If a couple is eager to have children despite a known hazard artificial insemination by donor might be acceptable to them and would eliminate reasons for concern.

Sterilization is legally required in Denmark on women known to have an IQ of less than 75. And North Carolina began, twenty-five years ago, requiring sterilization of residents seriously deficient mentally. So far, nearly 100,000 North Carolinians have been sterilized. To the extent that genes are responsible, sterilization is a negative eugenic measure. But it is doubtful that it has had much effect on the gene pools of Denmark or North Carolina. People who are seriously retarded mentally have a dramatically lowered capacity to reproduce.

A second approach to weeding out genetically defective humans is to prevent the union of people who might breed defective children. The state legislature of Illinois has recently had before it a proposal to hold up, at least temporarily, marriage licenses on genetic grounds. It was drawn up by the Chicago Bar Association. The wording was vague. All applicants for licenses would be required to submit not only to the usual test for venereal disease but for "any other diseases or abnormalities causing birth defects."[11]

Supposedly, people with long toes would be included. One of the drafters explained: "We are going to have to try to reduce the number of nonproductive members of our society." There was a further provision for mandatory counseling before a marriage license would be issued if test results indicated that the marriage would run an unusual risk in producing children with birth defects. The proposals were attacked as being so vague as to frighten away many people who would otherwise apply for marriage licenses. The proposals were finally dropped.

The theologian Paul Ramsey has suggested that the marriage-licensing power of the state might well be used to help prevent the transmission of "grave, dominantly inherited diseases." He doubts that the right to bear children is an unqualified right.

Perhaps the marriage-license bureau will become a checkpoint at which normal young persons carrying dangerous recessive genes can be identified. Then, at least, measures to discourage the births of defective children can be taken.

It would be heartless, however, to turn away two normal young people deeply in love, simply because tests show they both have the same unrealized problem. Of course, another complication in such a plan is that the bride-to-be is often already pregnant when the decision to marry is made.

The challenge is to keep them from falling in love. The biochemist Linus Pauling has proposed a forthright solution. Small avoidance signals should be tattooed on the forehead of every young carrier. That measure would certainly be resisted. However, a voluntary program of putting the tattoo in an inconspicuous place — behind the left earlobe, for instance — might work. Then, before young people decide to marry, they can check tattoos to see what is what.

Or in high school health classes each student could be encouraged to fill out, and keep, a simple medical form. It would include a section of genetic data as well as a check-off list of common ailments, like whooping cough, that the student had already had. Young men and women who were considering marriage could compare their genetic records. The problem of recessive genes should be explained in health classes. Test programs should be available.

A good example would be a test for the recessive sickle-cell trait. Among black Americans it is too common for comfort when marriage is contemplated. If the girl has the trait she has a real reason to be careful in mate selection. About one out of ten men she might consider will also have the trait, according to Robert F. Murray, Jr., a geneticist at the Howard University College of Medicine.[12]

A third proposal that has been made for weeding out genetically defective humans is to set an age limit on childbearing. Mongolism is only one genetic defect that becomes more common with maternal age. *Theological Studies* carried a proposal for "proscribing childbearing in women over thirty-five." The word *proscribe* can mean either "denounce" or "forbid." With more and more women having serious careers and postponing childbearing, "forbidding" childrearing would be unreasonably harsh. A simpler solution would be for all doctors or technicians performing Wassermann tests on marriage-license applicants to hand out a matter-of-fact booklet that would explain the genetic facts to keep in mind in childbearing. And it would contain a section on the extra hazards that older wives should keep in mind.

We have seen that negative eugenics, if approached compassionately and without coercion, can have some happy results. Happy for the family, happy for society. Genetic counseling, fetal monitoring, and voluntary, large-scale screening should, I believe, be encouraged. On the other hand, something uncomfortably manipulative looms when enthusiasts of eugenics start talking about laws regulating who procreates with whom. All mandatory regulations pose an unnecessary infringement on our personal freedom. Uncomfortable problems then arise of deciding who shall weed out whom.

There is the danger, too, that enthusiasm for the weeding-out can get out of hand. Are we going to become somehow unpatriotic — perhaps second-class citizens — if we give birth to less than "normal" children?

Are we going to be subtly encouraged to lose our sensitivity to the problems of people who are in distress because of ailments with a genetic origin? Stigmatization can be unspoken.

Let us not forget that Lord Byron had a clubfoot. Dostoevski was an epileptic. Woody Guthrie had Huntington's disease. The inventive genius Steinmetz was grotesquely deformed. Abraham Lincoln had a congenital ailment that made his fingers and toes abnormal in size and caused a number of other abnormalities.

Would the state of Illinois, where he grew up, now discourage the marriage that led to his conception? (As mentioned previously, Illinois is the state that considered holding up the marriage of anyone possessing "any diseases or other abnormalities causing birth defects.") Many people, like Lincoln, unquestionably are motivated to compensate for defects by excelling in what they can do.

In this chapter we have been exploring "negative eugenics" on an individual and mass scale. Sir Julian Huxley predicted: "Negative eugenics is of minor evolutionary importance and the need for it will gradually be superseded by efficient measures of positive eugenics."

I suspect that this is an overstatement. But let us turn to positive eugenics.

Packaging Superior People 18

Modern genetics is on the verge of some truly fantastic ways of "improving" the human race.
— James J. Nagle, geneticist

In concept, positive eugenics is as old as the selective breeding of animals. As for humans, Plato argued that human baby production should be limited to people selected for desirable qualities. At the turn of the twentieth century Sir Francis Galton tried to make a scientific movement of eugenics (he coined the word). "It has now become a serious necessity to better the breed of the human race," he said. "The average citizen is too base for the everyday work of modern civilization."

He was vague about what constituted baseness. But to him "civilized man" was epitomized by the English upper classes. They were the repository of "virtually all that is biologically precious in the English nation and possibly in mankind." Biology was then a rudimentary science and offered him few techniques for launching his movement.

A couple of decades later an American, Herman J. Muller, proved that genes could be modified and spent the rest of his long life ardently promoting positive eugenics. He argued for large-scale programs by which mankind could be transformed genetically. He contended that Man was now in a position, if only he would, to "transcend himself." Muller was promoting sperm banks as one possible vehicle long before they proved to be practical. The banks would be used to store sperm contributed by society's most superior and eminent males.

He argued that in the future, when sperm banks became an actuality, the "best genes" should be used. Children then would blame their parents if the parents, instead of obtaining these "best genes," had sought simply to procreate themselves. Seeking immortality through passing on one's own genes could become an act of "heedless egotism."

In the late 1930's Alexis Carrel, a French-American physiologist who had won a Nobel Prize twenty-seven years earlier, made the statement: "Eugenics is indispensable for the perpetuation of the strong. A great race must propagate its best elements. . . . Eugenics may exercise a great influence upon the destiny of the civilized races."

During the World War II period the Nazis made eugenics into a dirty word for a while. They issued harsh edicts on sterilization and elimination. They put girls to work as breeders. They talked of building a super-race.

By the 1950's and 1960's, new biological discoveries made mild, seemingly benevolent eugenics programs feasible. General betterment of mankind became the goal. Support or fascinated interest came from such eminent biologists as Sir Julian Huxley, H. Bentley Glass, and Robert L. Sinsheimer.

One of the rationales used by scientists in pressing for positive eugenics came from the threat of overpopulation throughout the world. (A number of non-Western governments are finally becoming serious about trying to slow down the population explosions in their respective countries.) If you are going to try to control the quantity of population, why not also control the quality? In 1974, at a meeting of the Eugenics Society in London, the British geneticist J. A. Beardmore of the University College of Swansea took up that idea. He suggested it would simply be a matter of years before "some formal consideration of population quality as well as of quantity will need to form a part of any governmental policy on population."

What about the preferences of the people themselves? Will ordinary people always prefer procreating themselves if offered the beguiling chance of creating children genetically superior to themselves? The argument may be made that they simply are starting with superior seed. They will still be responsible for the important environmental influences that shape children. What about adults who feel desperately inadequate or homely? Would they like something different or *extra* for their children?

It is quite probable that in a matter of years both parents can contribute their genes and still have a choice of additives from other humans. Multiple parenthood may well be in the offing.

It is also probable that authoritarian countries, which can pretty well control their populaces, will embark on large-scale programs of positive eugenics before Western countries do. They can make massive propaganda appeals for volunteers. Also, they can if necessary make regulations about who is given preference in procreation. And they will probably have clearer ideas of the traits they want to breed toward, such as strength and tractability.

A person breeding German shepherd dogs knows exactly what to breed for in trying to achieve the ideal for that breed. The long list of traits that constitute the ideal is printed in manuals. The list for each breed is used as a checklist by judges in dog shows. It includes both structural and behavioral traits.

Muller, Glass, Huxley, and other scientists have, perhaps fortunately, come up with a mixed assortment of human traits to promote. And they haven't pretended that the ones they chose constitute perfection. They come closest to agreement on superior intelligence, which is partly inheritable. (Inheritance apparently plays a slightly larger role than environment in determining IQ scores. About fifty studies in various countries lend support to this.) All three scientists mention a cooperative or adaptable disposition. And on one or more lists appear sound health, beauty, rugged physique, expressiveness, vigor, longevity, and reduced need for sleep.

Muller, in particular, included some traits that quite certainly have little or no genetic basis. He mentioned moral courage, integrity, progressive spirit, eagerness to serve mankind, compassion, and appreciation of nature and art. In his future society such traits presumably would be left not to eugenics but to euthenics, the movement to improve the human species by the control of environmental factors. Euthenicists shape people by shaping environment. Their number would presumably include the operant conditioners and the imprinters.

For the present, the biological technologists offer us four principal ways to work toward engineering a superior species of Man:

- Enriching the human seed
- Regulating who can produce babies
- Making biological alterations
- Reshaping family composition

A fifth approach, that of duplicating humans of a desired model, is still some years off. We will get to it after considering the others.

ENRICHING THE HUMAN SEED

In their scenario for our future — starting, say, a few years from now — the positive eugenicists expect to convert sperm banks into seed markets. These will handle not just frozen sperm of known quality but also frozen female eggs and frozen embryos. The larger markets would have their own laboratories for uniting selected sperms and eggs, and for cultivating the resulting embryos for a few days before freezing. Such prepackaged embryos of ready-to-go humans would be defrosted before being implanted in women.

The current pattern of a flat price per ejaculate would disappear as the emphasis shifted to quality. There would still be the staple items of semen provided by nearby, money-hungry college students. But the stress would be on suppliers of seed and embryos who have, as Muller puts it, "outstanding gifts of mind, merits of disposition and character or physical fitness."

If you are going to use donor seed to have a baby, they reason, why not get the best? Semen and eggs of healthy men and women with a guaranteed IQ of 130 or more could certainly command a $100 premium over the base price.

Athletic stars with magnificent physiques and famous movie stars with glorious physiques often supplement their incomes by endorsing products. Would some of them be willing to endorse themselves in a sense by selling a product created by themselves, their semen or eggs? It could be very attractive financially, especially for those whose incomes had started declining, as they usually do for such people. Men, for example, could easily add up to $50,000 annually to their income.

Perhaps some former Mr. Universes and some former Miss Americas could be induced to supply their seed, at say $500 per specimen. If the personage was not only physically outstanding but had outstanding talent the going price per specimen might rise to at least $1,000. Such a personage might be a modern-day star of the Marilyn Monroe or Johnny Weismuller type.

What about great athletic stars who were valedictorians of their college classes, or later became professors or business executives? Their semen might command a premium of $1,500. The frozen sperm taken at the prime of life from a man who became a national cabinet member or a distinguished scientist or the president of a major corporation surely might command a $2,500 premium. What about the semen of a modern-day Cellini, Beethoven, or Mark Twain who was running low on funds? Certainly there would be people willing to pay a premium of quite a few thousand dollars.

If the notable individuals were reluctant to be identified by name, they could still command an attractive premium for their seed. The seed-packager could identify the person as a former Academy Award winner. Or a Nobel Prize winner. Or the winner of four national golf tournaments. Or a self-made millionaire. Or a man who had been to outer space.

And what about the price for a frozen embryo that combined the seed of a certified genius and a well-known actor or actress?

In general, a female egg would cost substantially more than a thimbleful of male sperm. Eggs are harder to obtain. And even with chemically induced superovulation a woman would be hard put to supply the market with more than about 130 eggs a year. A single ejaculate of a male genius could be stretched to fertilize several dozen female eggs if the fertilizing were done with an eyedropper in a laboratory. *In vitro* fertilizing requires only a drop containing a few thousand sperms. The long, exhausting race up the female's reproductive tract, which leaves behind at least 99 percent of the sperms in an ejaculate, would be eliminated.

Presumably, each seed bank would issue a catalogue describing the various donors by traits, with prices.

Bentley Glass points out that donors would not have to be real, live people. They could be free-floating organs. He has stated that he expects techniques to be developed "for the cultivation in the laboratory of portions of human ovary and testis permitting successful continuous production of mature ova [eggs] or sperms." Glass explained that successes in producing mature female eggs "from cultured mouse ovaries lead me to expect that only persistence by a sufficient number of skilled biologists is needed to attain successful cultivation of human reproductive organs."

These organs presumably could come from well-endowed young people cut down by accident or disease before their prime. The organ cultivation, he reasons, would permit a "continuous production of eggs and sperm, and formation by fertilization in the laboratory of as many human embryos as may be wished."[1] Wow!

Speaking of mice and men, the most dramatic possibility of seed enrichment was discovered by Beatrice Mintz of the Institute for Cancer Research in Philadelphia. She wasn't — and isn't — interested in human-seed enrichment. But she perfected a research technique with mice that permits multiple parenthood! Wow again.

Her strategy, in studying how organisms develop through cell division, was to produce "markers" that could be traced. She mated two purebred black mice. Simultaneously, she mated two purebred white albino mice. When the fertilized eggs of each mating had grown to the eight-cell stage they were placed side by side in a culture dish. The outer covering of each embryo was dissolved so that cells from the two embryos could be intermingled. She put together an eight-cell embryo composed of cells from both sets of parents. In short, four parents. This new embryo proceeded to grow. It was implanted in a prepared female mouse and grew to birth.

The offspring showed dark and light markings, not only in the coat but even in the eyes.

She told me that she has put together more than a thousand "handmade" mice born of more than two parents.

Could human individuals with multiple parentage be created? She said she was "quite sure" they could be. Robert Sinsheimer agrees that it probably could be done. He has speculated on what characteristics would emerge, especially the psychological and behavioral ones. His conclusion: "No one knows." He imagines that the human result would be exotic.

Mintz takes a dim view of most of the ideas that excite the positive eugenicists. She thinks that creating multiparented humans would be irresponsible. Why? Because the results aren't predictable. Unlike laboratory mice, people aren't purebreds. The child could turn out to be a hermaphrodite, unless chromosomes were carefully checked for sex or unless the two-sexed fetus was aborted after being spotted in fetal monitoring. Also, unless there is a careful matching of pigmentations in the parents the child could have a vaguely striped appearance.

REGULATING WHO CAN PRODUCE BABIES

Until well after the dawn of history Man commonly arranged for the most talented males to procreate more than their share of the society's babies. This was done through the widespread institution of polygamy. The rulers, chiefs, and other males who could demand or afford it surrounded themselves with wives. The embryologist Robert Francoeur points out that King Solomon could pass on his famed intelligence via the offspring of six-hundred-odd wives and more than three hundred concubines.

Today's enthusiasts of quality control contend that medical technology and the looming population problem warrant regulations favoring the well-endowed. Roger W. McIntire, a psychologist at the University of Maryland, contends: "With the population problem now upon us, we can no longer afford the luxury of allowing any two people to add to our numbers whenever they please." Enter social manipulation.

In the January 15, 1971, issue of *BioScience,* the biologist Carl Jay Bajema summed up four ideas for regulating the size of the population, with eugenics in mind:[2]

1. Grant two *marketable* baby licenses to each family. You would need a license to have a baby. The ostensible aim would be to achieve zero population growth. But since many families have no children, or only one, they could sell their unused licenses. This idea of a marketable baby license was apparently first advanced by the noted American economist Kenneth Boulding.

Bajema figures that this arrangement would not only control population growth but work to improve the human stock. The achievers of the world tend to be the most affluent and would therefore be able to purchase licenses for extra babies from those who are less successful "because of their genetic limitations." (I know humans who are "unsuccessful" for other reasons.) Bajema concedes that for this plan to work well "societies would have to make sure that achievement and financial reward are much more highly correlated than they are at present." That would take some massive manipulation. And incidentally, what is achievement? In 1976 the *average* player in the U.S. National Basketball Association was making $114,000 a year.

2. Grant each family two nonnegotiable baby licenses and periodically create a pool of licenses equal in number to those that had been unused. People desiring more than two babies could compete regularly for licenses from this pool by winning contests. The contests might be tests involving "mental ability, sports, music, arts, literature, business, etc."

3. Administer a "simple eugenics test" to any couple desiring more than two babies. The test would be an examination of the first two children born of the couple. The children would have to be above average physically and mentally for their parents to qualify for additional licenses. This proposal was apparently first advanced by Bentley Glass.

4. Require all married couples desiring children to "pass certain tests" before they can qualify to have *any* children. This solution might be recommended if the first three proposals did not seem to be tipping the genetic scale favorably. Couples who failed to pass the tests would be granted a conditional baby license. The condition: they must go to a human seed bank and follow prescribed instructions. They would have their child "by artificial insemination and/or artificial inovulation." In this they would utilize "human sperm and eggs selected on the basis of genetic quality." The seed of the husband or the wife or both would not be used to create the baby they would be permitted to have.

The psychologist Roger McIntire, cited above, has been proposing compulsory licenses for parenthood for some time. He is an enthusiast of behavior-modifying methods of the Skinner type. His concern is less with eugenics than with eliminating parental incompetence. He would use child abuse (plus overpopulation) as the reason for making parents take lessons in parenthood in order to get baby licenses.

McIntire is not alone in being interested in a parental competency requirement. This is indicated by the fact that he was invited to present his ideas to the 1974 meeting of the American Psychological Association. He had earlier unveiled his ideas to a meeting of the Eastern Psychological Association.

Eugenicists would certainly applaud his ideas on who could qualify to have babies. The would-be parents couldn't be dummies. His system would be far tougher than the tests devised in the Old South for screening would-be voters, black and white. McIntire suggested that to qualify, couples would have to pass a course in child rearing at a local community college. The course would cover not only principles of nutrition, health, and learning, but behavior-modification principles, such as modeling, imitation, and reinforcement. There would also be lessons on "extinction procedures and adjunctive behavior."

Let's assume — and it's a big assumption — that these principles could be presented to typical would-be parents in understandable terminology. Since a depressingly large number of Americans have trouble filling out a job-application form despite ten or twelve years of schooling, it is surely reasonable to assume that millions would flunk the examinations in the child-rearing course. And failure to pass would presumably disqualify them for parenthood. Hence we would have eugenics entering the picture.

In McIntire's scenario as presented to the APA, Congress would set up a federal regulatory agency to administer a national parenthood-licensing program. He explained that it would be "similar to driver-training and licensing procedures." Similar? Applicants for driver's licenses don't have to go to college and take courses.

McIntire would draw upon medical and computer technology to make sure unenlightened people didn't get away with illegal parenthood. He points out that scientists are perfecting long-term contraceptives. One is a three-year capsule, implanted under the skin. Soon, he expects, semi-permanent contraceptives will be available that will remain in effect until an antidote is administered.

In his scenario he calls the contraceptive and its antidote Lock and Unlock. He envisions that the federal regulatory agency's local commissions would control the distribution of both Lock and Unlock. Presumably Lock would be, by law, administered to all females (or males) of reproductive age. Later, anyone who wanted an Unlock in order to have a baby, would first have to obtain a certificate showing that he or she had passed the test for sound parenthood. McIntire added in his scenario presented to the APA: "Since the records of distribution are stored in federal computer banks," the identification of any illegal babies made possible by the unauthorized use of Unlock becomes a simple matter. "Parents convicted of this crime," he projected, "are fined and required to begin an intensive parenthood-training program immediately."

In his project all this would occur within the coming decade. Knowing how Congress acts, and the Catholic Church reacts, I imagine it may take a little longer.

He suggests that the government could make a plausible case for intervening in child-rearing decisions within the family. After all, it is now saddled with substantial burdens in the rearing of children. It has to run schools, medical programs, youth programs, crime-prevention programs, colleges, parks, and so on.

McIntire has received hundreds of letters from people who have read about his Lock-Unlock proposal. The comments range from "Right on!" to "Sieg Heil!" About half, he told me, were flatly disapproving. He might find cheer in the fact that all really revolutionary ideas take a few years to catch on with the public.

MAKING BIOLOGICAL ALTERATIONS

Direct genetic engineering with humans is not likely to be a possibility that can fascinate or worry us before, say, 1984. But before the year 2000 it will probably be possible to modify physical traits and bodily processes if only one, two, or three genes are in control. With fruit flies it became possible several years ago to create, by gene manipulation, changes in the color of the eyes, the structure of the body, the number of bristles, and the like.[3]

Genetic diseases caused by the failure of a single gene to order up a critical enzyme will be coming under control.

Apparently, in some of these diseases, such as a certain liver disorder, the cause is not a missing gene, but rather the failure of the gene in question to be switched on.

Although every cell in the body has the same set of genes, most of them get switched off as they move into specialized areas like the eyeballs, the toes, or the liver. Some get switched off by mistake. Scientists at the universities of Cologne, Harvard, and Cambridge have learned the mechanism of this switching-on and -off process. They are starting to learn how to influence it.

The ultimate height of the growing bodies of boys and girls is starting to be manipulated by hormones. Mothers who feel that their daughters will be social misfits if they are too tall — or their sons misfits if they are too short — are consulting pediatric endocrinologists. Heavy doses of estrogen taken daily for a couple of years starting at age ten can cut two or three inches off the projected adult height of a daughter. Growth hormones have been used with some success to add an inch or two to short boys. Some scientists now believe that in the future a far more effective promotion of growth may lie in a component of growth hormones called somatomedin.[4] Some youngsters can go the full route with it without difficulty. Others, especially girls, have been afflicted with such side effects as chronic nausea, hypertension, and darkened nipples.

Speaking of nipples, there has been some talk of the wisdom of debreasting women as a part of overhauling people. I heard the idea first advanced by a medical philosopher at a conference on medical ethics near New York City. What fascinated me was that none of the fifty-odd scientists, physicians, and philosophers present arose to shout their consternation. Both men and women were present. Should we become the first mammals without mammae?

The idea was cited by H. Tristram Engelhardt, Jr., of the University of Texas Medical Branch, Galveston. It turned out that he had discussed the matter in two publications and that a New Orleans medical researcher had presented a paper referring to the idea at a Cambridge University symposium.[5]

Engelhardt pointed out that millions of newborn girls in the Western world are condemned to suffering or death because of breast cancer. Enormous sums of money are being spent on research and treatment of the disease. The whole problem could be eliminated simply by nipping off a bit of breast tissue at birth. Such an operation would be about as complicated as circumcision. Baby bottles have largely removed the functional need for breasts, although some argue that early breast feeding has advantages over bottle feeding. There is little evidence that breast size is significantly related to human milk production. A woman who wears a size AAA bra can nurse effectively. The anthropologist Robert Ardrey points out that female chimpanzees have very meager breasts even while nursing. He suggests that large breasts were developed in human females as a sex enticement. Humans, unlike the great apes, normally engage in frontal copulation.

For the modern liberated woman, breasts are often an encumbrance on the job, certainly so for foresters, jockeys, acrobats, soldiers, and mechanics. And they are also an encumbrance in such sports as tennis, golf, swimming, high-jumping, and the hundred-yard dash.

But what would breastlessness do to women's sense of identity? As for the use of breasts to attract men, padded bras and synthetic breasts held in place by suction could perhaps be substituted until the world got used to breastlessness. A poll on the subject among women in the Western world might be interesting. Perhaps a substantial response would be a hesitant yes, but only on condition that everybody else be breastless too.

RESHAPING FAMILY COMPOSITION

Evidence is piling up that size of family and the spacing of children have a lot to do with the intellectual development of the children. The known facts are being discussed by eugenic and euthenic enthusiasts. This new knowledge perhaps could be used to produce a general rise in intellectual level.

The approach would be through mass-education programs, through counseling provided by family-planning clinics, and by programs to limit family size, especially of the less-endowed. Mothers with only a few years of schooling tend to have more than twice as many children as mothers with college educations.[6] Dobzhansky calls that situation "undesirable."

Large-scale studies in France, the Netherlands, Scotland, and the United States all point to a relation between intellectual growth and the size and shape of families. Scholastic test scores were examined as a function of family size and birth order. Scores declined with increasing family size. Scores declined within the family. First children generally showed up among the brightest.[7]

A University of Michigan psychologist, Robert B. Zajonc, won a prize from the American Association for the Advancement of Science for helping develop a theory to account for this.

The intellectual growth of a child, he contends, is greatly affected by how much intellectual input the child gets from other members of the family. When there are two adults and one infant there is a lot of input. When there are two adults and five children, all under the age of eight, there is often a serious dilution of intellectual input per child. But in a large family this decline can be counterbalanced by spacing the children three or four years apart. Then the younger can get significant educational input from the older ones.

Under his theory, any movement to encourage zero population growth will produce a somewhat higher level of intelligence.

Zajonc adds, however, two important points: One, the tendency for a decline in intellectual input per child in a large family can be counterbalanced if the parents spend a good deal of time with each child. Two, children in a sizable family may have a better chance to develop social competence, ego strength, and moral responsibility than children in a small family. In short, higher intelligence gained by small family size may have its price.

Human improvement is an attractive goal. When it is sought through the volunteer programs aimed at specific objectives, such as better mental or physical health, or becoming better parents, we can all applaud.

What about a *movement* to realize some preconceived ideal human? This understandably can, at first glance, fascinate us, particularly those scientists who believe they have the insights and technology to help bring it about.

In savoring the possible gains we must also start looking at some probable social costs. For example:

• Many of the proposals impinge upon freedom of choice. Glass asks: "Cannot the substitution of a greater freedom of choice in new respects compensate for the restriction of some time-honored privileges?" Any trade-off of freedoms we now have for the freedoms in "new respects" should, I suggest, be viewed warily.

• The whole notion of an "ideal" human prototype to strive toward is worrisome. The notion would push us toward standardizing humans. We should cherish our present diversity, I believe. It makes life interesting for us. It stimulates our creativity.

- A higher level of intelligence is one of the generally agreed-upon ideals to strive for. Yet even here social questions arise. Government figures indicate that only 20 percent of the available jobs in this country really require a college education. Yet a sizable majority of American parents already want their children to go to college. Overeducation has become a national problem. Can we become too bright as a society for the society's well-being?

  Another point: If we raise intelligence levels without also increasing training in altruism and social responsibility we can end up as a society of sharpies.
- The use of donor seed to create "ideal" children raises several social questions. The purchase of the "ideal" seed of notables is not only a bit fatuous but belittling. It puts a dollar value on human life. Even assuming that prospective parents simply buy certified Class A seed, problems will arise in parenthood. Raising children who obviously are genetically like you not only can be gratifying but is actually easier than raising genetically strange children. That at least has been the experience of parents who adopt.

  Charles Fried, a professor of law at Harvard and an investigator of bioethics, states: "The child from random egg and random sperm grows up in a family situation which is in some sense wholly the result of chance, and this . . . may be a disturbing and unsettling fact."
- Finally there is the hazard of outright State control of the "ideal" characteristics we should possess, as in the Lock-Unlock situation. Harriet Pilpel, an authority on the legal aspects of the new reproductive technologies, told me that she is appalled by the idea of seed banks collecting human seed with "desirable" characteristics. It could provide the State with a technology for imposing its own standards.

The geneticist James Nagle who, as previously cited, foresees "fantastic ways" emerging for "improving" the human species, is uneasy about how some of these "ways" will be applied. He says the possibility that the application of these technologies will come about "under the direction of an elite, dogmatic leadership creates the greatest social anxiety" about them. We could end up, he suggests, under the dictatorship of "hereditary endowment manipulators."

And now we turn to the fifth way in which the biological technologists may be able to engineer superior people.

Duplicating Humans
of a Desired Model

19

Man is indeed on the brink of a major evolutionary perturbation . . . vegetative propagation.
— Joshua Lederberg, biologist

To biologists, vegetative propagation is the phrase sometimes used for any kind of nonsexual reproduction. In short, it is virgin birth. Or to be scientific, it is parthenogenesis (from the Greek *parthenos,* meaning "virgin," and *genesis,* which of course means "origin").

But how can it be? The wisdom of history from long before the Ark is that it takes two, a male and a female, to procreate. The birds do it. The bees do it. The fish do it. The bunnies do it. We do it. Even the buttercups do it. Each does it in his own fashion, but does it. Almost all visible lifeforms begin with the commingling of two differing seed materials.

All this was indeed true before the new biological micromanipulators got into the picture. But it now appears possible, in principle, to produce a dozen or a thousand identical twins of a valued racehorse or a prize athlete or a noted genius or a vain dictator. Or you could have an identical twin made of yourself in a way that would involve no mingling of seed.

Since word got out about this possibility a few years ago, there has been jocular speculation of mass-producing Albert Einsteins, Artur Rubensteins, Mao Tse-tungs, Margot Fonteyns, Mickey Mantles, Woody Allens, Lillian Hellmans, shahs of Iran, or queens of England.

The process for duplicating or mass-producing copies of an individual is best known today as cloning. In 1932 Aldous Huxley in his fantasy *Brave New World* dreamed up the name Bokanovskification for some such process. His scientists of six centuries from now could by some wizardry grow ninety-six human beings where only one had grown before.

Titillating or not, the possibility of genetically replicating individuals without sexual congress is taken seriously today by many prominent scientists besides Lederberg. Among them are James Watson, Robert Sinsheimer, Robert Edwards, and Jean Rostand.

The *Southern California Law Review* has taken cloning seriously enough to publish a 104-page assessment of the legal problems that would arise from the technology of cloning.

The main argument is not over whether parthenogenesis of humans can be done, but when, and how best to do it.

A few contend that it may never become feasible and that therefore it is preposterous to get worked up about the subject now. I have before me thirteen estimates by well-known biologists. One, Marc Lappé, calls it a distant possibility. Scientists at the Rand Corporation arbitrarily estimate that human cloning will be widespread by the year 2005, or twenty-eight years from now. All the others give a closer date. Some say within fifteen or twenty years. A few say it will be possible within ten years. The embryologist Leon Kass, who is no enthusiast of human cloning, is in the last group. He reasons:

"Given the rate at which the other technical obstacles have fallen, and given the increasing number of competent people entering the field of experimental embryology, it is reasonable to expect the birth of the first cloned mammal sometime in the next few years. This will almost certainly be followed by a rush to develop cloning for other animals, especially livestock. . . . With the human-embryo culture and implantation technologies being perfected in parallel, the step to the first clonal man might require only a few additional years."[1]

A lot may depend on whether one or another nation decides to make a major effort to achieve the cloning of either livestock or humans. Certainly, unless human cloning is banned, it is likely to become a reality within the lifetime of the great majority of people now alive. Since it will indeed represent "a major evolutionary perturbation," as Lederberg put it, I feel it deserves our thoughtful if brief inspection. It could become a momentous venture in attempts to manipulate Man.

The theoretical attractions of being able to create one or more identical twins of an individual human at will by nonsexual — or asexual — reproduction are several.

Such a capability offers the chance to preserve and expand more precisely than sexual reproduction can the special talents of individuals insofar as those talents have a genetic basis. Sinsheimer has stated: "Cloning would permit the preservation and perpetuation of the finest genotypes that arise in our species — just as the invention of writing has enabled us to preserve the fruits of their life's work." And Lederberg argues that if you have a superior individual — and its genetic blueprint in every cell — "why not copy it directly, rather than suffer all the risks, including those of sex determination, involved in the disruptions of recombination." In the traditional mating of humans — who of course are not yet near to being purebreds — you inevitably get a good deal of random assortment of genetic materials.

Thus, asexual reproduction of a human musical genius would probably offer a better chance of ending up with one or more musical geniuses than leaving the matter to romance. This assumes that the asexual process works well.

The sports farms of professional athletic organizations might well be attracted to the idea of making direct multiple copies of their 6'10" basketball stars. Since they already pay these stars more than $200,000 a year, building copies might seem like an attractive investment. Coaches have pretty precise ideas about the physical configuration of good football or soccer players. What professional ball team wouldn't want to produce for the farms a dozen copies of Pelé or O. J. Simpson? Some marriageable girls might feel honored to offer their wombs for the nurture of these prospective superstars. In any case, surrogate mothers could be hired.

Similarly, military leaders with billions of dollars available for research, might well be interested in mass-producing humans high in endurance, strength, or proneness to obedience. Dictators might justifiably see universal cloning as a way to achieve a more efficient, predictable, regulatable populace. Here are some other theoretical attractions:

- If you had a hundred clones or a thousand clones in a community there would be one clear medical advantage. Their organs could be freely transplanted among fellow clones without the still very bothersome problem of graft rejection.
- If a husband and wife were in deep distress because a greatly beloved child was dying, they could arrange to create another child that would be genetically identical.
- A woman who wanted a baby but had not come across a satisfactory mate could have by virgin birth a baby of her own flesh. The baby, though younger, would be the mother's identical twin.
- People interested in personal immortality could assure themselves of at least a start. They could arrange, through cell banking, to have persons of their exact genotype live on. Not for eternity, perhaps, but for at least a couple of hundred years.

Still other arguments are offered. Asexual reproduction would permit sex determination because same sex would be assumed. And it would offer still another way to bypass infertility in either sex.

Animal breeders are showing a lively interest in cloning. Those already involved in breeding purebred strains probably would not achieve much improvement in such characteristics as milk production. But they could assuredly determine sex. A countryside full of identical cattle, however, might be more vulnerable to a new virus. So you shouldn't have too many twins. Breeders of purebred racehorses and dogs have rules requiring old-fashioned one-to-one mating. The rules probably would have to be bent before they could turn to cloning.

SPONTANEOUS VIRGIN BIRTHS

To comprehend how cloning is accomplished, a little background may help — or at least be of interest.

Reproduction by nonsexual means does occur naturally in some life-forms, especially plants. Certain plants just keep growing vegetatively from cuttings of slips or shoots. Take the leaf of an African violet, put it in a pot, and it will grow an entirely new plant. Bamboo can be grown from cuttings, also English ivy and a number of grasses.

In higher life-forms there are recorded cases of female eggs spontaneously dividing and growing. Scientists at the U.S. Department of Agriculture spotted a strain of domestic turkey hens that has a tendency to lay hatchable eggs without benefit of semen from a tom. Over several years thousands of these eggs were identified. A few dozen hatched infant turkeys. Some grew to maturity. At least one has actually sired offspring.[2]

Virgin births in human females almost certainly have occurred spontaneously, which may be a comfort to some protesting maidens.

The obstetrician Landrum Shettles, in his years of plucking hundreds of eggs directly from women's ovaries, has found several eggs that had already undergone cleavage. These eggs could not possibly have had contact with a male sperm.

Some years ago there was an uproar over a talk given in England by Helen Spurway, a geneticist at University College, London. On the basis of animal research she cited figures suggesting that every year several dozen human virgin births probably occur somewhere in the world. Other estimates have since placed the estimate higher and lower. As a result of the commotion over her remarks, specialists at Queen Charlotte's Hospital agreed to make serological and other tests of mother and child when the mother claimed that no male had been involved in her child's conception. A number of mother-child pairs showed up to be tested. The doctors announced that one mother and daughter had passed every test they had performed that would indicate virgin birth.

So much for *spontaneous* nonsexual reproduction.

CONCEPTION IN A BROTH

In recent years Man in his laboratory has been learning how to manipulate reproduction in species that otherwise depend upon sexual reproduction. Take most plants. Almost all depend on sex pollination to reproduce.

The carrot is one example. Fourteen years ago the world of biology was electrified by news of a feat performed by a cell physiologist at Cornell University, F. C. Steward. He had grown a mature carrot — complete with flowers and seed — from a single body cell that he had scraped from the tip of an ordinary root.   ·

A body cell, unlike sex cells, already has its complete double set of chromosomes, which are needed for cell division. But as indicated, during growth only the genes needed for the specialized functions throughout the plant or animal remain active. All the unneeded ones somehow get switched off. Another phrase to explain this is that they get "masked off." Steward had switched on, or unmasked, the entire complement of the carrot's genes. It was like a born-anew fertilized egg. How did he accomplish this miracle? He did it by soaking the cell in a nutrient bath consisting mainly of coconut milk!

Since then, scientists have reproduced nonsexually a number of ordinarily sexual plants, including endive, parsley, and aspen trees.

The business of reproducing orchids has recently been revolutionized by a technique first perfected in France by Georges Morel. If you as an orchid breeder come up with a uniquely beautiful specimen and want to reproduce it exactly, you can't depend on sex pollination. Instead, you now have it cloned. This can make it worth tens of thousands of dollars. A large number of body cells are taken from the very tip of its stem, put in a chemical bath in rotating test tubes for several weeks. Thousands of exact copies of a highly prized plant can be raised from the emerging material. They may sell for $50 each, with a warning that by patent, "Asexual Reproduction Is Forbidden." A nursery in Azusa, California, reportedly is having success in cloning several flowering plants.

Now let's move to techniques for inducing asexual reproduction in animals.

REPRODUCTON BY SHOCK

For some years scientists have been using shock to make the eggs of sea urchins and other simple animals start to divide. And development has continued until baby sea urchins emerged. The scientists have successfully used strychnine, heat, needle pricks — literally hundreds of forms of shock.

An unfertilized egg has only half the normal set of chromosomes. But with such simple animal forms, having only half the normal complement doesn't seem to make too much difference. With higher life-forms the absence of a second set of chromosomes — which comes from the male in normal fertilization — apparently prevents full-term growth. However, Anna Witkowska of the University of Warsaw reported that she had by electric shock triggered unfertilized mouse eggs into dividing and growing embryos as far as midterm.[3]

CLONING BY CELL MANIPULATION

Now we come to the two known techniques that could lead to the systematic nonsexual production of exact copies of humans. And the reproduction could be done in large numbers.

The first technique involves replacing the nucleus of the woman's egg. This nucleus, as mentioned previously, is very much like the yolk of a chicken's egg. The "white" surrounding the yolk is called cytoplasm. Also as previously mentioned, an egg, before it is fertilized, has only one set of chromosomes. Union with the sperm provides the second set, which permits the growth, by division, of a new human being.

The strategy of cloning is to replace an uncompleted nucleus with the nucleus of a regular body cell. (A body cell, it will be recalled, already has a full set of chromosomes, and most of its specialized genes have been "masked off.") For cloning purposes the body cell can be taken from almost anywhere in the woman's body. Or it can be taken from the body of another individual, male or female. It will come from the person whom you want to copy.

The nucleus of the body cell, when substituted for the egg's nucleus, can be tricked into sensing that it is a newly fertilized cell. It will start growing by division into a new human. Apparently, all the masked genes of that body cell, taken, say, from an earlobe, will be unmasked, or switched on, by a chemical reaction in the egg's cytoplasm. Its full set of genes become operative again as if off to a clean start!

That is the strategy. Executing it is still an enormous challenge, at least for humans.

The first evidence that it might be possible came in 1966. J. B. Gurdon, a cell biologist at Oxford University in England, began trying the technique with frogs. He destroyed the existing nucleus of a sex cell with a tiny beam of ultraviolet radiation. Then he obtained the nucleus of a body cell that he had taken from the gut of another frog or tadpole of the same species. After putting it in a tiny hollow needle, he gently inserted it into the burned-out spot of the egg cell. The egg cell — now with a full set of chromosomes — reacted as if it had been fertilized. It began dividing.

The whole genetic system of the borrowed nucleus formed an embryo that grew to be a normal frog with all the characteristics of the donor, not of the mother who supplied the egg. It was an identical twin of the donor.

Gurdon spent a lot of time experimenting to achieve those first normal cloned frogs. Altogether, 707 attempts produced 11 clones.

Around the world scientists began attempting to duplicate the technique with mammals. And quite probably they have attempted it with the cells of that mammal most interesting to us, Man.

The leap from frog to mammal, however, has turned out to be gigantic. For one thing, the frog's egg is very big. It is an eighth of an inch in diameter. And it is fairly easy to manipulate. It has about seventy-five times the mass of a mammal's egg, which is a speck. There is also some evidence that mammal eggs are more fragile and that the masking of genetic material in the body cells of mammals is more complicated. Still, scientists are perfecting the necessary micromanipulative tools.

The second technique, which opened up as a possible strategy a few years ago, was the discovery, as mentioned in Chapter 16, that cells can be fused. James Watson refers to the technology of fusing animal cells as relatively "simple." The fusing procedure, he suggests, "has raised the growing possibility that further refinements of the cell-fusion method will allow the routine introduction" of the nucleus of a human body cell into a human egg cell. This would occur after the egg's nucleus had been removed.

In the cell-fusion method there is an actual blending of two cells into one, with the egg nucleus removed. It occurs when they are placed together in a laboratory culture. Viruses provide the stimulation to help the blending occur.

At this writing cells of mice so fused have been divided and grown to the embryo stage. A number of these early efforts have shown defects. So there would seem to be quite a way to go before human cloning becomes a reality.

A modified form of cloning would be to make ten or fifty copies of a fertilized egg. Such a copy would include the combined inheritance of mother *and* father, not just one individual. For a few days after conception, as cells divide and form into a tiny ball, cell specialization does not yet begin. Each cell still carries the unmasked blueprint of an entire new being. These cells can now be separated from the ball with little damage. Biologists believe it should soon be possible to plant such cells in a number of different wombs and grow what amounts to a flock of identical twins. It probably will have greater appeal in livestock breeding than human breeding.

IF CLONING OF HUMANS BECOMES POSSIBLE

If and when cloning does become a reality, James Watson foresees that there will be "a frenetic rush" to do experimental work with human eggs. He adds: "The growing up to adulthood of these first clonal humans could be a very startling event."

How long would it be before we would see movie-production empires ordering up hundreds of facsimilies of sex goddesses of the Raquel Welch type?

There would be problems right from the start. Who would decide what individuals were to be mass-produced by cloning? Would it be left to the free-enterprise market mechanism? Or would the State take over? If so, there would probably be black-marketing. Or would a nervous world set up the International Commission for Genetic Control to license clonists? That has been proposed by the English geneticist Lord Rothschild.

And again, what would we do with the mishaps, the monstrosities bound to result, at least in the early days? What if only ninety percent of the genes get switched on? Remember, Gurdon with his frogs averaged one success out of seventy tries. He, of course, was still learning.

Lederberg is afraid that social policy on cloning would not be set after the principles had been thoroughly debated, but rather by the "accident of the first advertised examples." Therefore he is anxious to hold off on humans until fetal monitoring is substantially more advanced than it is today.

A new set of human ethics would seem to be required. These New Ethics would permit scientists to experiment with latent humans through gestation, and would permit destruction of the fetus until at least twenty-four hours after birth. Otherwise, we would be confronted with an unacceptably high proportion of defective humans.

Leon Kass quite flatly puts the situation this way: "There is no ethical way for us to get to know whether or not human cloning is feasible." Are we ready for the New Ethics?

If cloning were accomplished, the predicted, initial "frenetic" rush might occur. But I suspect that public interest would end as abruptly as the public interest in moon-walking did once the initial curiosity was satisfied. Cloning would be reserved for special situations, special needs. Within two decades there would be disappointment that adult clones were not as precise copies of their donor parent as had been projected, even though they were genetically identical. The big successes in duplication would come when only identical physical configuration was sought, as in football stars and buxom movie queens. Disappointment would be more likely when duplications were made of people with special talents, high intellects, or outstanding personalities. Environment, starting right in the mother's womb, would still play a major role in development. A child's environment, including vivid life experiences, certainly has at least as much to do with aspiration, character formation, and disposition as heredity. And it has a good deal to do with intellectual development.

In discussing the idea of mass-producing an Einstein, the science writer

Horace Judson noted humorously: "Einsteins? But mathematical geniuses and violinists, we know, are the offspring of Jewish mothers — and you can't get them from a jar."

Cloning presumably would be done primarily with very successful people. What about the tendency of sons born in the shadow of very successful men to try to carve out lives of their own in some quite different field?

All the clones of a batch would, of course, be of the same sex. An occasional narcissist might wish to grow an exact copy of himself. But the vast majority of married couples would be repelled by the idea of having a baby that would be a copy of only one of them. Not only do most couples like the idea of conception by sexual union but they want the ensuing child to be a personification of their love.

If cloning became general it would represent one more assault on the already badly undermined role of parenthood. Many men would feel castrated. Many women would feel that they had been reduced to incubators. And under cloning, what would happen to child rearing? One of the most fascinating aspects of child rearing is to watch resemblances, talents, and personality emerge. Under cloning there would inevitably be a drop in interest in what would emerge. More would be foreknown, especially about physical appearance.

And what would be the identity problems of a child who was not only an identical twin — which is hard enough — but an identical twin of a parent? Kass suggests that the result could be psychologically disastrous.

As a major method of reproducing, cloning would be a disaster too. It would be a form of inbreeding that would head humanity toward a dead end. There would be a sharp decline in genetic variety. Diversity, it is widely believed, promotes human progress.

Cloning of animals, particularly livestock, may on balance be a worthwhile idea deserving thorough investigation. But for humans, cloning would be, on balance, unjustifiable. The hazards, the ethics, the meager chances for any significant gain for mankind, should impel us to reject it as anything but a challenging stunt.

All research concerning it should, in each country, come under the control of the national commission on reproductive research, proposed earlier. I do not believe that these commissions should ban laboratory research on the kinds of cell manipulation that could lead to nonsexual reproduction. Cell fusion, for example, can be an important research tool in understanding the genetic basis of such diseases as cancer.

A ban should begin, however, on research with embryos after the time they are ready for transplant into the womb. I believe that the ban should last indefinitely. At the same time I recognize that enormous pressure might build up from the scientific world to attempt cloning. Scientific curiosity is a powerful — and ordinarily a highly commendable — force for progress. And discussion of the scientists' ambition would be reported by the media and would excite public curiosity.

If such pressure does develop, then the ban should continue, at least until a number of primates have been cloned successfully. By successfully I mean with a mortality rate and an abnormality rate not significantly higher than that prevailing in sexual reproduction. This measure would put off any decision on human cloning for at least a dozen years. Probably more. And if a decision is then made to attempt to clone humans it should be done only as an experiment. It should be totally controlled by the national commissions, just as the National Aeronautics and Space Administration controlled manned landings on the moon. For the cloning experiment, medical ethics and the law would need to be specifically amended in order to permit the disposal of seriously defective clones that were not detected until after birth.

After a dozen successes in a row had been achieved, the goal might be set to try for six sets of clones. Each set might have six individuals. Once these six sets had been created, observed, and tested, the project should be closed down, just as the Apollo program was closed down after six manned landings on the moon. In the years since the final landing public interest in sending off other moon-walkers has declined sharply. And I suspect that public interest in building additional clonal colonies would decline just as sharply.

In general, cloning is an evolutionary perturbation that humanity doesn't need, and shouldn't want. How the clones turned out as they grew up might be watched for a while with as much curiosity as the early years of the Dionne quintuplets were watched. But the successful rearing of the quintuplets did not set off any broad surge to find ways to have more quintuplets.

After the official closedown, there would be no more cloning unless, on appeal, the commission made an exception for some extrordinary reason.

The Work on Man-Animals and Man-Computers

20

Can we tolerate the manipulations that blur the gap between animal and man?
— Gerald Leach,
author of *The Biocrats*

We come now to two particularly exotic concepts for modifying Man. They involve the physical union of Man, in varying degrees, with animals and computers.

I will avoid science-fiction projections and speak only of laboratory developments and discussions by scientists in good standing.

Science, the journal of the American Association for the Advancement of Science, recently had an article entitled "A Computer Under Your Hat."[1] It was based on a report to the 1976 annual meeting of the AAAS. As for Man-animal mixtures, they now exist as very small living creatures in many laboratories. Joshua Lederberg has observed that "there is enormous scientific interest in organisms augmented by fragments of the human chromosome set."

A hybrid of Man and animal is often called a chimera. In Greek mythology a chimera was a fabulous creature with the head of a lion, the body of a goat, and the tail of a serpent. It spat fire. In modern biology a chimera is any living structure in which the chromosomes of different species are combined. Often, there is a human component.

The biological chimera doesn't spit fire, but it has a tendency to inflame imaginations. This is especially true if the Man-animal hybrid is called a subhuman.

Its creation only became possible — like that of the artificially fertilized embryo and the clone — with the development of the new techniques for manipulating tiny fragments of life under a microscope. Manipulations requiring a steadiness of movement beyond the capability of a human hand usually are handled by powered micromanipulators.

Why would anyone, you may wonder, want to mess around combining elements of Man, that glory of Creation, with those of lesser animals?

Well, there is that compelling factor, scientific curiosity. Could primates be made even brighter if their chromosomes were mixed with human ones? There is the already real value of using hybrid cells as tracers in studying genetic defects and very early human development.

Some scientists feel that Man and beast could benefit mutually by swapping desirable traits. James Nagle mentions as a theoretical example the possibility that a liver trait found in animals might improve human liver function. Certain elements of human body chemistry might help one or another domestic animal perform more efficiently or produce longer. The general concept is that each might move toward perfectibility with help from the other.

Scientists at present can do a great deal of exploring of Man-animal combinations as long as they stay away from producing subhumans. Any man-made creatures that looked or acted like humans would probably create an uproar. But gradually, it is speculated, people would get used to them.

A panel of scientists convened by the Rand Corporation, a research organization, guessed that the routine production of specialized human mutants might come by the year 2025. That would be within the lifetimes of many of us. Such "parahumans" might be designed to serve a variety of functions. They might be trained to be the sewer workers and stoop laborers that are now becoming increasingly hard to procure in the Western world. They might be kept in enclosed areas as sources of organs for transplants. If we come to using animals as surrogate mothers during pregnancy, as previously discussed, they might be modified with human input to function better. Man-animal hybrids might be given a body chemistry more congenial to the normal gestation in a human womb.

The technology for mixing different species of animals, including Man, is still mostly in the test tube. As of this writing the following achievements (among others) have been announced in the literature:

• Man and mouse cells have been fused, with the help of the Sendai virus, as already indicated. The mouse nuclei tend to take charge because of their faster maturing rate. There are studies in which the human nucleus has taken charge, but only because the nucleus of the mouse cell was removed beforehand.[2] In plant life, hybrids created by cell fusion have grown to full maturity.

- In a number of instances the body cells of Man and ape have grown together in cell cultures.
- Human cells have been combined with the chromosomes from a large number of other animals.[3]
- At Erasmus University, Rotterdam, a series of time-lapse photographs have shown the early embryos of a mouse and a rat emerging into a large, ball-like blastocyst (young embryo).[4]
- Tiny human organs from fetuses — lungs, kidneys, testes, ovaries, and the pancreas — have been inserted into the abdominal walls of mice and have quadrupled in size. This occurred at the Pathological Anatomical Institute in Copenhagen.[5]
- The biologist Jean Rostand has cited work being undertaken to incorporate human tissue into an ape.[6]

Obviously we are still some way from the predictions of reputable scientists that living specimens of genuine Man-animal hybrids will be achieved. We are apparently quite a way from a donkey with a vaguely human facial expression. We are quite a way from an ape-Man that can turn bolts on an assembly line for seven hours a day.

If a generation from now those scientists enthusiastic about the idea have indeed engineered Man-ape hybrids, how are we going to treat them? Will they be entitled to the same legal protection we are? Or will we treat them in the way some societies have treated human slaves, that is, as less than human? In Zaire, for example, certain Africans are having trouble accepting the official view that the Pygmies of Zaire are human beings. Gerald Leach has summed up neatly the dilemma we may face, with this prediction:

"If a man-animal hybrid is produced we will keep it in a zoo or laboratory, not our homes, and pay to see it and embarrass ourselves. But . . . it might embarrass us more than we can stand. Biologists working in this area should perhaps consider the myth of the Minotaur, the bull-man that had to be locked in a labyrinth because it was too awful to gaze on."

A large part of the embarrassment might arise in trying to decide what really is "human." We think we know, but do we? A number of human conceits might be shaken. What are the boundaries of humanness? What characteristics make the human species unique on this planet? Remember, the gamut of humans runs from aborigines to atomic scientists. I have collected from a number of published lists traits that are supposed to be exclusively human and that set us apart.

Man is the only animal with a brain capacity of more than 800 cubic centimeters. This superior brain capacity, it is thought, accounts for man's wit and wisdom. The statement is untrue. Many dolphins have bigger brains than many humans, and seem to be remarkably fast learners. John Lilly, a longtime student of dolphins and a founder of the Human/Dolphin Foundation in the United States, contends that large-brained dolphins are "quite as intelligent as humans" but by necessity exhibit their intelligence in different ways. Within their society, he says, they "are more intelligent than humans." Lilly finds that they have a philosophy, an ethic, and a social sensitivity to each other and to other species (humans included) that "we may not achieve in 1,000 years."

Man is the only animal that makes long-term plans. What about squirrels gathering nuts for the winter? So what, if they are more instinctively driven than humans. And what about the fact that billions of humans seem to live only from day to day?

Man is the only rational, reasoning animal. This assumes an ability to analyze and think out problems systematically. What about the raccoons that can escape intricate puzzle boxes invented by psychologists? They have escaped from boxes requiring the mastery of seven different kinds of precise actions. Simple instinct? Or what about rats that have outperformed college students in solving maze problems? Rats, like wily, ambitious corporate middle managers, seem to have a special maze brightness because they spend so much of their lives figuring out ways to get to the cheese.

And what about the beavers that cut down specific types of trees and lug them or their branches hundreds of feet to streams or rivers. They must battle rushing currents to put the branches in a position exactly right for dam building. Some beaver dams are 15 feet wide, some 250 feet wide. Thus the beavers can't just be working by rote. And the dams themselves are of no interest to the beavers per se. The beavers build them as long-term projects to force a rise in the water level. The water level must rise high enough to permit them to have secure, moated homes behind the dams. Just programmed behavior? Perhaps so, but a lot of decision making would seem to be required.

Man is the only animal that can make and use tools. Chimpanzees have some capacity to do both. They break branches to the right size and use them for tools or weapons. They throw stones at enemies. They can sweep floors, hammer nails, drive minicars accurately to destinations hundreds of yards away.

Man is the only animal living in a context of time. This one is ridiculous. Regardless of the time of year, my family's dog knows when 3:45 P.M. arrives. That is one of his feeding times. And he has a fit if we are ten minutes late. And what about some humans? Many peoples in the world seem, in Western eyes, to be unconcerned about time and to be undepend-able about being "on time." Their efforts to adopt Western ways have become symbolized by the wristwatch.

Man is the only animal that uses symbols. Chimpanzees may not have invented symbols but they certainly can use them. The token-economy system, which behavioral psychologists have developed to motivate hu-mans, was derived from token systems developed by animal experimenters back in the 1940's. Chimps quickly comprehended that if you work hard for a few hours and earn ten chips, the chips can be turned in for a bunch of grapes.

Man is the only animal with the potential for communicating within the species. Dolphins seem to have an intricate communication system based on sounds. And it has now been discovered that chimps can learn to com-municate with humans in the sign language deaf-mutes use. To do this, they have learned hundreds of human words or word symbols. The sign language is necessary because chimps lack a human voice box and therefore can't speak the words they have learned. A female gorilla at Stanford Uni-versity has a sign-language vocabulary of three hundred words.

Man is the only animal with a sense of responsibility. What egotistical non-sense! What about the mother moose, a terror to men wandering in northern woodlands when she is guarding her calf? What about German shepherd watchdogs?

Man is the only animal that exhibits awareness, self-consciousness. It is true that Man is strong on this. Lately he has exhibited an orgy of aware-ness. Guidebooks to self-understanding and autobiographies full of meth-ods for self-examination have in recent years glutted the best-seller nonfiction lists. But we don't really know that awareness is unique to Man. Aging elephants certainly show awareness that death is approaching. They travel miles to be near elephant graveyards.

Man is the only animal with imagination. A man or woman, unlike animals, can form mental images of what is not present or has never actually been personally experienced. Possibly so. We just don't know enough about what goes on in the minds of bright animals — or of many primitive people.

I won't get into the contention that Man is the only animal with a soul, since the existence of the soul is hotly argued even among humans.

There are some things we can say for sure about the uniqueness of Man. He is the only animal that at present clearly looks like a Man. He is the only mammal that is a full-fledged biped and has an efficient opposable thumb. He is the only animal that usually wears clothes instead of hair, feathers, scales, a leatherlike skin, or a protective shell. He has a unique complement of forty-six chromosomes. However, even here, Lederberg points out, "no more than a small percent" of the total genetic units in a human egg "differentiates the human being from the ape." The line between primitive humans and neighboring apes is real, but not always awesome.

We are on much safer ground if we define Man in terms of what *certain* of his representatives are *capable* of doing. Some humans can compose symphonies, invent pocket calculators, write books of poetry, paint great murals, keep records, invent clocks, build cathedrals and great bridges, fly spaceships, make nine-layer cakes, design skyscrapers that won't fall down, or invent strapless gowns that won't fall down either.

Best of all, we have some wonderful humans with a deep capacity for caring about their fellow humans, and for working with dedicated selflessness to improve the condition of the species. That indeed is a distinctively human capability.

The only point I am making in all this is that chimeras offer the possibility to erode further Man's sense of dignity and uniqueness. This process of erosion, which started with the findings of Copernicus, has continued down through Darwin and Freud to the most recent findings or proposals of B. F. Skinner and Joshua Lederberg. Shall we, through our ingenious manipulations, continue the erosion by creating subhuman chimeras just because it is probably possible?

Nagle suggests, correctly I think, that true subhumans with visible human characteristics would create moral and ethical questions of enormous magnitude.

The prospect is that we are going to see more and more chimeras created with a human component. Some may be man-sized. That is still another reason why it is urgent that each modern nation set up a commission on human reproductive research. The only possible justification I can see for creating subhumans is for medical research and organ transplants. Subhumans should never, ever, become a part of the public human scene.

ON CREATING THE MAN-COMPUTER

The advances of medical technologists in mechanizing the human body will be examined shortly. Here I will focus on one development so spectacular that — like chimeras — it could make a fundamental change in human identity. I refer to plans for wedding the living human brain and the man-made computer. This involves something quite different from building a mechanical robot that walks, lifts, and (for decoration) has a metal head with vague human features, or robot arms strapped to human arms. It also involves something quite different from the handsome man in the TV series *Six Million Dollar Man*. This fictional character has been engineered, in vaguely explained ways, to perform superhuman muscular feats. It also involves something quite different from the comparable female in the TV wonder-person spinoff, *The Bionic Woman*.

The Man-computer, in short, involves modifying and enhancing the brain by computer. Reading the mind electronically by computer will be easier than feeding the mind electronically. Some computer linkups with the brain will be less formidable than others. A small computer that can recognize abnormal brain-wave activity, as noted, is already being perfected. It may be implanted inside the head. When it recognizes an abnormality it triggers a reaction that smooths out the brain waves.

MIND READING, OF SORTS

With a computer that records brain-wave patterns you can know *in advance* how a person is going to react to a question or a problem, and take action. John Hanley of the Brain Research Institute at the University of California, Los Angeles, found that such mind-reading was possible with chimpanzees. A chimp named Jerry was taught to play tick-tack-toe. (The game can usually be mastered by humans only after the age of four or five.) Once Jerry had mastered the game, Hanley's group planted electrodes in his brain. These were to record his brain-wave activity while he decided what move to make next. They restricted their observations to the times when he had a chance to make a winning move that would complete a row. To their surprise, and delight, they found that they could tell at least fifteen seconds before he made a move whether it would be correct or incorrect. Their forecasts were right 99 percent of the time![7]

All aspects of Jerry's brain-wave pattern showed more activity if he was about to make an error. Even when the brain-wave recorders were simply placed on the outside of Jerry's scalp the predictions were 70 percent right. More sensitive instruments presumably could improve the rate of correct predictions. Hanley suspected that Jerry's brain had to be in a certain state for him to arrive at the correct move.

This work of Hanley's has led to surmises that in the future airplane pilots should wear brain-wave attachments that would ring a bell if the pilot's mind was not in the right state to start a critical maneuver. How many times have we read reports that an accident was due to "pilot error"?

Another suggestion is that students at the consoles of computerized teaching machines could wear brain-wave recorders. These would give the machine 10 seconds' notice when the student was about to make a wrong answer. In that time a lightning-swift computer could easily rephrase the question or switch to a slightly easier one.[8] This, in turn, would give the student a euphoric sense of brightness and perhaps help motivate him to be an enthusiastic student.

MINDS THAT MOVE MATTER

Lawrence R. Pinneo, a forty-eight-year-old neurophysiologist with a bushy white beard at the Stanford Research Institute, had long yearned to be able to write as fast as he could think. Could he do it by thinking into a computer? Some idea! He still cannot write as fast as he can think, but after a two-year experiment he has proved that you can think into a computer, and that the instructions you think can cause the computer to activate and move remote-control cameras or other machines. In short, the machines obey your mental instructions.

Pinneo started with the motor theory of thought. This holds that verbal thinking is nothing more than subvocal speech. With a number of subjects he attached electrodes to the area of the scalp near the region where speech originates. On command they were to think of a word, such as "schoolboy" or "start" or "left." They were to repeat the word in their minds ten times. All this thinking of words was being registered by a computer. It averaged out a recognition pattern for each word. He proceeded to build up a vocabulary of fifteen unspoken English words that the computer could recognize. He trained the computer to recognize actually spoken words (overt speech) as well as think words (covert speech). They came out much alike in the word patterns that the computer stored away.

In his pioneering effort he was successful more than half the time in getting think words classified properly into the computer. With some subjects the success rate ran as high as 70 percent. With spoken words he had 100 percent success.

In his preliminary report Pinneo stated: "We conclude that it is feasible for a human verbally to communicate both overtly and covertly with a computer using biological information alone, with a high degree of accuracy and reliability, at least with a small vocabulary."[9]

This is interesting as an exercise in scientific versatility. But what would the practical applications be, assuming that 100 percent accuracy is achieved with a much larger vocabulary of words that were only thought, not spoken? There seems to be very little chance of improving upon the tape recorder and typewriter in recording thoughts. It could reduce the crews needed to man the cameras and lights in television or motion picture studios. But what could it do in activating most machines that cannot already be done by pushing buttons on a panel? There is speculation that people could drive cars or airplanes just by thought control. I would still feel better if someone had his hands on the controls.

Perhaps the best practical use would be in surreptitious situations. Bank tellers ordered by robbers not to make a move could think the word that would activate the alarm. An executive who wanted to end a wearisome interview — and could not reach the hidden button — could mentally summon a secretary. She would accept the signal as a cue to remind him about some other appointment or commitment.

MACHINE-TO-MIND EDUCATION

Or indoctrination. Is it possible that eventually the Pinneo procedure could be put into reverse? Perhaps the same electrophysiological signals that form words in the mind could somehow be programmed into a computer and fed back to the mind. There are hundreds of thousands of words, but all can be broken down into a few dozen units of sound.

Memory seems to be acquired electrically even though it apparently gets stored chemically. Is automatic learning possible without conscious participation if signals are fed into specific neuronal structures? Pinneo has at least speculated on the possibility of using cybernetic communication to help a child's learning process.

Imagine the revolution in education if students could learn without the enormous bother of studying! Some informed people talk of the possibility of computerized machines that could imprint in our brain in hours information that otherwise might take days or weeks of learning. Science-fiction stuff?

Of course you might end up with even more Johnnies than we already have who, after a dozen years of schooling, can't write two successive coherent sentences. Learning to write correctly and clearly is much more than a memory exercise. Throughout the world it is an exercise for disciplining oneself to think clearly. Writing is a method not only for the expression of one's thoughts, but for the formulating of them as well.

THE COMPUTER AS A PART OF THE BRAIN

The 1976 meeting of the American Association for the Advancement of Science had, as mentioned at the beginning of the chapter, a session on the possibility of incorporating a memory-enhancing computer into the brain. It was described as a fascinating session. A compact computer would serve as a booster to the brain of Man. It would be tuned in to the brain's own electrochemical language and would both receive and send out information.

As projected, it would help the mind function faster and greatly expand the brain's capacity for reliable memory.

The discussion was launched by Adam V. Reed, a young psychologist at Rockefeller University. He has been doing work on a first step toward this dramatic innovation. That first step is deciphering the brain's internal language. It seems that animal studies are beginning to yield clues on neural coding and on the brain's pattern of processing information.

The goal, Reed said, would be a tie-up of mind and computer that "will have the information right in your head instantly . . . whenever you think you want to know something." The computer in the head would become a natural part of the person's brain. Physically, he said, it would require making contact with natural neurons for both input and output without interfering with their normal operation.

Achieving such a brain-booster, he conceded, will take some doing. He thought it could be done within fifty years, or within the lifetimes of many millions of people now alive in the Western world. Five formidable problems must be solved:

1. We must refine at least tenfold the already amazing miniaturization of thin-wire electrodes.

2. We must learn vastly more about how to get computerized information into the brain.

3. We must locate the neurons in the brain that are the best ones to use in making hookups with electrodes.

4. We must learn the brain's internal language so that it can be used by the computer.

5. We must unravel in detail the secret of how information gets stored as memory and is later retrieved.

None of the scientists present hooted at the feasibility of Reed's ideas. Some expressed skepticism, or thought such a tie-up to a computer would take a very long time. However, John McCarthy, a computer expert at Stanford University, said: "If fully, successfully done, this would constitute a complete evolutionary jump in the species."

He seemed to like the idea, for he added: "Fundamentally it is something that can't be suppressed."

Reed was questioned about the possible abuses that might arise from such a mind-computer marriage under a totalitarian regime. Would it open up possibilities of thought control, indoctrination, even memory tapping? Reed dismissed these speculations, I think rather naïvely, by saying, "Those who work in the field would simply shut off the availability of that technology." Many scientists under Stalin and Hitler may have been unhappy and may have dragged their feet, but most did not "simply shut off" their know-how on atomic energy, rocket building, eavesdropping, or brainwashing. They tried to reach accommodation with the regime. Sometimes they succumbed to the threat of imprisonment; sometimes the lure of favored treatment seemed too strong to resist.

On the other hand, a populace even under a dictator probably would not put up with any universal program to put computers in their heads. Almost certainly they would not permit it unless it was done under the promise of brain enhancement.

Reed said that a brain-computer hookup would be "a great thing to have as long as it was under one's own personal control."

As a practical matter, the cost of constructing and implanting the computers — if they are ever perfected — would be so high that only the rich or the government could afford them. The government might allocate them to people who work under pressure or who must have many facts available for instant recall: astronauts, for instance, or secretaries of state or managers of atomic energy plants. Any computers left over might go to sportscasters, who could brighten their chatter by mentioning how many times Johnny Unitas ran the football on fourth downs in 1966.

In terms of manipulation, ethics, and social hazard, the Man-computers seem substantially less worrisome than Man-animals. There wouldn't be the monstrosities to cope with. The computer linkup would be made after birth, not before. Presumably, its installation would require the consent of the "wearer," who would also have the option of deciding when he wanted it removed.

If computers that feed the mind ever become widely available, they could conceivably contribute to the serious depersonalization of Man that we are already witnessing. Individuality might suffer because of problems of evaluation. Professors wouldn't know if they were grading a man or a machine. And there would be problems of sizing up new acquaintances. If at a cocktail party an attractive new neighbor alluded to Akbar, the Mogul emperor, or cited Cato, what would we conclude? Was she unusually well read? Or was she simply trying out some new information fed to her the night before while she was asleep?

Resetting the Clocks in Our Bodies

21

*By 1990 we will know of an experimentally tested
way of slowing down age changes in man that offers
an increase of twenty percent in life-span.*
— Alex Comfort, British gerontologist

Comfort pointed out that the rate of aging in laboratory animals "can be altered by relatively simple manipulations." Looking slightly further over the horizon to the year 2000, the California gerontologist Bernard Strehler states that by then the human life-span may be increased by ten to thirty years if sufficient money and talent are devoted to reaching that goal. His maximum figure would take the average life expectancy beyond one hundred years.

Manufacturers of light bulbs know almost exactly — to within a few hours — how long a light bulb will burn. Similarly, manufacturers of radios and other products can gauge how long the product will last with average use, give or take a few weeks. In the trade these expectancies are referred to matter-of-factly as "death dates." Even the best product cannot last forever if it depends on moving parts or energy. It usually has a known death date. Death dates of products can be, and have been, manipulated for maximum profitability.[1]

Living creatures have death dates, too. There are maximum expectancies and average expectancies. Few reach maximum expectancies for a variety of reasons. Golden hamsters never live longer than three years. Dogs never live longer than thirty years. Horses have a maximum of fifty years. The champion in longevity is not the brainy one, Man, but the giant tortoise. It sometimes lives 180 years.

Man comes in so many varieties, and lives in such varied conditions of environment and wear-and-tear, that it is misleading to give one death date for him. As of today, the human biological clock or clocks appear to be set in such a way that human life ends before 120 years have elapsed. Usually, much before. A few humans have apparently lived longer than 120 years, though verification is difficult because birth records were not kept in many parts of the world until a century ago. But the practical maximum of 120 years is thirty years longer than all but a few remarkably hardy people can expect to live.

Since we are dealing with billions of individuals it is probably better to think of human death dates in terms of average life expectancies. Even here we get gross variations. In the United States in 1976 the life expectancy of males at birth was approximately sixty-eight years. Women lived about eight years longer. Their expectancy at birth is now almost seventy-six. The figure for males is about the same as that for males in Germany, Canada, and Great Britain, and slightly above that for Russian males.

Note some of the extreme variations in life expectancies by countries as given in the list below. The figures are based on United Nations data of life expectancies for *males,* at birth:

	Years
Guinea	26 (the lowest reported)
Afghanistan	38
Burma	46
Turkey	54
France	68
Sweden	72 (the highest reported)

Presumably, the prevalence of disease and malnutrition and severity of environment account largely for the great differences in expectancy.

Of the seven countries listed in which the life expectancy of males at birth exceeds seventy years, only two are outside northern Europe: Japan and Israel.

For Americans the 20 percent increase in life-span projected by Alex Comfort would take expectancies up to about eighty-two years for males and ninety-one for females. That would be quite a jump, especially since the life expectancy of Americans has been at a standstill for a quarter century. Expectancy may even be declining now because of pollution, unsound nutrition, and the increase of smoking among women.

For hundreds of years individual humans by the billions have yearned for an increase in longevity, not for other people necessarily but for themselves. They have invested in magic elixirs, in pulverized bull testicles, in monkey glands. In recent years one of the more popular alleged regenerators goes by the name of cell therapy. It has been widely tried in Europe. Patients, including world celebrities, have paid hundreds of dollars to have a fluid based on cell extracts from unborn lambs pumped into their buttocks. Some claim they felt peppier afterward. The explanations advanced for how or why the treatment should work have often been vague and contradictory. Few scientists take it seriously. Comfort blasted "pseudomedical techniques based on cell extracts."[2]

Another alleged life-expander that is popular in some foreign countries and has several scientific supporters in the United States is a Rumanian drug known as Gerovital. It does not at this writing have the blessing of the U.S. Food and Drug Administration. Many of the world's wealthy people journey to a Rumanian spa for a course of treatments. The main ingredient is procaine hydrochloride. If Gerovital does perchance have any life-expanding effect it would apparently be through its action in inhibiting an enzyme believed to influence the brain's hypothalamus, a major regulatory center.

WHY HUMAN BODIES WEAR OUT

Age, Comfort says, "is the only disease we have all got." But scientists cannot wholly agree on what we have got. The once-popular "watchspring theory" — that the body just runs down because our energy decreases — is generally rejected. Some think aging is caused by the massive destruction of cells. Certainly that occurs. Some think the major cause is that genetic material starts wearing out or becomes blurred. Others believe immunological defenses start crumbling; still others, that the endocrine glands get a drop-off in instructions from the brain and start slowing down their functioning. And surely for individuals, genetic inheritance is at least partially involved. Longevity does tend to run in families.

Whatever the specific causes, aging in biological terms is, in Comfort's words, "the increasing inability of the body to maintain itself and perform the operations it once did."

With age, waxy deposits resulting from the degeneration of bodily tissues start accumulating between the cells. And fatty deposits start showing up inside the cells.

After the age of thirty our muscle cells start dying off. If we reach ninety years of age they have usually decreased by one third. Muscle cells, like brain, kidney, and nerve cells, do not renew themselves. In many people, at around the age of thirty, the brain's nerve cells start dying off at about 100,000 a day. This is not as ominous as it sounds because at that rate it would take 273 years for the supply of nerve cells to be exhausted. By the age of seventy the average person has lost perhaps 14 percent of these cells. But he still has about 8½ billion cells left, far more than he needs in order to function.

By the age of seventy a person has lost perhaps a fourth of the tiny urinary tubes of the kidneys that carry away waste from the blood. And the seventy-year-old has lost at least half of his taste buds.

One of the more startling discoveries about aging was made by the microbiologist Leonard Hayflick in 1961, while he was at the Wistar Institute in Philadelphia. Up to then, scientists had commonly believed that most body cells (except those of the brain, nerves, muscles, and kidneys) would double by division indefinitely if kept in a proper culture. It turned out that the only cells with this capacity are cancer cells. Hayflick found that there is a maximum number of times that ordinary body cells will double by division in culture. For humans, it is about fifty times.

He took cells from tissue of a four-month-old aborted human fetus and put it in culture. The cells multiplied about fifty times and then died within a few months. The cells divided vigorously at first, then gradually slowed down, and divided very slowly as the fiftieth division approached. (Inside a living body, division is at a much slower pace. The figure of fifty divisions may apply only to cells in culture.)

Hayflick later took cells from the lung of a human twenty years old. In culture these cells divided about twenty times and stopped. He found that if he froze cells at, say, the fifteenth division, action stopped until they were thawed. When thawed, the cells went on and divided about thirty-five more times. Each species of animal he tested had its own maximum number of divisions in culture. Mouse cells divided about twelve times and stopped.

The scientific world at first took an incredulous view of Hayflick's reports. Skeptics can still be found. However, his findings have been replicated in other laboratories and are now widely accepted as a rough norm for the division of human cells in culture under normal conditions.

Perhaps the cells contain a kind of clock that runs down at a predetermined rate. And perhaps the cell clock is only one of several kinds of clocks in the body. In 1975 *BioScience* carried opinions by some leading researchers that the body might have a master clock or pacesetter in the brain, perhaps in the hypothalamus. If there is a master clock, it would appear to have the power to override the cell-division clock. Evidence of this is what happens to the "age" of ovaries when they are transplanted to younger or older females. Experimenters have taken from old female rats ovaries that had stopped cycling and have transplanted them in young rats. The ovaries then began cycling again. And young ovaries transplanted to old female rats stopped cycling.[3]

One of the problems scientists have to cope with in their efforts to lengthen life-span is time. They don't like to wait fifteen or forty years — that is, until their subjects die — to get verification of their theories. Ways are being found to assess success or failure on the basis of a segment of a subject's life-span. To develop valid data that will enable such segments to be used, scientists at the Gerontology Research Center of the U.S. National Institutes of Health are establishing baselines of normal aging rates. About 650 healthy men of various ages have over the years been given periodically about eighty tests. These are biochemical, psychological, clinical, physiological. They measure variables known to reflect aging — grip strength, skin elasticity, heart and kidney function, amount of cholesterol, maximum breathing capacity, fluid pressure in the eye, problem-solving ability, memory, and the like. Computers can tell whether any one individual in the experiment is remaining more or less youthful than would normally be expected.

Scientists are investigating several possibilities for making significant changes in the human life-span. Some of them would be more pleasant than others. Here are several approaches being made to lengthen life in a major way:

LOWER FOOD INTAKE

Scientists are finding that in a variety of animals a restricted diet can lengthen life by as much as 50 percent. The animals in some of the experiments tend to be smaller than normal. That might suggest that if we applied such dieting to humans, the average height might drop a few inches. But what would be wrong with that, if the drop was widespread? Large body size is no longer particularly functional except in athletes and steelworkers. Rats given normal diets but not fed every third day extended their life expectancy by about 20 percent.

But wouldn't reduced food intake leave us feeling chronically hungry? Apparently not necessarily. Two researchers, one from Rijks University, Netherlands, and the other from the Fox Chase Cancer Center in Philadelphia, let 121 rats eat all they wanted. The rats were divided into three groups. Each group got a diet with a different protein-calorie balance. But they were all offered heaping amounts.

It turned out that some rats just naturally ate a lot less than other rats throughout life. Some lived three times as long as others. There was a clear tendency for the lean eaters to live longer. Overeating during adolescence particularly shortened life. If someone wanting a long life were to alter his diet according to these test results, he would eat a diet strong in proteins early in life and a diet light in proteins after fifty. Whether the results will, in fact, apply to humans has not, at this writing, been fully tested.

THE COOLING APPROACH

There is considerable evidence from animal studies that a slight lowering of body temperature greatly extends the life-span. The studies suggest that lowering the body temperature even a few degrees Fahrenheit might in humans add decades of healthful life. Cooling slows down the rate of chemical reactions. But can cooling be done with humans — and without making us shiver perpetually?

To slow down bodily processes during and after surgery, hospitals occasionally lower the body temperature a few degrees. They do it by wrapping the patient in a hypothermic blanket with circulating cold water. Patients can tolerate the blanket for a few days but it is not something practical or desirable to use indefinitely.

But perhaps the "normal" body temperature can be changed. Scientists say there may be nothing immutable about 98.6° F. — or 37° C. The dangling human testis — the organ that helps assure the continuity of our species — requires a temperature several degrees below the normal body temperature if the sperms are to function properly. The body temperature of hibernating mammals drops down to just above freezing. All the change does is put them into a deep sleep. The heartbeat slows to about six strokes a minute. What exactly happens is still not clear. It is known that if you inject the blood of a hiberating squirrel into the bloodstream of a housed squirrel that is not hibernating, the housed squirrel will soon fall into a deep sleep resembling hibernation.

Two studies of body cooling are of special interest. Both involve efforts to reset the thermostats of mammals to produce a cooler body without discomfort. And both are based on the probability that thermostatic control resides in one small region of the hypothalamus in the brain. As indicated earlier, the hypothalamus may contain the body's master clock.

At the Purdue University Laboratory of Neuropsychology, Robert D. Myers and colleagues have succeeded in producing prolonged rises and falls of as much as ten degrees in the body temperature of monkeys and cats. To obtain the temperature rises, they perfused the critical area of the hypothalamus with sodium ions. To obtain the cooling, they used calcium ions. Apparently they achieved a resetting of the set point of the body thermostats. The animals did not appear to suffer either acute fever or chills.

The problem of this approach from the standpoint of widespread human applicability is that it involves implanting a slender tubular shaft into the brain through which to deliver the ions. And the flow of the ions must be regulated by a tiny pump.

The second approach is being made at Michigan State University by a team of biophysicists headed by Barnett Rosenberg. This group first had considerable success in greatly prolonging life by cooling the body temperature of so-called cold-blooded animals. In these animals body temperature tends to coincide with the temperature of the air outside their bodies. Now the team has turned to warm-blooded mammals, which of course have thermostats determining a fixed body temperature.

Rosenberg told me: "We have found a drug which in our preliminary testing is capable of turning off the 'thermostat' for significant periods of time without deleterious side effects in mice. These chemicals can be given orally or injected into the blood." The drug used, he said, is so common that he will not at this time reveal it out of fear that people would start using it incautiously. He revealed the name to me and I can understand why he is cautious. He indicated that in due course the name will be published. The drug turns off the thermostat instead of resetting it, so that bodily temperature drops to that of the surrounding air. If a human whose thermostat had been turned off were in an outdoor temperature below about eighty degrees, his heart would start fibrillating.

At present, Barnett's group has colonies of mice living in incubators where temperatures are fixed at various levels. The mice get an injection every other day and it turns off their thermostats for about twenty hours. The animals have become lighter in weight than those in the control group, and if earlier findings hold they will live longer.

In Rosenberg's opinion, for complete success with humans the resetting of the thermostat would need to be permanent and at any desired level. He is planning work in that direction.

He has provided me with charts showing what might be achieved in lengthening the human life-span by body cooling if findings from animal tests prove valid for humans.[4] If a white American male with a median life expectancy had his thermostat reset downward soon after birth by approximately four degrees Fahrenheit he would probably live to be a hundred. If the thermostat were reset downward by about eight degrees he would probably live to be 140.

Neither Rosenberg nor Myers indicated that the cooling feats they have performed had any serious side effects on the animals treated. The drops in temperature were too small to approach hibernation. Still, until tests have been performed on humans one must wonder whether long-term cooling might make the human subjects languid, drowsy, or even, if the drops were big enough, stuporous.

STRENGTHENING THE BODY'S IMMUNE DEFENSES

Killer cells that guard the body against viruses, bacteria, and other foreign invaders seem to become less dependable with age. Because they are processed through the thymus, a small gland under the breastbone, scientists call them T cells. (They originate in our bone marrow.) The thymus shrinks with age and as it shrinks it processes fewer of these killer cells.[5]

The defenses also become less reliable in destroying invaders that can cause disease. Furthermore, sometimes the killer cells seem to get confused and attack the body's own cells, and thus create ailments.

Researchers at the U.S. National Institute on Aging have succeeded, with genetically inbred animals, in transplanting bone marrow and thymuses from young to old. The results were dramatic. They were comparable with giving a human oldster the system of a youngster. Mice in which such transplants were made have been living up to a third longer than normal. One institute researcher thinks that it may eventually become feasible for young people to put some of their killer cells in deep freeze and draw them out in their later years. Using one's own cells would eliminate the compatibility problem.

The thymus, incidentally, may become involved in another strategy — a hormone strategy — for lengthening life. A team headed by Allan Goldstein at the University of Texas Medical Branch, Galveston, found that a hormone called thymosin does wonders in raising the vitality of laboratory animals. It is produced by the thymus. When injected into mice, thymosin increased their vigor and disease resistance, and consequently their lifespans. Goldstein said: "We have good reason to suppose the hormone will do the same in man."[6]

HELPING THE GLANDS SECRETE BETTER

Scientists seeking the causes of premature senility and the degenerative diseases have found that depletion of a leading transmitter of signals between brain cells seems clearly to be implicated. That transmitter is dopamine. Depletion can be greatly eased by giving the person large doses of L-Dopa, a building block of dopamine. Prematurely senile patients who have taken L-Dopa have thrown away their canes. (Readers may recall that a reported side effect of L-Dopa, which was accidentally discovered, was increased sexuality.)

Recently, researchers at the Gerontology Center of the University of Southern California have found that in mice dopamine levels in the hypothalamus consistently decline with age. The hypothalamus governs the body's crucial pituitary gland, which in turn sends out hormones to regulate growth, reproduction, metabolism, and so on. With a decline of dopamine in the hypothalamus the whole chain of endocrine gland action — including perhaps that of the thyroid — becomes erratic. George Cotzias at the Memorial Sloan-Kettering Cancer Center in New York has found that mice fed on a heavy dosage of L-Dopa live about 10 percent longer than average.[7]

RESETTING THE CELL DIVISION CLOCK

Vincent Cristofalo of the Wistar Institute, where Hayflick did much of his work on cell division, has increased the life-span of certain cultured human cells by 40 percent. He did it by adding hydrocortisone. In late life cells seem to have a harder and harder time dividing. They can't synthesize a certain protein that powers the division. Cristofalo reasons that the hydrocortisone may do the synthesizing and therefore enables cell division to continue.[8]

Perhaps a human time clock based on the fifty cell divisions in culture that Hayflick reported isn't immutable. The recent burst of sophistication in manipulating genetic material suggests that it may be possible to change the instructions given to certain genes. (Aging has some genetic basis.) Perhaps viruses can be sent in with new instructions. At least, gerontologists are talking about this as a possibility.[9] One way to spot the genes involved might be to trace them through the enzymes associated with aging, enzymes that the genes control. A few of the enzymes have tentatively been identified.

If a human cell has the capacity to divide fifty times in culture because of genetic instruction, possibly in a few decades genes carrying instructions for seventy divisions can be substituted for those we now have.

But perhaps we won't have to wait for a genetic breakthrough to reset the clock of human cells. Two California physiologists reported in 1974 that they had more than doubled the normal fifty divisions of human cells in culture. The two are Lester Packer of the University of California, Berkeley, and James R. Smith of the Veterans Administration Hospital in Martinez. They added Vitamin E to a culture of human embryonic cells obtained from Hayflick. In Hayflick's experiments the cells had, it will be recalled, died after about fifty divisions. With Vitamin E added they kept on dividing for 120 population doublings! Vitamin E protects living cells against damage from a class of chemicals that combine with oxygen and are known as oxidants. Oxidants are hard on cells.

Even if Packer and Smith's finding stands the test of many replications, this does not mean that we can live twice as long by eating more Vitamin E. Some experts assume that most people, provided they are not unduly exposed to gas fumes, radiation, or high oxygen levels, and the like, already eat enough Vitamin E to combat most of the oxidants within their bodies. And cells in culture presumably are much more vulnerable to damage than cells in the body. But Vitamin E may well help fight off the ravages of some cell irritants, and thus perhaps extend life. For more on Vitamin E and even stronger anti-oxidants, keep reading.

THE PRESERVATIVE APPROACH

Of all the biological approaches to increasing the human life-span this one seems most immediately applicable. People seeking longevity may quite soon be encouraged to add chemicals to their diet, under a doctor's guidance. The chemicals would smother what are called "free-radical re-actions."

Some of these chemical inhibitors already are being used commercially to preserve the freshness of cornflakes, shortening, potato chips, and many other shelf items in grocery stores. They protect the foods from free-radical reactions. These chemicals also go into chicken feed. Vitamin E is just one such inhibitor. There are more powerful ones available.

Denham Harman, a biochemist at the University of Nebraska Medical Center, has pioneered in testing free-radical inhibitors. He told me: "It seems very likely that we could now, on the basis of what we already know, increase the average life expectancy of man by as much as five to ten years with a corresponding increase in our years of healthy functional life."

That seems like a modest claim when he adds that with one inhibitor he tried the average life expectancy of male mice was increased by 44 percent. Ultimately, he believes, the use of inhibitors could raise the average life expectancy beyond eighty-five years and increase significantly the number of people who would live well beyond one hundred.

Free-radical reactions occur, for example, when you combine oxygen with gasoline in an engine. They occur when butter becomes rancid, when smog forms, or when linseed-oil paints go hard in the can. They occur in the human body, notably when certain fats oxidize by coming in contact with oxygen. Free-radical particles pervade nature. Apparently, oxidants contribute heavily to their creation in the human body. The free-radical particles are incomplete molecules — sometimes called "chips" — looking for something to bind to. And when they find that something the ensuing reaction tends to create a commotion.

Alex Comfort compares a free radical to a "convention delegate away from his wife; it's a highly reactive chemical agent that will combine with anything suitable that's around." In the human body free-radical reactions tend especially to damage blood vessels and brain cells over one's lifetime.

One of the better inhibitors of free-radical reactions is BHT (butylated hydroxytoluene). Harman is also particularly impressed with Santoquin and 2-mercaptoethylamine. And he has gotten modest results with Vitamin E.

It remains to be seen whether the inhibitors have a life-lengthening effect on humans and if they can be taken without harm. (When and if the stronger ones do become available, they will need to be used only under medical supervision — most inhibitors are powerful.) Harman says it would never be too late in life to start, but the younger one starts the better. And on the dictum that it could "do no harm" he says anyone might — while waiting for human test results — step up his weekly intake of the well-tested Vitamin E by 300–500 milligrams. That is a moderate dosage. He adds that not enough experimentation with Vitamin E has been done to know whether there would be a beneficial effect on the human life-span.

THE LIFE-STYLE APPROACH

Medical people are becoming aware that increased life expectancy can be achieved by changes in life-style.

In the United States, for example, several decades of increases in life expectancy ended in 1954. The curve flattened out and hasn't changed much since despite an outpouring of medical advances.

Why? Perhaps because 1954 was about the time a host of social changes got into full swing that may work against long life. A huge number of pension plans came into effect, and with them a great expansion of compulsory retirement by the age of sixty, sixty-two, or sixty-five. There is some evidence that people who keep active live longer. Perhaps one reason for the big and increasing lead wives have over husbands is that homemaking never ends.

With affluence, diets became richer, fatter. With automobiles and an array of laborsaving innovations, jobs and housework began requiring less exertion. The long-distance transfer of corporate personnel became commonplace. The disintegration of the family accelerated. These and other factors brought a sharp upswing in social fragmentation. People had less chance to enjoy a sense of community or continuity.

Network television started having an impact on most of the homes in the land. A whole population became more sedentary as people began spending three to four hours a day seated in front of television sets, often watching vigorous athletic contests.

During this time anthropologists roaming the globe began discovering pockets of people who seem to live extraordinarily long lives. Genes might play some part. But most of these people lived in rugged terrain, seemed to lead hard lives. Noted especially were the Abkhasians, a proud, clannish people in Georgia between the Black Sea and the Caucasus Mountains. The anthropologist Sula Benet of Hunter College helped focus attention on them. Medical people journeyed to Georgia to take a puzzled look.

An Abkhasian's chances of living past ninety are six times greater than the average American's. The Abkhasians have a large, jaunty, uniformed chorus whose members are all men over one hundred years of age. A photograph of the group suggests that virtually all are robust, erect life-enjoyers.[10]

The Abkhasians eat a frugal, low-fat diet. Benet noted some other interesting clues. They have a strong sense of identity with their village that seems to give each one "an unshaken feeling of personal security ar d continuity." She found that they didn't have a phrase for "old folks," only for "long-living people." And these long-livers were not a retired group apart, or seen as a social burden. Abkhasians never "retire." They continue all their lives working at tasks geared to their energy level. With increased age, status rises rather than decreases.

Any systematic biological program to prolong life would be of dubious social value if it simply increased the number of senile people in the population. We all know remarkable people who are over eighty-five and yet are mentally sharp and creative, and have a zest for life. We have the examples of Pablo Casals, George Bernard Shaw, Winston Churchill, Konrad Adenauer.

But at present, the great majority of people in that 1 percent who live past eighty-five undergo mental impairment. This impairment is particularly apparent whenever they try to shift to unfamiliar tasks requiring eye-limb coordination.

Not only do the brain cells decrease in number, but with prolonged age the nerve cells may lose some of their ability to transmit messages.

The neurologist David A. Drachman of Northwestern University believes this loss to be a major cause of senility. Since transmission of images is at least partly a biochemical process, Drachman reasons that senility can be countered biochemically. Memory loss is one aspect of senility. He has helped people with serious memory problems by injecting a drug called physostigmine.[11]

Meanwhile, scientists at the University of Göteborg, Sweden, have been working on techniques for increasing protein production in brain cells as a way to counter aging in the brain.

Even if the medical technologists can keep our mental vigor in pace with our physical vigor as they prolong human life, do we want the elongation?

That may seem a rude, ungrateful question. Our value system has always accepted long life as a major desirable goal. But now we should be starting to face the question, How long is long enough?

The yearning for immortality is a deep one. Individually, most of us would say we want to live to be 120 or even more. But many of us would specify that we wanted such longevity only if we were assured of good health, mental vigor, financial security, challenging things to do, interesting friends to know.

Most modern societies would have to be substantially revamped to provide for those ifs. The concept of retirement would have to change. Who would want to spend sixty years riding around a retirement community on an adult tricycle? We would probably have to have a society where multimarriages and multicareers were taken for granted. "Children" might have to wait until they were seventy to receive inheritances.

But let's accept the proposition that individually most of us yearn to postpone personal death, or a terminal illness leading to death, as long as possible. What about the proposition that it would indeed be desirable if people lived twenty years longer, with no average increase in senility? Would such a revolution be looked upon as *socially* desirable? Probably not, if that revolution was not accompanied by economic, demographic, and social revolutions. Most of the world is already haunted by overpopulation. Society is already groaning under the burden of high medical costs. Making such a transformation would certainly add greatly to national medical costs.

Many societies with new productive technologies are already anxious about their unemployment rates. Would the retirement age be postponed twenty years? That would aggravate unemployment unless we drastically shortened the workweek. On the other hand, if present retirement patterns were maintained, the people under sixty gainfully employed would have to start putting at least a third of their income into retirement plans. Retirees certainly would become a dominant political force in every country. They tend, for one thing, to be niggardly about supporting schools. An indication of the problems that might arise can be seen in the fact that already in the United States there are thirty Social Security beneficiaries for every hundred working people.

Already retirees are among the most alienated people in many societies. They more than most suffer from loneliness, a sense of uselessness, the sense of being a burden, fear of impoverishment, loss of dignity, a disgust that their once-unique selves are becoming blurred by the ravages of age. In the United States, a Senate committee recently heard a report that alcoholism has been rising among the elderly.

In February 1974 the *Gerontologist* carried a guest editorial that concluded: "We may all recognize that life will be less pleasant once we choose to extend life." It suggested that the situation could become "ominous."

From the view of both personal and social well-being, is it possible that people in the Western world already live long enough? Is it possible that we would have a more pleasant world if the medical profession paid less attention to strategies for promoting longevity by manipulating our biological clocks and more attention to preventive medicine?

I suggest that this approach would permit more of us to enjoy robust health as long as we do live and at substantially less medical expense. We would spend less money on expensive treatments and less time feeling miserable. As a Census official put it: "More can be gained from saving the life of a forty-five-year-old man from heart attack than prolonging the life of a sick seventy-five-year-old." Here we are weighing lives. But that would seem to be a reasonable social view.

If, as a result of focusing on preventive medicine, we happen to live a little longer, that would be socially tolerable. But I do not think greater longevity for the population should be the objective.

Now let's turn to another area where we are seeking an enormous growth in the manipulation of the human body. That is the imminent revolution in human spare parts. While the primary objective of this revolution is to correct certain bodily malfunctions, it too may have the overall effect of lengthening the life-span. And it raises a whole set of other questions.

The Human of Totally Replaceable Parts

22

Our bodies are becoming more like cars, and surgeons like mechanics.
— *Behavior Today* (November 1975)

We come finally to the grossest *physical* manipulations of Man now being attempted: the substitution of worn-out or defective parts. A leading Chicago anesthesiologist, Vincent Collins, says: "We have the capacity to make a composite man."

The providing of parts for humans already constitutes a multibillion-dollar industry. In the decade ahead the production, sale, installation, and servicing of human spare parts is likely to become the fastest-growing industry in the modern world. Certainly in dollar volume it will rival the automobile spare-parts business, conceivably the entire automobile-building industry. Hospitals will have spare-parts departments just as garages do. These will be supplied by area warehouses.

As technology develops, the parts will come mainly from four sources:

1. Other people — alive, dead, or living dead. One technical name for the living dead is "neomort." Garage mechanics matter-of-factly refer to the process of removing desired parts from other cars, often immobilized, as "cannibalizing." *Medical World News* has suggested this as a reasonable term to use when speaking of obtaining human organs for transplant. The people from whom organs are taken often don't have to be of the same sex as the recipient.

2. Artificial spare parts. The technical name for these is prosthetics. They are created by specialists called bioengineers.

3. Animals. For some sophisticated purposes it will be important to use primates or a new kind of subhuman that is a man-ape hybrid close to

man. For other purposes, when only size and shape are important, a calf or pig may be better and cheaper. Some of America's foremost transplant experts have been working on cross-species transplant research.[1] They have had limited success in using the organs of chimpanzees, rhesus monkeys, and baboons in transplants to humans, especially kidney transplants.[2] The first efforts have met with difficulties mainly for the same reason that man-man transplants have been troublesome: graft rejection. This problem is starting to yield.

4. Clones. This is a bit over the horizon and is mentioned only in passing. Joshua Lederberg cites as one of the attractions of cloning that it opens up the possibility of "free exchange of organ transplants with no concern for graft rejection."

How this ability to draw upon the bodily resources of an identical twin would work out in terms of interpersonal relationships has been left vague. Ideally, if you want a clone twin just as a source of spare parts, it would probably be better not to get to know him or her too well.

One report speaks of "clonal farms," where anyone could keep a deep-frozen twin available for organ withdrawal. But there does seem to be a psychological problem. How do you get a twin of yourself grown big enough to be useful for transplant — at least thirteen years — without becoming personally involved with this creature you are about to put into the deep freeze? If the organ desired happened by chance to be something that could be spared — a kidney, a part of a lung, a patch of skin — it could be a loving gift between twins with no need for keeping things impersonal by deep-freezing.

Before full utilization of spare parts can be achieved with humans, two problems need to be solved better than they have been so far. One is technical. The other is semantic.

THE PROBLEM OF REJECTION

To install a new part in any human body from any source other than a twin or a clone, the obstacle of graft rejection must be overcome. This is particularly true if the part consists of living tissue.

Graft rejection occurs as a result of the body's fine, built-in defense system against alien intruders. Alien organisms are quickly identified and attacked by lymphocytes. The graft withers. If it is a vital organ, the recipient dies.

Scientists have been experimenting with a number of strategies to cope with the body's defense or immune system. The strategy most commonly used today is to overwhelm it with drugs. Some fairly effective ones have been developed. But a dangerous price is paid. By knocking out the natural defenses, the body becomes dangerously vulnerable to invasion by aliens: bacteria, fungi, viruses. The current strategy is to use a variety of drugs on the recipient to minimize vulnerability. Most of them suppress cell division.

Another strategy is tissue-typing, to seek a compatibility of antigens. Antigens stand as outposts on the perimeter of every cell in the body. The aim is to try to get as good a match as possible between the antigens of the donor and the antigens of the recipient. That way the implant may not attract quite as much hostile attention. Until recently, the problem with this approach has been that it has taken a few days to complete the testing. In that period, the desired organ of a dying donor might become unusable. Now, a team at the University of Wisconsin has cut the testing time down to a few hours.

Several other possible strategies are under investigation. There is some evidence that soaking an organ to be grafted (such as skin) in a tissue-culture solution for a few weeks can help make it acceptable to the recipient's defense system. And massive irradiation of an organ prior to implant seems to help.

THE PROBLEM OF DEFINING DEATH

When a critical organ is involved and the spare part is to come from a human, that human is often near death. He has agreed to surrender the desired organ after his death.

But what is death? The definition can be important to the possible success of a transplant.

Until the 1960's, few had bothered to raise the question. Death — absence of life — came when the heart stopped beating and couldn't be induced to start beating again. You could check it by a finger to the pulse or, if a doctor was around, by a stethescope pressed to the chest. If there was any question you could also confirm that the person had stopped breathing. You held a mirror in front of his mouth. All vital signs disappeared fairly quickly. Within a few hours rigor mortis set in, if still further confirmation was needed.

Since the late 1960's hundreds of articles on how death should be redefined have appeared in medical, legal, and ethical journals. The reason is that the medical technologists have developed machines to keep the heart beating and perfusing the body's organs with blood. Machines also keep lungs pumping the breath of life — oxygen — into the body.

This can be done with bodies that otherwise seem quite dead. Many vital signs are absent. The eyes are fixed and dilated, and show no response to light. No amount of probing can produce reflex actions or signs of pain. The tongue muscles seem paralyzed. For more than ten minutes the brain has shown no electrical activity. The electroencephalogram's once-bouncing recording now shows a flat line.

Or there may be twilight situations. The animal part of the brain — the brain stem and midbrain — shows some electrical activity, but there is very little electrical activity in the "human" or thought-processing part of the brain. That apparently was the situation in the celebrated right-to-die case of Karen Quinlan. She was kept alive by force feeding, massive doses of antibiotics, and respirators for more than a year while she was in what was described as a hopelessly "vegetative state." When court permission finally was obtained to disconnect the respirator and let her die, it turned out, apparently to everyone's surprise, that she still had a feeble capacity to breathe and pump blood.

In an obviously hopeless situation, one in which a patient has a terminal illness or is in an irreversible coma, some doctors are reluctant to stop using all the miraculous life-sustaining technologies available. They are afraid of malpractice suits, or worse, charges of being a party to a homicide.

In 1977 California became the first state to permit a person who knows he is dying to make a "living will." He can authorize or order the withdrawal of extraordinary life-sustaining procedures. The ailing person himself must make the directive. (That would rule out Miss Quinlan, who was found in a coma.) And there must be two witnesses who could in no way gain from the death.

The main reason for the push to redefine death, however, is not simple uncertainty because of the new life-supporting systems. The reason is the desire to keep the organs of an obviously dying person fresh. Organs inside the body of "persons" who would be obviously dead without life-supporting systems can be kept fresh for days or weeks. They can be kept fresh until arrangements can be completed to open up a desperately ill recipient-to-be and until permission to remove organs of the neomort have been received from next-of-kin. Is such a donor dead or alive?

Awkward situations have arisen. In Birmingham, England, surgeons had cut open an elderly victim of an accident who had been declared dead. Their objective was to pluck a live kidney. They stopped when he showed vital signs. He died for good a few hours later.

In general, transplant surgeons want an obviously dying person to be declared dead as quickly as possible. That gives them more time to make arrangements for taking the organ needed, while machinery keeps the donor's heart and lungs pumping. When they have a choice, they prefer donors who are slowly dying of a disease that does not affect the organ they want to take for transplant. That vastly simplifies all necessary permissions and preparations.

The above considerations have by now brought about a fairly broad consensus among medical people that death occurs not in the heart but in the brain. Surgeons want to keep their options open on the heart. Courts are starting to concur.

There are plausible medical grounds for making this shift to the brain as the best indicator of "death." With modern resuscitation techniques a heart that stops beating can be reactivated for nearly ten minutes after it stops beating. And an enfeebled heart can be kept beating effectively with mechanical help for months.

In contrast, the thinking part of the brain is substantially wrecked if the flow of blood is interrupted for more than three minutes. And it can never be resuscitated. The entire brain is virtually destroyed if the interruption continues for eight minutes. An overly conservative statement of brain death was made by an official of the U.S. National Institute of Neurological Diseases and Stroke. He said that if a patient's brain shows no electrical activity for half an hour the patient is "as dead as he ever will be." The Supreme Court of New York has now recognized that a person can be declared dead on the basis of brain-related criteria.

In 1975 the *British Medical Journal* carried an article that offered a dramatic reason why death should be defined in terms of the brain. It stated: "Though the concept of brain death is not easy to explain to the layman, the extreme example of the victim of the guillotine is perhaps helpful. Nobody would consider the body, after the head has been severed, to represent an individual, living being, yet the body could be resuscitated and the organs kept alive for a considerable period."

Some medical people even argue that great gains could be achieved in transplanting organs and in research if death were defined as the irreversible loss of cortical function. If the cortex stops functioning its owner has lost all the characteristics of being a live human person. The body — even part of the brain — may still be "alive," but the body has lost its humanness. If it is no longer human, they ask, why should not organ borrowing be as permissible as it is from any other living creature?

Others argue that in today's world transplanters should feel free to take organs from any certified corpse if there is no evidence that the former person has specified otherwise. Finding next of kin and getting permission often produce serious delays and frantic operating room scenes. It is now a common practice to carry in your wallet a card stating that you wish your body to be given to a certain institution for medical research or that you give permission for certain of your organs to be used for transplantation. Many thousands of people carry these cards.

When needed organs become a matter of life and death, money may enter the picture. In May 1974, the *Michigan Law Review* devoted more than one hundred pages to an analysis of issues arising from "the sale of human body parts." An examination seemed pertinent, it said, because "virtually every part of the human body is now, or will soon be, reusable."

Sections were devoted to such topics as the "routine removal of organs in the absence of objection," the "living source as seller," and the "sale of cadaver parts." The study explored conditions under which the contribution of a body could result in a tax deduction or a no-premium life insurance policy. And it raised such interesting questions for the future as these: "Will a debtor be able to put up his kidney as collateral for a loan? May a person be forced to sell a kidney in order to satisfy a money judgment?" The organ market, the author suggested, should be regulated to protect the public against such abuses.

Here is a sample of how organs are put up for sale. In July 1975, the *New York Times* carried this item in its commercial notices: *"30 year old man in excellent health desires to donate a kidney for fee. Negotiable. 274–2750."* According to the *Hastings Center Report* the people who have called the National Kidney Foundation offering to sell one of their kidneys now number in the hundreds. The asking price in one newspaper ad was $5,000.

In both the Soviet Union and the United States the possibility that future murder rings might thrive by bootlegging body parts has been discussed in the press. The parts would go to fences who could find people so interested in saving a life that no questions would be asked about where the body parts came from.

When a transplant is to be made, hospitals should have review officers to verify the source of the organ to be transplanted, and also to verify that the patient whose body is about to be invaded for a badly needed organ is legally dead.

The problem of defining death is eliminated if the part to be substituted comes from an animal or is manufactured. And in the latter situation the problem of graft rejection may be reduced since alien organisms are not involved.

WAREHOUSES FOR THE LIVING DEAD

The amiable and highly respected New York psychiatrist Willard Gaylin has explored at length the possibilities of "harvesting" the living dead.[3] He imaginatively extrapolated the logic of known facts about technological advances in modern medicine and came up with this vision for tomorrow:

Special "bioemporiums" would house row upon row of "living cadavers" attached to respirators. Since their brains had stopped functioning they would be legally dead. But they "would be warm, respiring, pulsating, evacuating, and excreting bodies." They would require feeding and maintenance and "could probably be maintained so for a period of years."

These banks of neomorts, he suggested, could serve science in dramatically useful ways. For example:

- As a source of spare parts. Major organs are difficult to store. In a bioemporium, neomorts would be maintained with body parts computerized and catalogued for compatibility. This would make for a much more efficient system of storing for future transplant such organs as lungs, kidneys, hearts, ovaries. Blood could be drained regularly. Bone marrow, cartilage, and skin could be harvested. The neomorts could also be used as manufacturing plants to make hormones, antitoxins, and antibodies.
- For the training of physicians. Interns and residents could get practice in doing spinal taps, making skin grafts, operating on eyes, removing kidneys and testicles, and performing major plastic surgery on the face.
- For medical experimentation. The neomorts could provide the missing link in moving from animal studies to testing drugs and new surgical procedures on humans. With such bodies available, we would no longer need to use "prisoners, mentally retarded children and volunteers."

Gaylin tried to make a cost-benefit analysis, in which he clearly sounded sardonic. Then he got to what seemed to be his major thesis: that there would be costs in terms of human values. Would repugnance fade with time and education? Perhaps so, but would that be good? Some claim, he suggested mildly, that repugnance is a "quintessentially human factor whose removal would diminish us all, and extract a price we cannot anticipate in ways yet unknown and times not yet determined."

He suggested that revulsion may be "one of those components of humanness that barely sustains us at the limited level of civility and decency that now exists." His conclusion was that measures to sustain life may not be justified if those measures destroy "those very qualities that make life worth sustaining."

BODY PARTS THAT ARE NOW REPLACEABLE

There are few parts of the human body that cannot already be replaced, at least partially. The brain is one, but as we will see, the scientists are working on it.

We can expect that within a decade a number of wealthy individuals with once-ailing or battered bodies may be carrying around as many as a dozen body parts they were not born with. And the substitute parts will not include wigs, cosmetic breasts, or false teeth. (Even today some false teeth are actually implanted surgically and anchored to the living jawbone.)

The structural parts of our bodies are now largely replaceable by transplant or manufactured substitutes. Just about any bone or tendon can be replicated by fabrication. The jawbone, fingers, wrists, bone for the inner ear, a part of the skull, a complete shoulder or hip — you name it. The artificial hip is a godsend to otherwise healthy people over seventy-five years of age. Perhaps their greatest vulnerability to disablement and dependency is a hip that has become fragile. One of the most difficult parts to replicate was the total shoulder mechanism. That is now available.

Skin, arteries, nerves, corneas, hair, can all be transplanted, either from self or donor. It is now, at least technically, possible to transplant a whole eye. For some reason artery, bone, and cornea transplants rarely excite a destructive immune response. We can get blood changed almost as easily as we can get oil changed in our cars, although it is more expensive. During the 1976 Olympics there were rumors that some long-distance runners were getting a competitive edge on energy by having their blood changed on their days off between races.

Much more challenging is the replacement of the vital functioning organs of the body. I refer to such organs as the liver, spleen, kidneys, heart, lungs, pancreas, brain, thymus, ovary and testicle. These are all organs in which there is processing activity going on. But progress is being reported on almost all of these. Catalogues of the spare parts presently available in stockrooms have now been issued.

The first big success story in replacing a vital, processing organ involved the kidneys. A kidney is a washing machine, with a little chemistry thrown in. It filters our blood to clean out the wastes that constantly accumulate. The wastes go to the bladder as urine. Although the human body contains only about five quarts of blood, the heart pumps two thousand quarts of that blood through the kidneys every day!

The removal of wastes is done by millions of tiny filters and tubes. Wastes can seep out to the bladder, but the relatively large blood corpuscles cannot. If the kidneys become damaged by infectious disease, defect, or accident, the filtering becomes inefficient. Poisons from the wastes relentlessly and painfully build up. Until recent years, they usually killed the kidneys' owner by degrees.

Today, tens of thousands of people with failing kidneys have new ways to get the filtering done: either by an external, man-made dialysis (filtering) machine or by a fresh, transplanted kidney.

A crude artificial filter was first achieved in 1943. This was at a time when all transplanting was considered impossible because of graft rejection. The filter's ingenious pioneer, Willem Kolff of the Netherlands, has continued to be a leading designer of artificial human organs. He developed his kidney dialysis machine during the Nazi occupation of his country. Later he went to the Cleveland Clinic, now famed for its work on artificial organs. Most recently he has been exercising his genius with a team at the Institute for Biomedical Engineering of the University of Utah.

Kolff exploited two, then-recent inventions to develop his first dialysis machine. One was cellophane, a membrane just porous enough to allow the toxins of the blood to pass through to a cleansing fluid. The other was heparin, the blood-thinning drug that would prevent the blood from clotting during the washing process.

Today in America alone, at least twenty-five thousand people spend a substantial part of their lives attached to dialysis machines. They may go to a treatment center three times a week to get a five-to-six-hour cleansing. The cost, at this writing, is at least $450 a week — or about $22,000 a year. Or they may have their own $5,000 home installation with a twenty-five-gallon tank of cleansing fluid. In the early 1970's the federal government enacted loosely written legislation to provide 80 percent of the cost of these treatments for people on Medicare. Costs are now double those expected and may in the next several years approach a billion dollars a year. The legislation is particularly lax in its provision for payment of retainers to doctors. The fee specified is about $200 a month for each patient, just to be in charge and to be on call in case anything goes wrong. A draft report by the Oversight Subcommittee of the House Ways and Means Committee noted that a doctor "on call" to one hundred patients might be receiving under the program a retainer of $240,000 a year.[4]

The thrust of current research is to make the machine portable, to free people from being immobilized so long. A machine-in-a-suitcase has been developed that will permit a person to take a trip for a couple of weeks. Kolff, who has been striving for an artificial kidney that can be worn like a hiker's backpack, now has an eight-pounder that will at least permit the patient to walk around occasionally during treatment.

Another serious shortcoming of the dialysis machine is that usually it does not clean the blood as well as a real kidney does. Many patients on the machine complain of being weak and listless. Their complexions may be grayish.

Thus, the more desirable solution, when it works, is a kidney transplant. The mechanics of the transplant primarily involve hooking up the replacement kidney to a vein or an artery, and to the tube leading to the bladder. But keeping the patient alive with his new kidney is a big problem. How do you find someone else's kidney that will not be automatically rejected by the body? How do you keep it from deteriorating until implanting can be done? (Packing the new kidney in ice water helps.)

The pioneer work on kidney transplant was accomplished during the early 1950's at the Peter Bent Brigham Hospital in Boston. And the first successful human transplant was performed there in 1954 by a team led by an old acquaintance of mine, Joseph Murray. He is a lean, warm-hearted, soft-spoken man. At that time, 1954, there was little knowledge about strategies to combat graft rejection.

Murray's team made the plunge with identical twins. Because identical twins have identical genes, he thought there would be no serious rejection problem. One of the twins was dying. The other was willing to surrender one of his two kidneys.

Everyone is born with two kidneys. Why do people have two kidneys, two lungs, and two testicles or ovaries, but only one heart? The paired organs — all highly vulnerable — were presumably built by Nature to provide a better chance of survival in case of accident. The heart may have required too precise a synchronization to have permitted a paired arrangement. At any rate, a human can get along quite well on one kidney. Usually all he loses is a safety factor.

The transfer operation of the Murray team on the twins was a spectacular success. Soon the twin who had received the transplant married his nurse. He became a father and lived for eight years before dying from heart failure. His heart had been weakened by his original kidney disease.

Murray and his team continued to make transplants, mostly successful, with identical twins. When they shifted to fraternal twins, they used irradiation to reduce the tendency to graft rejection. It was seven years after their first transplant that they were able to use newly discovered drugs that work to suppress graft rejection. The drugs opened up for the first time the possibility of extending vital organ transplant to people other than twins.

Today it still helps if there is a blood relationship. A sister may give one of her kidneys to a brother, or a father may give one of his to his daughter. A wife may give a kidney to her husband, but since there is no blood tie the chance that the transplant will succeed is no more than average.

But the "average" for survival has been improving. In the world today there are more than twenty thousand people with transplanted kidneys. Some of the transplants were done more than eighteen years ago. If the substitute kidney is from a person who has died very recently the chance that the recipient will go on living has risen to better than 70 percent. If the kidney is from a relative the success rate is approaching 90 percent.

Many countries in Europe have united to form Eurotransplan, which regularly issues a large directory of people seeking transplants and possible donors. The total cost is still high, in the United States well over $10,000.

Almost all of our other vital, functioning organs have proved to be more difficult to replace than the kidney.

The world now has several centers where a broad assault is being made on the problems of inventing artificial organs, including a mechanical heart. One of the most famous centers is at the Cleveland Clinic.

A VISIT WITH A LEADING BUILDER OF ORGANS

History is being made in a modest, sprawling annex behind a gasoline station across the street from the huge Cleveland Clinic. The building includes one of Cleveland's few barns and contains the world's best library on artificial organs. It also houses the Clinic's Department of Artificial Organs, which Willem Kolff established. The present head is Yukihiko Nosé, an organ expert who was brought from Japan to succeed Kolff.

Nosé is a youngish, tall, bespectacled, usually beaming man who speaks fluent, colloquial English. His associates include not only medical specialists but mechanical, electrical, chemical, and biomedical engineers as well as biochemists and polymer chemists. About one third of the thirty-odd people on the staff are from outside the United States, from such countries as Austria, Italy, Chile, Brazil, Japan.

The department concerns itself only with artificial substitutes for the vital, functioning organs in the trunk of the body, the biggest challenges in the field of artificial body parts.

Nosé took me on a tour of the center, which has substantial federal funding. There were the twin operating rooms separated by a monitoring center. One of the operating rooms is devoted exclusively to operations on animals that are performed weekly to test artificial hearts or heart parts and their functioning. "Tomorrow," Nosé said, "we will remove the natural heart of a calf and put in an artificial heart and keep it living as long as possible. We are testing out a new design."

The other operating room is used for surgery connected to the development of artificial substitutes for other vital organs such as the liver, lungs, pancreas, and kidneys. Yes, kidneys. Nosé hopes to see the time when a victim of kidney failure can actually carry an artificial kidney inside his body, or at least wear a fully portable one on the outside. The staff has recently been working with a lightweight glass tube filled with charcoal and other substances. Such a device would not only eliminate the bother of being tied to a dialysis machine fifteen hours a week, but bring down considerably the cost of kidney substitutes.

From the operating rooms we went into the barn, where there were mostly calves and a few sheep. Many had bandages on their necks or on their sides. One calf with a big incision had been living for some months with an artificial aorta. In another pen was a contented cow that had had an artificial heart passively beating alongside her regular heart for three years.

Since my visit, Nosé and his associates have made medical history (in 1976) by keeping a calf alive on a totally artificial heart for 145 days. The calf's natural heart had been removed.

The reason so many calves are used is that they have a chest cavity of approximately the same size as the human one. Nosé would prefer to use large dogs like Saint Bernards, but is restrained by public opinion. The use of dogs for experiments offends many people, he explained. "They don't seem to care about calves but they get upset about dogs. This creates a problem because dogs would be better. They can be trained to behave as we wish them to after the operation, and calves are not cooperative. Also, calves increase their body size and weight after surgery."

Next, we went into a kind of museum, where all the various artificial organs that had been used in animal experiments at the center are stored in liquid-filled jars. Most of the organs are made of materials similar to plastic or rubber. In one jar was a natural human heart valve sewn inside an artificial heart.

Lying on a table was a transparent plastic container that had a rubber heart inside it. An aide demonstrated it for me. He blew into one of the "ventricles" and you could see through the plastic that the artificial heart inside was throbbing as a natural heart would throb.

Nosé held up a part of an artificial heart. He said there was a plastic film on the outside, but on the inside was a "biological" material. Much of their research on artificial organs that must process blood focuses on materials for the "heart's" surfaces. They had assumed correctly that there would not be a really serious problem of graft rejection if artificial organs were used. But when blood is involved, a serious problem does arise. As indicated, blood tends to coagulate or become damaged when it is outside living tissue. How do you avoid on the one hand graft rejection of living tissues and on the other the tendency of blood to coagulate on artificial materials?

The answer, Nosé hopes, is in "biological" materials comparable to tanned shoe leather. He added: "With biological additives — not living biological material — we find we are getting better results." Blood is less apt to coagulate.

Next, we came to a device that looked vaguely like a glass toaster. It was an artificial liver that works outside the body. At present, the experimenters use slices of very fresh liver from animals to make the "liver" machine perform its metabolic duties. Nosé said that the liver is the most difficult of all organs in the trunk of the body to simulate artificially. It receives the food we eat after the food has been pretreated in the stomach. In the complex factory of the liver, the food elements are converted into a form the body can use, especially fuel for energy. A part of the problem of simulating the liver, he said, is that "we know so little about how it functions." Even without full knowledge, his group can now maintain most of the "critical functions" of the liver in an artificial liver.

The liver is also one of the most difficult live organs to transplant. Interestingly, the problem of graft rejection is not as big with livers as with many other organs. But the liver deteriorates faster after death than just about any other organ. To cope with this, some of the early attempts to save humans with liver failure involved making a connection to the liver of a *freshly slaughtered* pig. The first successes in using natural human livers were at the University of Colorado. Throughout the world there have now been perhaps three hundred natural liver transplants. Some of the recipients have survived more than five years.

Scientists at the Cleveland Clinic are also working on artificial lungs. I saw several specimens that had been tried out in animals but with little success. More progress has been made in developing lung machines outside the body.

Lungs absorb oxygen from the air and expel the leftover carbon dioxide. Lung machines have been used for some years to keep patients alive for several hours during heart operations. Nosé hopes his group will soon have a device that can keep a patient alive for at least several months.

A lung is even more difficult than the liver to transplant. Only a few dozen transplants have been achieved and most of the recipients have survived less than a year, and with considerably less than 100 percent lung efficiency. The operation itself is difficult. But the main problem is that the lung is the one internal vital organ that is always vulnerable to infection from the air we inhale.

The pancreas is almost as difficult to transplant as a lung. In the whole world the patients with successful transplants are numbered in the dozens. On the other hand, the pancreas has been fairly easy to imitate artificially. It is the organ that controls the sugar level. When it malfunctions we get diabetes. The Cleveland Clinic has devices that work quite well outside the body. Scientists at various centers are beginning to try out models inside the body.

The clinic has also done some preliminary work on an artificial uterus, an organ Nosé worked on while he was still in Japan.

It is the artificial heart, however, that now is absorbing most of the attention of Nosé and many of his colleagues.

TOWARD THE TOTALLY IMPLANTABLE ARTIFICIAL HEART

Readers are familiar with the fact that after the first wave of attempts to transplant actual human hearts, starting with Christiaan Barnard's attempt in 1967, there was a great discouragement. Only one recipient in ten, on the average, survived for a year. Most died of infections or graft rejection. Interest dropped off.

But one man kept at it. He had played an important role in developing the technique that Barnard used. The man is quiet, publicity-shy Norman Shumway of the Stanford University Medical Center. In 1968 he began doing one transplant each month. He concluded that many of the early failures were due to the very poor physical condition of the patients. They were so near death from heart problems that their kidneys and lungs had already started failing because of the ailing heart. He began restricting transplants to people under fifty-five years of age who had a reasonable chance of withstanding the shock of a major operation. Also, he developed a technique for snipping out tiny samples of an implanted heart to check for the first signs of graft rejection. How do you snip into a heart that is inside the body? He did it by placing a snipping device in the jugular vein and working it down into the heart. He also began using stimulants to help the new heart. Gradually, he began using the new drugs that help control graft rejection, and the new techniques to store hearts until he was ready to implant them.

The worldwide rate of successful heart transplants still remains at about one in ten recipients who survive for a year. In contrast, the long-term survival rate of Shumway's patients is running better than one in three.

Still, barely three hundred persons have undergone the transplant of a donor heart since 1967. And hundreds of thousands of people are dying each year of heart failure.

This has led to an enormous interest around the world in developing artificial hearts and artificial parts for hearts. These could be mass-produced. Graft rejection would be minimized. And there would be no agonizing waiting around to find a blood-compatible donor who was at the point of death. The greatly increased interest in the artificial heart has led, for example, to the joint American-Soviet venture to develop a totally implantable artificial heart. Experts from the two countries have visited each other's laboratories. The Germans and Japanese, to cite two other conspicuous examples, are also deeply involved in the same effort.

The development of an artificial heart, it appears, is essentially a plumbing problem. The heart — long thought to be the center of human emotions and the soul — is, let's face it, essentially a pump.

Fixing ailing human pumps with artificial parts or devices is already close to being a billion-dollar industry. This industry includes the artificial-valve jobs, which cost about $3,000 to perform, plus hospital costs. It includes the eighty thousand or more pacemakers, which are installed in Americans each year. These cost about $2,000 per installation, plus hospital costs. The industry also includes lung oxygenators, or artificial lungs.

Still another major industry that has just sprung up involves a new way to relieve heart problems by surgery. There has in the past two years been an explosion of operations to bypass damaged or clogged blood vessels within the heart. In 1976 about seventy-five thousand such operations were performed in the United States at a cost ranging from $10,000 to $15,000 — or a total medical cost of about a billion dollars for bypass operations alone.

The technique was first developed at the Cleveland Clinic. Typically, an unnecessary vein in the leg is taken and grafted around the damaged or clogged portion of the blood vessel in the heart. It becomes a new channel through which the blood can flow.

Some authorities — Donald C. Harrison of Stanford is one — contend that a significant number of these operations have been needlessly performed and could have been avoided by the use of such simple methods as losing weight, quitting smoking, and taking drugs that slow the pulse and decrease blood pressure. An official of the American Heart Association, on the other hand, argues that the bypass "is just a good operation that is going to be used a lot."

But what about the *total* implantable artificial heart? Is it a farfetched idea? Where would the power come from to make it pump?

Nosé does not see it as farfetched at all. His group has taken the natural hearts out of calves and replaced them with plastic hearts having an external power source. The calves, as noted, have lived for five months. The power source was still external, but mechanized power sources now are almost small enough to be implanted along with the artificial heart if desired.

In one surprising way operating the implanted artificial heart is apparently easier than had been expected. It had been assumed that a human with an artificial heart would need to regulate the flow of blood by turning a knob on the outside of his chest. He would need more blood when he walked up hill than when he was sitting watching TV. Some talked of the possibility of trying to turn this regulatory job over to a programmed computer. Nosé is now pretty well convinced that no control mechanism will be needed, either manual or computerized.

"The artificial heart is designed to pump whatever comes in," he explained.

When the body demands more blood it will push more into the heart. Even if an additional control mechanism is needed after all, Nosé thinks it will be a relatively simple system.

One of the most difficult challenges in developing a long-term artificial heart is to find a material tough enough to stand the wear and tear that a human heart undergoes, and at the same time one that does not damage the blood's properties. Regarding the problem of wear-and-tear, Nosé pointed out:

"The human heart flexes fifty million times a year. You can wear out a lot of material in a year. What we want is material that will flex at least two hundred million times, or for about four years. Most rubber starts cracking at about five million flexes, or a little more than a month. Most of the materials we have tested thus far rupture within a year, but some are more promising. We have found one that could take two hundred million flexes."

As for a power source, that too would seem to be a formidable problem if it is to be implanted within the body. The requirement is for enough power to pump several thousand quarts of blood a day through the body's circulation system with its thousands of branch lines. More than a gallon of blood a minute is pumped in the human of average size.

The U.S. National Institutes of Health and the Atomic Energy Commission are spending millions of dollars each year, largely through contracts to aerospace companies and grants to research centers, to develop an implantable power source. Three inside-the-body possibilities are being explored:

• An electric motor powered by rechargeable batteries
• An electric motor powered by a biological fuel cell (the fuel cell would draw its energy from the food we eat, just as our natural heart does)
• A nuclear engine powered by several dozen grams of plutonium 238

Nosé showed me a mockup of a miniaturized nuclear power plant that was about two thirds the size of a quart bottle. The Westinghouse Electric Corporation and North American Philips Corporation, with Atomic Energy Commission financing, have been working on a bench model for an implantable nuclear engine in the body. It is hoped that radiation will not be a serious threat. After ten years or so, radiation would probably sterilize the person in which the power source was implanted. And it would also sterilize any constant bed companion over that period.[5] But by then presumably the couple would have conceived any children they desired anyhow. (Or they could sleep in separate beds.) The initial cost of the plutonium would be high. But when people with artificial hearts die — as we all must sometime — unused nuclear fuel could be recovered and transferred to devices under construction.

In the first efforts to install artificial hearts in humans the power source will probably remain outside, in a backpack. Partly this will be because of the size problem. But also it will enable the design engineers to make adjustments and repairs before they go for the totally implantable heart.

But the goal is to get the power source inside the body. Nosé explained: "Penetration of the skin always creates infection problems."

CAN THE BRAIN BE REPLACED?

Nosé pointed out that much of the cortex of the brain has some resemblance to a computer. Conceivably, a number of its functions could be duplicated. Getting the computer down to pack size would be a big, big problem. And reproducing artificially the special organs inside the brain such as the hypothalamus would be, he acknowledged, "most difficult." At best, only parts of brain function could be reproduced artificially on the basis of present knowledge.

However, within walking distance of the Cleveland Clinic is a laboratory where the world's pioneers in transplanting live brains are at work. This audacious undertaking is going forward at the Neurosurgical Research Laboratories of the Cleveland Metropolitan General Hospital. Its leader is Robert J. White, a tall, good-humored, forceful, offhand man who can make brain transplant sound simple.

White is professor of neurosurgery at Case Western Reserve Medical School. About fifteen years ago he concluded pragmatically that the best way to transfer the brain of one human being to another human being would be to transfer the entire head. And that is what his group has been doing in animal experiments.

One reason he prefers the whole-head approach is that it is vastly more simple. Lifting a living human brain out of a skull and depositing it in another skull without interrupting circulation or damaging the parts is difficult in itself. The brain is, as he puts it, like a ball of jelly. And it has billions of nerve-cell connections. More important is the fact that in preparing to lift out the brain he and his colleagues would have to cut a great number of connections to the rest of the body. The most significant are those leading to the brain's information-gathering outposts: the eyes, ears, nose, tongue, fingers, face, toes, and so on. Trying to restore all those connections within the recipient would be a monumental task. And a brain without any contact with the five senses would be a useless brain.

By simply switching heads, on the other hand, only a few connections need to be severed and then restablished in the neck of the recipient body. The brain remains operational as far as the eyes, ears, nose, face, and tongue are concerned. The recipient's body will remain out of contact with the brain until scientists master the problem of regenerating severed spinal cords. But they are working on that!

There is, of course, the problem of keeping the head and its body-to-be alive while the switch is being made. White has demonstrated that both can be done.

He prefers to call the procedure "a cephalic transplant." He found that the phrase "head transplant" seemed to give some people the shivers. And in his writings it is always a "cephalon" his group is transplanting, not a head. But in casual conversation with people he often does say "head."

The laboratory's menagerie of animals that were about to have something done about their brains included eighteen monkeys, six rabbits, twenty rats. An operation is performed at least once a week. Monkeys have brains pretty much like Man's structurally, if in miniature. But monkeys are so small that making hookups of arteries and veins takes especially nimble fingers. Transplanting human cephalons, White assumes, would be easier. His plans called for work with chimpanzees. They are far bigger than monkeys and come closest to having brains and nervous systems like ours. There is, again, a problem of expense. Chimps cost about ten times as much as monkeys.

Usually there are also in residence at the laboratory a number of "isolated" animal brains. Some are in cold storage and presumably are not doing much thinking. Others are functioning, warm brains that are being kept alive by hookups to blood machines or to live individuals of the same species. What is going on mentally inside those disembodied brains is a matter of speculation. They appear on the basis of brain-wave and chemical tests to be functioning normally, in fact somewhat better than before they were "isolated." They are feeling no pain because nerves were severed while the animal was under anesthesia during the operation.

Probably these brains in warm storage are bored and lonely, living on old memories. At any rate, White's group is learning a great deal that is of interest to medicine about the physiology of healthy and sick brains.

He showed me the machines used to keep life-giving blood flowing into the brains. They varied in size according to the size of the brain to be fed. The one for a monkey was considerably smaller than the one for humans. Another device, which his group has used successfully, keeps a human brain alive for several hours after all the other parts of the patient's body are unquestionably dead.

The machine — it is partly glass — performs for the brain what the lungs and heart ordinarily do. Rotating disks stir up films of blood over which oxygen is passed. The oxygen is absorbed into the blood and pumped to the brain.

White's group began making medical history in the 1960's, when they first lifted out and preserved, alive, the brains of monkeys and dogs.

Other investigators, too, began having some success in replacing *parts* of animal brains. The front of a lizard brain has been successfully switched at the Weizmann Institute in Israel by David Samuel. The untrained recipient lizard that received the new frontal part remembered a trick for finding food that had been taught to the debrained lizard. The fronts of lizard brains and human brains, however, are different by a magnitude of millions. The Soviet Public Health Minister has disclosed that Russian scientists have switched certain parts of dog brains and that the dogs have survived for more than a week.

White indicated to me that if he could cool the human brain as he cools monkey brains, diseased or ailing parts of the human brain could probably be replaced. Successful cooling of the human brain may occur within the next few years. At present, a surgeon going in to remove a tumor, for example, has only three minutes in which to operate while the blood supply is cut off. But if the brain can be cooled down toward freezing he will have a full hour in which to work. If the brain is frozen for longer-term storage you must add antifreeze because, White said, ice would wreck the brain.

While the White group was isolating brains, a Russian scientist, Vladimir Demikhov, managed to graft the upper body of a dog onto a second dog, and maintain the graft's life for several days. In effect it was a two-headed dog, and sometimes the grafted head would nip at the host's head. White said he had been to Russia five times for consultations, and has twice visited Demikhov.

The White group's next achievement of note was grafting an isolated brain onto the throat of a living dog. This grafted brain did no thinking for the host dog. It simply proved that the piggyback brain could be kept alive and functioning from the blood of the host.

The successes of the White group in actually transplanting heads — or if you prefer, cephalons — of monkeys began in the early 1970's. Since then, White and his colleagues have been building up a broad body of knowledge about brain function — and how human head transplant might be handled. That is, how it might be handled if they decide to try. Public opinion and scientific justification will determine if and when they make the attempt.

White has on the wall of his office pictures of monkeys with transplanted heads, and I have before me as I write this, one such picture on short loan. The monkey has a support collar about his neck, with a number of tube connections showing. He looks angry, but monkeys commonly look angry. White advised me that if a monkey was pugnacious and irritable before his head was transplanted he tends to remain pugnacious and irritable. On the other hand, if the monkey's original disposition was pleasant or tranquil that disposition seems to continue. White added:

"It's disturbing, I will admit. The head continues to function just like it was on its own body."

Most of the heads that have been transplanted show every indication of being alert to what is going on. The naturally angry monkeys will snap at your finger if you give them a chance. The monkeys with transplanted heads will make sounds. They will accept and chew food. Their eyes will track you as you move.[6]

The White group has kept monkeys with transplanted heads alive for nearly a week. There is evidence that the brain could be kept going indefinitely. The problem is that the other organs of the head start to deteriorate because of graft rejection. White explained that when you transplant the whole head of a monkey "the brain does all right." But the tongue and the tissues of the face undergo swelling as a part of the rejection process. Failure, however, does not involve the brain. It does not seem to suffer graft rejection. But other parts of the grafted head are rejected by the body. He commented, "It is an interesting thing." His evidence is that rejection of the brain is at the very least substantially delayed. "We don't even bother to do tissue typing for the brain. I have never seen the brain rejected. Possibly it is because lymphatics are much involved in the immune process, and brains have no lymphatics. Also, the blood vessels in the brain have a unique barrier."

The group usually closes down an experiment in five days because of failure of the organs and tissues of the head. He said: "The brain is still going on happily and looks fine when we take it down."

So in that one sense, the brain is one of the easiest of all organs to transplant!

White is acutely aware that his findings may force us to reconsider our ideas of what the essence of our being is and where it resides in the body. He is a Catholic and has talked with the Pope about his work. Some years ago, when he first perceived the implications of his findings, he said: "Now that we are able to take someone's head, which contains the mind, we must ask: Where is the soul: in the entire body, or in the mind only? When we place the head of one man on the trunk of another man does the new creature have one soul or two souls? I say he has only one soul because the soul is in the head, or better, in the central nervous system. But theologians will answer that the soul must be considered to be in the totality of the whole individual, that you can't confine it to a single organ."[7]

White said to me: "What makes up you and me is our mind, not our hearts. All other organs supply parts to keep the brain alive." Or perhaps to obey the brain's commands. Glands outside the head can have a profound influence on our behavior; but most seem to be masterminded by organs within the brain, such as the hypothalamus and the pituitary.

Even some medical doctors have been slow to adjust their thinking to the primacy of the brain. One of them, in writing about organ transplant in *Mental Hygiene,* summed up the "staggering possibilities" of brain transplant by saying: "A man faced with cerebral arteriosclerosis could buy a new brain with nice soft arteries." But if the implications of Dr. White's findings are as they seem, why should he bother? If he arranged for the installation of a nice new purchased brain it would require removing his own brain (head) from his body and installing a stranger's brain in its place. Furthermore, he might be legally liable as an accessory to homicide. If under the new consensus death occurs in the brain, not the heart, obtaining the still-live brain of a dying man would surely involve killing the person's body, even if his body were on a respirator.

No, to make any sense it would seem to have to work the other way around. A person with a healthy, vigorous mind might want to shed a shattered, badly diseased body. He would wait until the local hospitals had a patient with an otherwise healthy body who was dying of a brain tumor or a brain hemorrhage or other affliction of the head. The body could be kept functioning by machinery while the brain died. Then a switch could be made.

There would be little point to making the switch, however, until science learns how to get the two severed spinal cords to grow together in an orderly way. The lower body of such a born-again human would be numb. The spinal cord is about the size of a broom handle. White indicated to me that his group had solved the problem of restructuring the bones of a severed spinal column but not the regrowth of the nerve tissue of the spinal cord. A considerable amount of federal money is going into experiments to regrow spinal cords, but any success for humans apparently will take some time. White said: "If we can grow spinal cords, we can transplant heads of humans with hope of achieving reasonably normal people."

Fifteen years ago most scientists concluded that regrowing a severed spinal cord was impossible. This doubt persisted even though in a few situations seeming miracles had occurred, with the help of a fever-producing substance called pyrogen. Also, it was known that nerve axons (connecting lines) have a capacity for regrowth.

In June 1976 the fourth biennial Conference on Regeneration of the Central Nervous System heard a report that *Science News* called of "monumental importance."[8] A Soviet scientist, Levon A. Matinian of Armenia, reported that after twenty years of effort he had succeeded in regrowing the spinal cords of rats. His rate of success was 40 percent: of 350 rats, 140 were now walking. Apparently one of the problems is to prevent scar tissue from forming while the severed nerves have a chance to regrow axons. He found that the enzyme trypsin and several other enzymes seemed to work best in stimulating regrowth. Some American investigators are wary; they would like more details.

AFTERTHOUGHTS

Most of the surge to develop the techniques for transplanting natural organs or for substituting artificial ones seems to represent a commendable application of technological ingenuity. Hundreds of thousands of people who would be healthy except for a damaged or diseased body part could be restored to a more effective life. But as we move toward the replacement of numerous complex body parts in innumerable people of all ages, some uncomfortable questions need to be asked.

The already-soaring health costs in most advanced nations may become a real social burden if we accept the thesis that anyone with an ailing organ has a "right" to have it replaced and to be kept alive at all costs. Some of the substitute organs, with the hospitalization and servicing required, would be enormously expensive. And some people could not be kept vigorously alive, just barely alive, and thus would become burdens on their families or on society.

Increasingly, the technologies of organ replacement are being developed primarily or almost entirely with public funds. Once developed, can private entrepreneurs with patents, medical or otherwise, feel free to build fortunes by reproducing or installing such organs? Will only the rich be able to afford them, like buying seats in the lifeboats of a sinking ship? Possibly not. Most advanced societies are moving toward national health care.

But that raises other questions, if we assume that everybody has a "right" to good health and to live as long as science can manage it. One question is that of supply. What if there aren't enough organs to go around? Who gets them and who doesn't? And even if there are enough, the question of cost arises. Installing and maintaining them in all the people who want them could become an enormous tax burden on the whole population.

Baffling problems of distributive justice are bound to arise. Consider the totally implantable artificial heart as an example of the problem faced with substitute organs in general. Many patients are already clamoring for such a heart. Once it is perfected quite a few years will go by before enough hearts can be produced to satisfy the demands of the hundreds of thousands of people wanting them. More than a half million Americans die each year of heart attacks.

The theory of distributive justice most cited by ethicists was developed at Harvard by John Rawls. Organs would go to those people whom a reasonable person would designate as candidates if that reasonable person were cloaked "in a veil of ignorance" as to his *own* social status, assets, prejudices, and shortcomings. That may be a good approach to pure, blind justice. But applying it to such a complex, massive, and politically loaded problem as distributing totally implantable artificial hearts seems impractical.

In the United States funds for the development of the totally implantable artificial heart come very largely from the federal government, so the highest-bidder approach to distribution also seems unlikely.

Some other basis is needed. But what basis? In 1973 the National Heart and Lung Institute had a panel to ponder this. The panel came up with $25,000 as the cost per installation, a figure that will probably prove to be conservative. Even at that, as the organs become widely used, the total costs could quickly mount up to billions of dollars. And these could be tax dollars. The panel acknowledged that once we were faced with these realities, we would probably be forced to rethink the long-held concept that life should be preserved at any cost.

But let's get back to the problem of choosing recipients. The panel settled on three basic alternatives for giving applicants priority:

1. "Decisions based on appropriate medical criteria, i.e., providing the artificial heart to those with the most urgent need." (As we have seen, Shumway succeeded in heart transplants only by *rejecting* the candidates with the most "urgent" need. These people had less chance of surviving the operation. However, "medical criteria" that determine who has the best chance of being restored to vigorous health would, in my view, be a good starting point.)

2. "Decisions based on estimates of 'social worth.'" The panel firmly rejected this sticky approach because it would be undemocratic. Political dissidents might be rejected, the famous and rich be favored.

3. "Decisions based on some form of random selection." The panel came out in favor of a lottery when medical criteria were not compelling.

The panel might have cited, I think, two other reasonable bases for choice of candidates: family responsibility and age. I would think that a thirty-three-year-old man with three young children should be given priority over a sixty-eight-year-old widower whose children had all married. The panel, in fact, seemed to assume that older people would need and get most of the artificial hearts. It alluded to the fact that much of the financial support in the United States would probably come from Medicare, which offers low-cost medical care for people over sixty-five.

I believe that putting most of the available artificial hearts in senior citizens — and this sounds harsh — would be the *least* justifiable course in terms of sound social policy. People who would — except for an organic defect — have the most years left of normal life expectancy should have a higher priority than those with an actuarially short life expectancy. This would seem to comply better with the requirements of both justice and social foresight.

The operating lifetime of millions of automobiles has been more than tripled by a generous use of spare parts. Automobiles are not people, and the situation is far from comparable. Still, if the bulk of our spare parts goes to people past sixty-five there will be for millions of people an increase in life expectancy without any present prospect of an accompanying prolongation of mental acuity.

Sir George White Pickering, professor of medicine at Oxford, some years ago viewed the broad-scale substitution of human organs as a "terrifying prospect." He foresaw that senile people might overrun the earth.

As far as possible, I believe that any national program for the substitution of organs should have as its goal helping people live out their normal life expectancy. It should not be designed to help aging people live longer.

Then there is the possible undermining of a sense of personal identity if we move in a major way into substitution of body parts. The complexities and social fragmentation of the modern world already have aggravated identity problems for millions of people. No one can seriously argue that a transplanted kidney or even an artificial heart would necessarily alter a person's self-assurance about his identity. But substitution of organs that affect appearance, sex drive, memory, mental capacity, aggressive drive, disposition, metabolism, feelings, emotional reactions, creativity, certainly would.

While switching heads of individuals is a fascinating exercise in technological virtuosity, I believe it should be considered for humans only in extraordinary cases. Even if the regrowth of the spinal cord can be achieved, the problems of getting a brain used to directing the thousands of activities of a strange body would seem to be formidable. They might be formidable enough to drive the brain crazy.

Finally, we come to a potentially worrisome development arising out of the industry that builds organ-assist devices. The Medtronics Company is the leading manufacturer of the electrical devices that serve as pacemakers for the human heart. In the United States alone about eighty thousand are being installed each year. Officials at Medtronics, however, do not plan to stop at pacemakers for the heart. Their researchers are working on electrical devices that when installed in the body can influence the central nervous system, including the brain. They are working on ways to modify behavior electrically. They also talk of devices that will, for example, modify antisocial behavior.

So we come full circle, by way of manipulating organs, back to the behavioral manipulations of Skinnerism and the mood manipulations of Delgado.

III

CONCERNS
AND COUNTER-
MEASURES

23

Second Thoughts by the Human Engineers

It is a hard thing for an experimental scientist to accept, but it is becoming all too evident that there are dangers in knowing what should not be known.
— Sir MacFarlane Burnet, Australian biologist

Only quite recently has it occurred to an appreciable number of scientists that their discoveries about modifying Man are not automatic boons to mankind. Some exceptional voices, as I have noted, have expressed deep concern.

Modern scientists have scored so many triumphs of ingenuity that most of them have learned to expect applause. And they have usually gotten it from an awed public (when the public could comprehend what had been achieved). They have tended to feel that their mission is to seek the Truth, whatever it may turn up. They are always to press on no matter where the findings may take them. If something can be done, it should be done. And anyhow it will be done. The operant conditioner Roger McIntire writes: "History tells us that whenever we develop a technology we inevitably use it."

James Vicary, who developed subliminal advertising, acknowledged that at first misgivings did occur to him. But he set them aside because, "as a researcher, I have always pushed on as far as I could."

Typically, biologists have wanted to be free of public guidance in deciding what innovations to attempt that could affect the human makeup. At the same time they have usually depended upon public funds for their work.

Lately, though, as ever more spectacular scientific feats are announced, the situation is changing. There is now a refreshing amount of discourse among scientists about the wisdom of what they are doing. At conferences they have been inviting the views of ethicists, philosophers, theologians, public servants. Some even accept the inevitability that the public deserves to have a say about what potentially human-altering projects they undertake.

The biologist James Watson, who helped lay the foundation for much of the current genetic manipulation, is one who has called for public guidance. He said of cloning:

"This is a matter far too important to be left solely in the hands of the scientific and medical community. The belief that surrogate mothers and clonal babies are inevitable because science always moves forward . . . represents a form of laissez-faire nonsense dismally reminiscent of the creed that American business, if left to itself, will solve everybody's problems."

One of the first hints of the new, troubled mood of many life scientists came in 1969 from a team of three brilliant young Harvard researchers, headed by Jonathan Beckwith. They had just isolated the first gene in history. Their feat won world acclaim. At a press conference, however, they were somber. One said, "We do not have the right to pat ourselves on the back" because in the long term the discovery might "loose more evil than good" upon mankind. They saw the possibility that other scientists, in building on their contribution, might develop genetic controls over behavior that some future dictator could exploit.

Some of the older scientists angrily told them to shut up and shape up or get out.[1] One of the three, James Shapiro, did in fact get out. He withdrew from a promising career in molecular biology at Harvard to become a social worker. Then, after a couple of years, he returned to science as a microbiologist at Chicago.

Another brilliant young scientist, Leon Kass, has shifted his main area of exploration and writing from biochemistry to bioethics. He first headed the Committee on Life Sciences and Social Policy of the National Academy of Sciences. Later he became a full scholar at the Kennedy Institute for the Study of Human Reproduction and Bioethics in Washington. Kass suggests that the new biology has gotten scientists into a situation comparable with that of the pilot in this familiar fable:

"Good afternoon ladies and gentlemen. This is your pilot speaking. We are flying at an altitude of 35,000 feet and a speed of 700 miles an hour. I have two pieces of news to report, one good and one bad. The bad news is that we are lost. The good news is that we are making excellent time."

In addressing his colleagues in the pages of *Science,* Kass admitted that no one likes regulations but he said that there is something disingenuous about a scientist who professes concern for the social consequences of science but who responds to every suggestion for regulation with insisting that research be unrestricted and technological progress be uncurtailed.

Kass worries about the fact that new technologies in the Biological

Revolution often become introduced as "the result of no decision whatso-ever." Or they are "the culmination of decisions too small or unconscious to be recognized as such."

The Kennedy Institute is only one of several institutions that have sprung up around the world to think about the implications of new developments in the life and behavioral sciences. In Great Britain, for example, the British Institute of Human Rights, and also the Working Party on Ethics of the British Association for the Advancement of Science, are carrying on large-scale programs of analysis and debate.

The world's best-known and most ambitious program is that of the Insti-tute of Society, Ethics and the Life Sciences at Hastings, New York. It stages debates between some of the more controversial experimenters and their critics. It holds conferences in which people from many disciplines discuss specific troublesome issues. It is encouraging colleges and medical schools to offer courses on the ethics of the new biology.

As this new mood of uneasy self-examination has taken hold, some of the leading proponents of human modification have shifted to less brash positions. Delgado has become less flamboyant in his observations. Sweet and Lederberg speak more softly (from Lederberg: "Scientists are by no means the best architects of social policy"). Even Skinner has modified some of his ideas. In *About Behaviorism* he sets out to provide "a good deal of clarification" and to "reinterpret the data" on some of his more con-troversial pronouncements. Few behaviorists, he now says, believe that human behavior is "endlessly malleable." (Yet another book has softened the arrogance of his public image: the autobiographical *Particulars of My Life,* in which he offers a frank and charming description of his early years in Susquehanna, Pennsylvania.)

The uneasiness of scientists centers on three areas: health hazards, ques-tions of individual ethics, and questions of social hazards to Mankind. I list them in that order because scientists seem to have considerably more understanding of health hazards — and concern about them — than they do of the ethical questions. And their awareness of possible social hazards lags far behind.

CONCERN ABOUT HEALTH HAZARDS

Hazards to health are something that life scientists find understandable. They have thought much more about them than about the social im-plications of what they are doing. Partly this is because in recent years they have seen health hazards leap out of unsuspected, inadequately tested scientific creation: aerosol cans, Thalidomide, DDT, vinyl chloride, as-bestos fiber.

Many reproductive biologists have backed off from efforts to implant a laboratory-conceived embryo in a woman and take it to birth because of an obvious health hazard. They could end us with a monstrosity that could not legally be killed. And its birth might reach public attention.

In 1974 molecular biologists around the world agreed to their historically unprecedented moratorium on certain kinds of genetic manipulation out of a health-related concern. By combining various animal genes with genes of certain bacteria they were creating life-forms new to the world. These opened up spectacular possibilities for genetic engineering. Some might end in modifying the human species. But by recombining genes they were creating life-forms that could multiply. Some of the new combinations were potentially dangerous and if they got loose could terrify humanity.

The man who was probably ahead of the rest of the world in this type of research was Paul Berg of Stanford University. He led the campaign to get a moratorium. And later he led the drive to establish ground rules on what could and could not be done.

At first, he and his associates in research had been reveling in breakthroughs using different life materials. One project of gene-splitting, at that time still in the planning stage, involved a tumor virus. One of his students, a girl, happened to mention — and defend — the project during a discussion after a lecture at the Cold Spring Harbor Laboratory on Long Island. The lecturer, Robert Pollack, had been discussing safety measures in using biological materials. The discussion became heated. Afterward, Pollack was upset enough to call Berg long distance and cite the hazards he thought might be involved. In the following months Berg became a convert to caution. Much later he told the science writer Horace Judson: "When Pollack called me, it was true that we had been heedless. We had not considered the possible hazardous consequences of what we were making. I do not see this as a great moral or ethical issue; I see this more as a problem of public health."

Berg's uneasiness continued to mount as other researchers began calling his group every day asking for biological materials with which to do similar experiments. "I would ask them what they wanted it for," he recalls. "Some of them had horror experiments planned, with no appreciation of the consequences."[2]

To head off some of these disturbing projects he began a crash campaign to get a number of distinguished colleagues to join him in writing the now-historic letter calling for a moratorium on certain types of experiment in which DNA molecules are recombined. It appeared in two of the world's most respected scientific journals: *Nature* (Great Britain) and *Science* (U.S.A.). The proposed moratorium was indeed respected, while it lasted, on a substantially worldwide basis.

CONCERN ABOUT INDIVIDUAL ETHICS

A physician's code of ethics requires him to do no harm and to do only what is best for the patient. The situation becomes somewhat blurred when the medical man is a researcher performing nontherapeutic experiments on groups of people who may or may not need medical attention. And the situation becomes even further blurred if the experimenters are physiologists or biologists or psychologists who are performing experiments on humans that involve or create medical problems.

Many dozens of health-related experiments on humans have been marred by dubious manipulative practices. A reexamination of ethics suddenly has become a major preoccupation. The uproar over experimenting with live fetuses would probably not have erupted in the 1970's if certain experimenters had not seemed so callous. There is still a lack of precise knowledge about when fetuses can feel pain. The Nobel laureate Frederick C. Robbins of Case Western Reserve University addressed a National Academy of Sciences symposium on ethical problems in experimentation. He said that some of the fetal experiments being conducted sounded "like a horror story." In one study the heads of eight live young fetuses were cut off and then injected with radioactive compounds to study brain metabolism.[3]

Revelations about the use of untreated control groups in human experimentation also has caused some squirming. Experimenters of course need to use control groups in order to confirm that the innovation being tested on patients really is responsible for change. But to do this, is it ethical to deceive people who urgently need medical treatment and think they are getting it when they are not?

For more than a quarter century hundreds of poor Alabama black men suffering from syphilis were given phony "shots" by doctors working for a federal health agency. The aim was to study the course of untreated syphilis. Others in the experiment were given the best treatment available, which in time came to be antibiotics. By 1952, twice as many of the untreated as the treated were dead.[4] It was only twenty years later, in 1972, after the Associated Press brought the project to public attention, that a recommendation came to close down the federal project.

Or consider the low-income Mexican-American women overburdened with children who came to a San Antonio clinic for contraceptives. Seventy-six of them, without their knowledge, were given dummy birth-control pills. They were a control group. Others were given contraceptive pills. All were requested to supplement their real or phony pills with a vaginal cream. The experimenters were seeking to determine if anxious women given placebos would report any of the side effects such as nervousness and depression that some women on genuine pills had been reporting.

Some did. But ten of the seventy-six who came in good faith for help and received the phony pills became pregnant.

Other ethical scandals have centered on administering dubious experimental preparations to chronic charity patients or retarded children. At a hospital for elderly chronic patients in Brooklyn, live cancer cells were injected into debilitated patients who did not suffer from cancer. A director of the hospital charged that the purpose was to determine whether cancer could be induced by the injection of live cancer cells. Experimenters claimed that this was not the purpose at all. It was well known, they asserted, that a person's immunity defenses reject foreign cells whether cancerous or not. They had found that the defense system of a person with cancer acts more slowly than that of a healthy person. Hence, what they were trying to establish was whether this slowness of action was due to cancer or just general debility.

The patients involved were not informed that the treatment was not for their own good. Nor were they told that they were being injected with cancer cells. No written consent form was obtained. If verbal consent was given, as claimed, it obviously could not have been informed consent.[5] When members of the hospital board voted for what one member called a "whitewash report," he challenged his fellow board members to submit to the same cancer injections. State officials, in reprimanding the doctors involved, held that they had no right to claim a doctor-patient relationship when they were acting as experimenters.

At an institution for mentally retarded children on Staten Island, some of the children were injected with live hepatitis virus. This was part of a federally sponsored program to develop a hepatitis vaccine. Professor Henry K. Beecher of the Harvard Medical School cited this in the *New England Journal of Medicine* as one of twenty-two published experiments he considered to be ethically dubious. He was denounced by many doctors.[6] The doctor in charge of the Staten Island hepatitis program defended his procedure by contending that hepatitis was already rampant at the institution. Therefore, he argued, it was safer for his subjects to get their hepatitis under carefully controlled conditions than to contract it naturally.

Similar ethical questions have arisen in connection with some types of psychological experiments. For example:

In a test of behavior under mental stress, each subject suddenly found a realistic dummy of a human body propelled in front of the car he was driving. Several of the subjects who were among the first to be tested expressed great distress. When they were informed that the accident victims were only dummies they were bitterly critical of the experimenters. Nonetheless the tests continued until thirty subjects in cars had struck what they thought were human bodies.

Military psychologists, in stress tests, took untrained soldiers, disoriented them, gave them false instructions, and then led them to believe that they had caused artillery to fire on their own troops. They were told that heavy casualties had resulted. Some of the soldiers broke down emotionally, some even went berserk. No account of debriefing could relieve the guilt of a number of them.[7]

Was the "knowing" attained in these tests worth the human damage done?

Today, scientists are vastly more aware of ethical issues in human experimentation than they were a decade ago. They are now, for the most part, agreed that any research project involving experimentation on humans should be reviewed and approved by a "peer review committee" before it is allowed to begin. A recent issue of *Modern Medicine* reported that most doctors favor experimentation on humans, even on prisoners, children, the mentally retarded, and other captive subjects, as long as there is peer review.

In practice, peer review committees consist overwhelmingly of fellow staff members. Few include lawyers, ethicists, ministers, or scientists in other areas of research. There is a need for the creation of a new profession: that of ethicist with medical or scientific training.

Today there is general agreement on the *principle* that human subjects of experimentation must give their "informed consent." This concept came largely out of the Nuremberg trials, in which Nazi medical experiments were described. Actually, the obtaining of "informed consent" is honored more in the breach. The June 1973 *Report* of the Institute of Society, Ethics and the Life Sciences stated: "The tragic fact is that less than twenty-five percent of the studies in our files claim that consent was obtained from the participants" and not a single paper documented the form in which the consent was given. Can a mentally disturbed person or a mentally retarded person actually give a truly informed consent? Consent to psychosurgery? The psychiatrist Willard Gaylin, president of the institute, put the problem neatly by saying, "the damaged organ is the organ of consent."

In all talk of ethics involving experimentation on humans scientists tend to specify that a third consideration be taken into account: the benefits must outweigh the risks. That is a highly subjective measure. Scientists are all too prone to assume that the benefits will always outweigh the risks. In view of their theoretical commitment to peer review and informed consent, the risk-benefit measure can come out as a loophole. The ends justify the means. The same sort of argument is used by policemen who defend searches without a warrant by contending triumphantly that after all they came up with useful evidence.

CONCERN ABOUT SOCIAL HAZARDS

The great majority of scientists who are busily engaged in projects to shape human behavior and development do not seem actively concerned about the social hazards of their work. At a 1975 conference on genetics and ethics in New York City, a Harvard biologist estimated that only about one scientist in a hundred was interested in the social and moral problems that might be generated by science. That was probably an overly pessimistic statement. However, it does suggest — and correctly, I think — that far more scientists should be concerned than are concerned, in view of the momentous changes that are being wrought.

As noted before, a number of eloquent voices from among both the behavioral and life scientists have expressed uneasiness. In biology some of the more forceful and persistent have been René Dubos, Sir Mac-Farlane Burnet, Leon Kass, Salvador Luria, Jonathan Beckwith, James Watson, Jean Rostand, James Nagle, Paul Berg, Robert Sinsheimer, and George Wald.

Among psychiatrists, psychologists, and sociologists some of the more thoughtful, cautionary voices have been those of David Krech, Carl Rogers, Stephan L. Chorover, Willard Gaylin, Amitai Etzioni, Perry London, Karl Pribram, Peter Breggin, Thomas Szasz, Richard Restak, and Elliot Valenstein, and the behaviorists Leonard Krasner and Israel Goldiamond.

They have focused their concern on the various aspects of man's remaking man. Most would agree that it is time, past time, for scientists, as well as societies in general, to arrive at value judgments on whether proposed innovations will contribute to or undermine sound social policy.

And most would agree that standards are needed for judging whether a proposed research project is socially hazardous or not. Most would also concur that the proper machinery needs to be set up whereby projects judged to be hazardous can be prevented from starting at all.

In making the necessary judgments, a number of questions need to be asked — and answered. For example:

What will be the impact on the quality of life, on the family, on population, on justice, on individual integrity, on individual dignity?

What is the prospect that the innovation could in the future strengthen the power of an aspiring dictator?

Would the innovation press us further toward a dehumanizing style of living?

On Controlling the Controllers

24

A decent society regulates all technology that is powerful enough to affect the general welfare.
— Perry London, psychologist

We have encountered in these pages several dozen possible innovations for reshaping and controlling our lives. Most involve an element of manipulation. Most assume a high degree of plasticity in humans. Some of the scientists who have developed these innovations, as we have just seen, are starting to become uneasy. The uneasiness is spreading to the public and its political leaders.

Consider the situation in the United States. We are still a long way from setting goals or priorities. We have barely started assessing the impact on humans of innovations that attempt to modify Man. We have done little to provide across-the-board laws and regulations that will protect us against abuse and folly. But we are making fumbling starts.

Many of the controversial developments we have been examining have come about, at least partially, through governmental funding. And the funding has often been legislated in mindless fashion.

In 1973 the U.S. Congress established the bipartisan Office of Technology Assessment, largely because of the efforts of Senator Edward Kennedy. Today "technology" is involved in most legislation. Inevitably, the OTA, spread thin, pays most attention to big-ticket items like aircraft carriers, spaceships, and solar power plants.

The Senate's Subcommittee on Constitutional Rights has made a study of federal involvement in behavior-control technology. It concluded that the government "through a number of departments and agencies is going ahead with behavior modification projects, including psychosurgery, without a review structure fully adequate to protect the Constitutional rights of the subjects." Partly as a result of this study the National Commission for the Protection of Human Subjects of Biomedical and Behavioral Research was established. Its function is to *advise* the Secretary of Health, Education and Welfare.

The Congress also established the President's Biomedical Research Panel, which was to report directly to the President, although there is little evidence that President Ford welcomed it.

It was also in mid-1976 that Senator Kennedy, chairman of the Senate Subcommittee on Health, declared himself dissatisfied, if not disillusioned, with developments. He said he was not impressed with the way the National Institutes of Health had been protecting the public's interest in biomedical and behavioral research. He began hearings that seemed likely to last for at least a year.

REGULATION BY THE SCIENTISTS THEMSELVES

Scientific goals can be shaped to some extent by controls that the investigators and practitioners exercise themselves. Each scientific society could, as the American Psychological Association has been doing over the past few years, update its professional code. By bringing in respected outsiders the societies could — and sometimes have — made peer review something more than a rubber-stamping by colleagues.

Scientists have also been debating the idea of setting up their own science court system. That would be one way to head off outside regulation. A scientific advisory group recommended in 1976 that a court be set up experimentally by the White House Office of Science and Technology Policy. The science court would try to establish — by adversary proceedings — what the known scientific truths are in a given controversy. The aim would be to try to weed out exaggerations, errors, and outright lies. The court would leave to other public bodies what is often more important: making value judgments on the particular question. But just getting the facts straight would help, particularly in the more esoteric fields like molecular biology.

As for the geneticists who accepted a self-imposed moratorium at the Asilomar Conference, they ended the moratorium only after they had themselves come up with a moderately tough set of guidelines. These guidelines were to cover certain kinds of particularly dangerous research in which genes from different species are manipulated in order to invent new creatures. The guidelines had only the power of moral censure, but there was an assumption, as *Science* put it, that the guidelines "will probably be followed closely by the national bodies in each country responsible for framing relevant regulations." A number of national governments did shortly enter the picture.

One of the most heartening aspects of the Asilomar Conference was that there were fifty-three delegates from countries outside the United States. The British were particularly active. And the Russians, as noted, voiced concurrence with the conclusions. The proliferation of nuclear materials in many countries is teaching us the hazards of allowing technologies to advance without international agreement.

The genetics group had proposed that the most hazardous experiments in recombining genes be banned. In very high risk experiments only organisms incapable of surviving outside the laboratory were to be used. Special "containment" laboratories should be constructed with low inside air pressure. Laboratory workers would be required to shower before leaving. A Boston-based scientific group, Science for the People, called the guidelines "mild."

PROTECTION BY GOVERNMENT RULES AND GUIDELINES

In the United States it was the National Institutes of Health, a federal agency, that took the geneticists' proposed guidelines under consideration. Hearings were held. Out of the new review came official guidelines for any research supported by NIH funds. These new guidelines were somewhat more strict and detailed than those reached at the Asilomar Conference. But they kept within the general pattern the geneticists had proposed. The NIH called them a "small" but "important" step toward protecting the public.

Some distinguished critics stress the "small"-ness of the protection offered. Robert Sinsheimer, chairman of the Biology Division at Caltech, warns that we are letting loose a rush of research that is irreversible. He stated that because of human frailty these new biological creations "will escape and there is no way to recapture them." Mark up another convert to caution. Sinsheimer has been a longtime enthusiast of the general concept of genetic engineering. Now he was concerned. He said, "Science has not taken so large a step into the unknown since Rutherford began to split atoms." Harvard's Nobel Prize-winning biologist George Wald went further. He called gene-splitting the biggest issue in the history of science, and called for a new moratorium on creating novel genetic combinations.

Both he and the biochemist Erwin Chargaff of Columbia University have been particularly disturbed by the continuing use of *E. coli* bacteria, an old lab favorite, as hosts in recombining genetic materials. *E. coli* come from the human gut. The reason they remain a favorite is that so much is already known about them. Chargaff charges that there is no practical way to know if altered *E. coli* bacteria are escaping. "You just have to sit and wait" until something odd starts happening to some segment of the human population. Both Sinsheimer and Chargaff believe there should be a ban on using any organisms found in humans or in the human environment.

Despite the guidelines, obstacles were encountered. The City Council of Cambridge, Massachusetts, objected. It called for a halt to such research in Cambridge until the councillors had a better grasp of the possible hazards to people. Cambridge is the home of Harvard and MIT. Both staff many genetic engineers.

By early 1977 the Council, after lengthy hearings, agreed to permit the construction of gene-splitting facilities in Cambridge, but under stiffer regulations than the National Institutes of Health had recommended. The chairman of the Cambridge citizens' review board expressed his gratitude to those scientists who had continued to call for more stringent control against the majority opinion of their colleagues. He felt that some of them had imperiled their careers by stating their views. On several campuses it was noted that faculty members who still lacked tenure had been pressured into keeping quiet for fear of losing jobs or promotions.

A year earlier the National Institute of Mental Health came out with a "policy statement" on the practices and techniques of behavior modification. The practices covered ranged from conditioning to electroshock. (The operant conditioners vainly claim exclusive right to the phrase "behavior modification.") The statement was called "Behavior Modification: Perspective on a Current Issue." It was more a lecture than a set of tough guidelines. Some forms of behavior modification were commended, others deplored. It warned of the need for "boundaries" to make sure a technique was never used by a totalitarian regime for "mind control."

Although units of the Department of Health, Education and Welfare finance much of the research on human behavior, other agencies and departments have been heavily involved. It is time that all government agencies financing research aimed at altering human behavior or development think about impact. They should require that applications for grants include a Human Impact statement. This would be comparable to the Environmental Impact statement now required by many agencies.

The pronouncements coming from agencies of HEW have varied in firmness. Some announce "criteria" rather than the more flabby and polite "guidelines." Or still more firmly they announce "regulations." The National Institute of Mental Health addressed itself to the commotion over psychosurgery by announcing mere "recommendations" for determining when and how psychosurgery should be performed. For one, the operation still should be treated as an experiment rather than as therapy. For another, it should not be performed on anyone involuntarily confined or incapable of giving informed consent. And it should not be performed without advance notice (to permit others to make an assessment).

These "recommendations," it should be noted, apply only to surgeons or facilities receiving HEW funding. They do not apply, for example, to surgeons in private practice.

Later, HEW's National Commission for the Protection of Human Subjects of Biomedical and Behavioral Research "recommended" that psychosurgery could be performed on involuntarily confined people after a court hearing was held and the operation was approved. In what seemed to be a major shift, the commission spoke of psychosurgery as having "potential merit."

PROTECTION BY LAW

A law is a far more certain bulwark than a guideline when there is agreement that safeguards are needed. Laws are comprehensive. They cannot be altered by administrative whim as guidelines can.

Let's return once more to the guidelines that the National Institutes of Health issued to geneticists on creating new life-forms. The guidelines don't cover any research being conducted or financed by the Department of Defense or other government agency. They don't cover research going forward in the laboratories of drug companies or of General Electric or of other corporations. They don't cover research financed by foundations. Most ominous, perhaps, they don't address the question of probable proliferation down to high school chemistry laboratories once the techniques are perfected.

The pages of *Science* have carried strong appeals by well-known biologists for Congressional action. One proposed that a *law* be enacted bringing all forms of "genetic engineering" under "federal control." That sounds reasonable to me. As I have previously advocated in this book, there should be created in each country a commission with absolute power over the creation of new forms of life. Along with it there should be an overall international organization to try to control the proliferation of new living creatures.

The International Council of Scientific Unions, which has representation from sixty-six countries, set up a committee on genetic experimentation. Its function is to "blow the whistle" on countries that do not comply with prevailing safety standards in recombining genes. What is "prevailing" is still not quite clear. Its only force will be moral persuasion.

In the meantime the prestigious National Academy of Sciences has urged that all American scientists, *however funded,* at least observe the NIH guidelines.

Many observers felt that guidelines were not enough, and that basic decisions on gene-splitting should not be left to scientist-dominated groups, including the National Institutes of Health.

In early 1977 the problem was becoming urgent. The *New York Times* reported that eighty-six universities in the United States alone were doing DNA research, and so were at least nine private companies.

A number of U.S. cities, in addition to Cambridge, that are sites of universities involved in gene research — including Madison, Wisconsin; Bloomington, Indiana; and Ann Arbor, Michigan — were hearing calls for municipal regulation of gene-splitting research. New York and California were drafting laws that could affect any gene-splitting research within those states. And at the federal level bills were introduced in both the Senate and the House to require federal control of all gene-splitting attempted anywhere in the land. Senator Kennedy of the Health Subcommittee began hearings.

Thus it happened that three years after the scientists themselves became anxious about gene-splitting the federal government accepted the need for laws — not guidelines — to control any research involving the recombining of genes. This occurred on April 6, 1977, when the Carter administration, through Joseph Califano, Secretary of Health, Education and Welfare, appeared before the Kennedy committee. He said that because of the potential hazards involved "there is no reasonable alternative to regulation under law."

Kennedy has offered the broader — and I believe warranted — forecast that "from a Constitutional point of view, the frontiers of law for the next twenty to twenty-five years will be in these areas of bio-ethics."[1]

CONTROL BY SPECIFIC ADMINISTRATIVE BOARDS

When the National Commission for the Protection of Human Subjects was established, it was assumed that it would regulate the use of humans in research projects. Instead, it turned out to be an advisory board that would merely make recommendations to the Secretary of HEW.

The federal government has two bodies that police some problem areas on the national scene: the Drug Enforcement Administration and the Environmental Protection Agency. As indicated earlier, I think there is urgent need for the creation of a third: the Human Reproduction Research Administration. It should have policing, not advisory, powers.

Many developments in reproductive research need policing because they can substantially affect the future of Man. At present there is virtually no public supervision of any of these activities:

Human seed banking
The creation of future humans in the laboratory
Predetermination of sex
The building of artificial wombs

> Attempts to clone humans
> The use of surrogate mothers
> The development of man-animal chimeras

Many of these developments should be halted or be put under tight control. All of them have the potential for shaping our evolution.

Since much of the competitive pioneering in reproductive technology is being done in other countries, an international search for consensus on policies is urgently needed.

To sum up, groping starts have been made to exercise some kind of social control over some of the activities aimed at manipulating human behavior and development. Other areas have been virtually neglected except for the influence of public opinion.

Public opinion, when informed, can be an important source of control. Fortunately, the publishing and broadcasting media and citizens' groups have been showing an increased interest in these developments that can affect Man in significant ways. It was public opinion that scared the psycho-surgeons underground for a while. It caused the U.S. Bureau of Prisons to rethink its use of operant-conditioning techniques on prisoners.

In England one concerned lawyer, Paul Sieghart, succeeded in establishing the Council for Science and Society. It has the specific purpose of making sure the public is informed. Its goal has been to stimulate "informed public discussion about the possible consequences of socially important pieces of scientific research, in each case at the earliest possible moment."[2] The primary focus has been on new biomedical technologies.

Every advanced society, I believe, should have a comparable organization.

New Trends That Can Enhance Self-Direction

25

Are we seeking the truth about man that sets all men free? Or are the truths we discover only making some men more free and powerful, while others become more vulnerable to manipulation?
— Sidney Jourard, humanistic psychologist

When Jourard asked these two questions in 1972 he had good reason to be worried. And much has happened since then to make the asking even more timely.

We have examined many developments that subject our bodies, our minds, our seeds, or our genes to manipulation. Is too much being done to make us predictable and manageable? Are we indeed becoming more and more malleable under the onslaught?

On the other hand, much has happened in the past few years to indicate that people will not surrender their integrity casually. In many ways our interest in being masters of our fate has heightened as we have perceived that we seem to be losing control over certain areas of our lives and our environment. Control of one's fate is, after all, deeply imbedded in Western thought.

Stephan L. Chorover, the socially concerned psychologist, does not see the human cause as a lost cause:

"With courage and insight we may still be able to seize back from technology both our freedom and our dignity."

In retrospect we see that the worldwide explosion, now in the early 1970's commonly called the Counterculture, was largely a revulsion. It was a revulsion against the way societies were drifting. It was, at least partly, a revulsion against feeling a loss of personal control, of having a say in important matters affecting our lives. It was a revulsion against a sense of depersonalization. It was a revulsion against being manipulated by militarists, by the advertising specialists who generate materialism, and by other skilled people-handlers.

During this same time we saw women by the millions vigorously object to devoting their entire energies to husband, children, and the home, and win recognition in a host of fields. They made great gains in equality and self-direction. And in the United States millions of blacks, male and female, were winning similar recognition and successfully establishing the same rights.

During these same past few years we have seen corporations around the world rushing to head off employee resentment by redesigning working conditions. Jobs by the millions have been "enriched" to provide more responsibility and autonomy. And there has been much less demand for nonessential conformity.

City bosses in vast, anonymous New York City have long been accustomed to treating the public as conglomerate blocs and pressing familiar levers to exercise control. Now suddenly they are dealing with people through more than sixty tough-talking community boards and through several thousand newly created block associations.

In the same past few years the American people for the first time in history cast off a President, one whose manipulations and thrusts for power became unmistakably evident.

Perhaps, too, the enormous recent growth of the Human Potential movement is significant. The movement, accompanied by experiments in expanding consciousness and by some exotic improvisations, tells us something. It demonstrates, at the very least, that a widespread quest for personal meaning and effectiveness is going on. Certainly, the movement springs in large part from a revulsion against a deepening sense of the loss of identity, a pervasive feeling that our lives are becoming less and less ours to command.

THE BEHAVIORISTS SHOW A NEW INTEREST IN SELF-DIRECTION

It also seems significant that some of the brashness and stress on mechanical behavior have gone out of the behavioral psychology movement. Albert Bandura of Stanford, the first true behaviorist to become president of the American Psychological Association, urged operant conditioners to behave with caution.[1] He has been a social learning theorist and a close observer of the human scene. In his presidential address, he spoke of the "external consequences" that operant conditioners manipulate in order to shape behavior. These consequences, he said, "influential as they often are,

are not the sole determinates of human behavior, nor do they operate automatically." At another point he said: "The most reliable source of opposition to manipulative control resides in the reciprocal consequences of human interactions. People resist being taken advantage of." Is this reference to "reciprocal consequences" an acknowledgment that Man does not like to be pushed around and in some situations can become quite intractable?

To his colleagues who function largely by manipulating rewards, Bandura had this to say: "Most people value their self-respect above commodities."

And he spoke of the possibility that people could, by themselves, apply behaviorist principles to solve their problems: "Applications of self-control practices demonstrate that people are able to regulate their own behavior in preferred directions by arranging environmental conditions most likely to elicit it and administering self-reinforcing consequences to sustain it."

No behavior shaper need be on hand to control the action! That was a fairly radical stand for a behaviorist to take. Except for Bandura, Israel Goldiamond of the University of Chicago had been just about the only prominent figure in the behavior-modification movement to stress the desirability of autonomy.

A year earlier Goldiamond said of his work with patients at his "self-control clinic" that the program requires self-analysis and self-management. "We train the patient to discover and change the contingencies, the environmental conditions, that govern his behavior." His group asks the patient to set the goals he wishes to attain, and to keep a daily log on his behavior.

A new handbook on behavior modification devotes a large section to four possible ways that people with problems can be encouraged to use self-management:

• Self-determination of goals
• Self-recording
• Self-evaluation
• Self-reinforcement

Apparently with a little counsel, any reasonably bright person can now become a modifier of his own behavior.

The behaviorists have also toned down some of their scorn for the humanists, whom they once dismissed as soft and unscientific. One noted behaviorist, Lloyd Homme, an early colleague of Skinner's, has written a book with the seemingly improbable title of *Humanistic Behaviorism.*

Leonard Krasner, another pioneer of behaviorism, has in his recent writings been trying to show that behaviorism and humanism are not opposite camps at all, nor mutually exclusive. He concluded recently that "the abusers of behavior modification are those who utilize the behavior-influence process for the 'good of others' without involving those concerned."

Behavior modifiers in a number of fields are finding that they have a far better chance of achieving *lasting* changes if the subjects themselves have a say in deciding what the goals of the therapy are, what rules to follow during treatment, and what rewards or penalties are imposed.

In reporting new trends in the treatment of juvenile delinquents, John Burchard and Paul Harig report that more success seems to be achieved if the behaviorists help the troubled youth change his environment than if they try simply to change his behavior. They explained: "The evidence seems to suggest that when a youth perceives himself as having produced a change in the behavior of others he is more apt to continue to engage in the behavior that produced the change than if he perceives that someone else did it for him."

The O'Leary team in its study of new trends in the use of behavior modification in schools reported in 1976: "Research on self-control is proliferating and . . . results are exciting."

Even in the matter of treating people for gross eating disorders, two behaviorists surveying the various approaches being tried concluded: "When a subject becomes his own therapist, self-regulated consequences provide a strategy which can be maintained long after termination of a formal treatment program."

This new stress on the possibility of self-management moves the behaviorists closer to the humanists. For many years the humanist Carl Rogers has stressed that the client must be central to the therapy and that the therapy must be nondirective. The therapist's main role, in his view, is simply to provide the feedback that will help the discouraged client improve his concept of himself.

Some of the younger humanistically oriented psychiatrists, far from being "soft," are quite hard-nosed in seeing to it that the clients assume responsibility for themselves. Peter Breggin, a psychiatrist from Washington, D.C., is an example. He believes that everyone has the potential to be free and self-directed. And he won't settle for less from his clients. This attitude has caused him to shun most of the ordinary strategies of psychiatrists. He states:

"I never give any drugs. I never refer anyone for electroshock. I never hospitalize anyone or talk to a client's relatives. If the person is suicidal or psychotic I don't bring in the authorities. I treat each human being as a totally responsible person."

Only in such an environment, he contends, can people wrestle with personal responsibility, find out who they are and who they want to be.[2]

The behaviorists have long held that their approach to the problems of people is superior to the humanists' because they come up with measurable results. But in the offices and factories of the world dramatic measurable evidence of results from the humanistic approach is now available. The evidence shows improvement in performance and morale after managers and employees were given more self-direction. The long trend toward increased control over employees, and manipulation of them, seems to be reversing. Corporate managements did not start instituting job-enrichment programs out of whim or a desire for popularity. They had a worsening problem as they kept moving toward further standardization, rationalization, and specialization. Humanism provided a solution.

Thirty years ago the revolutionary MIT psychologist Kurt Lewin certainly persuaded me, in the several days I spent with him, that an atmosphere of freedom and dignity brings out the best in people and makes them more productive. His early research findings have since been impressively confirmed and amplified by evidence that industrial psychologists like Chris Argyris and Warren Bennis have compiled, and by projects Abraham Maslow encouraged.

In industry the body of evidence shows that production and morale rise and complaints from customers about defects drop when employees are given more personal responsibility. This applies whether they are day laborers, salesmen, or managers. In pioneering tests, assembly lines and time clocks were thrown out. Women workers were told that now they would each have charge of the entire assembly of a product, whether it was a small computer or a radio. With guidance they drew up a procedure that seemed to them to be the most logical way to handle the entire assembly.

The women showed initial stress as they struggled with the new responsibility. But within a year they were making products better and faster than before, and were feeling much better about themselves.

An outstanding example of *removing* manipulation to achieve higher results occurred in Topeka, Kansas, at the Gaines Pet Food plant. The experiment began almost by accident. Sale of certain Gaines products had exceeded the production capacity of an existing plant in Kankakee, Illinois. The Kankakee plant had been plagued by labor trouble, absenteeism, goofing off, even some sabotage. In setting up the new Topeka plant the management was persuaded by humanistically oriented organizational specialists in the company to set up a workers' democracy. Some jobs were more boring than others. To get them shared, the workers were divided into teams to handle each operation. Each team selected its own foreman. At the beginning of each shift they met briefly to divide up job assignments, air grievances, and discuss the best way to meet the day's production quota. Each worker was gradually trained to handle virtually any job in the entire plant, even to monitoring the vast cooking machines.

Team leaders interviewed and hired new team members when there was a vacancy. And the team as a group could discipline any member who appeared to be goofing off. Even allowing for new technology, this work force at Topeka was soon outproducing the conventional force back at Kankakee by about 25 percent.

A number of team workers themselves expressed pride in the new way of producing the food. *Newsweek* quoted one worker as explaining: "It's different. I am still just a worker, but I have something to say about my job."

CONTROL BY THE SELF OF THE WAY THE BODY FUNCTIONS

The partisans of self-direction recently have been receiving support from some surprising sources: yogis, Zen monks, experimental psychologists, and physiologists.

It seems that an individual can create changes within his body that were assumed to be beyond the power of will. Some Skinnerians have suggested that purposeful will is a vastly exaggerated human conceit. Man acts primarily in response to stimuli.

Obviously there are some things Man can will to do without much stimulus beyond whim. He can wiggle his toe, roll his eyes, kick a tire, or climb a mountain because it is there. His skeletal muscles are under the direct, instant command of the brain through the somatic nervous system. But medical wisdom has long held that many of our crucial, soft visceral body parts — our internal organs, glands, blood vessels, all of which lack skeletal muscles — work more or less automatically. We can't modify them by any amount of will because they are controlled by the autonomic nervous system. That system, it was long believed, was under the control of the lower brain, not of conscious awareness.

Holy men in the Far East who seemed to perform remarkable feats in willfully controlling these organs were usually dismissed as benign fakirs. But a few scientists with their sophisticated new measuring instruments decided to check out some of these yogis and Zen monks. They came away amazed.

The conclusion had to be that these men were finding fulfillment through prodigious self-discipline and concentration. They sometimes used these skills to influence their bodies in astonishing ways. The author Gerald Jonas has reported on a number of feats that showed up on the instruments of the scientists.[3]

One holy man could, by willing it, race his heartbeat up to 250 to 300 beats a minute, a very dangerous rate. And he did it without moving a muscle! Another man could cut his heartbeat from sixty-three beats a minute to twenty-four in a matter of minutes. Yet another could, when he willed it, cut his body metabolism rate in half. This is pretty remarkable when you consider that a normal person's metabolism drops no more than 12 percent even at the deepest point of sleep. Another could, by willing it, break out in a sweat. He had learned to do this, he explained, in order to keep warm in a cold cave. Another could produce a ten-degree difference in temperature between his thumb and his little finger. The scientists concluded that this had to involve mental control over local blood vessels in his hand.

There have been numerous reports that when holy men meditate their oxygen consumption drops by as much as 20 percent. And their brain waves take on a distinctive configuration.

All this has profound implications for Western Man, beset by many ills of the viscera and brain that are caused by tension, anxiety, stress, and the sense of a decline in autonomy.

But how can Western Man learn to influence bodily parts without spending years practicing in solitude as the monks do? One answer lies in discovering the state of mind that governs a desired organic function, and then concentrating on achieving that state of mind.

It would help a person to achieve that state of mind if he could get constant guidance on how he was doing, some kind of feedback. Neal Miller, an experimental psychologist with behaviorist leanings at Rockefeller University, was a pioneer in this. He found that a subject could learn to lower his own blood pressure by watching a monitoring device, which showed at all times his blood-pressure level.

At about the same time a psychologist of Japanese descent, Joe Kamiya of the Langley Porter Neuropsychiatric Institute in San Francisco, was getting people to slow down their brain waves. They could learn to increase the proportion of the slow alpha waves, which are usually associated with a relaxed state. To help them do this he used an exquisite monitoring device, now famous as the biofeedback machine. His interest was in what goes on in the brain during different states of consciousness, such as bad dreams.

There are now many kinds of biofeedback machines. Which you use depends on what body activity you wish to monitor. Is it muscle tension, blood pressure, or brain-wave activity? A baseline of your normal pattern is established by computer. Then, when you succeed in altering the pattern in a desired way, the machine notifies you by a light signal or a soft-tone signal. The machine in effect reports: "Hey, that's right! Do it again!"

With such guidance from instruments people have trained themselves to lower their blood pressure as much as fifteen millimeters. They have eased the related disease of hypertension. They have increased or slowed down their heartbeats. They have eliminated or eased migraine headaches. Psychologists at the Menninger Foundation discovered that subjects suffering from migraine headaches found relief after concentrating on the thought that their hands were getting warmer. One housewife's hands did rise ten degrees in temperature, apparently because of a surge of blood into her hands. And her migraine vanished simultaneously.[4] Possibly it was related to a parallel easing of the constriction of the blood vessels at the seat of her headache. Other investigators have greatly reduced tension headaches in a large proportion of their patients by training them to concentrate on relaxing the muscles of the forehead.

One of the more intriguing feats of biofeedback has been training people to straighten out irregularities in their heartbeats by exercising the will. Bernard Engel, a physiological psychologist in Baltimore, was apparently the first investigator to achieve this. It has been duplicated by others. Patients can be weaned off the biofeedback machine as they learn to tune in on their circulatory systems. At home, by will power, many continue to straighten out irregularities.

The cost of spending weeks of daily training in a laboratory — and the waning effect that sometimes occurs when people stay home and concentrate — has created interest in buying one's own biofeedback machine. And an eager electronics industry has rushed forth with a host of do-it-yourself, popular-priced models. Prices of biofeedback machines range from $98 to $10,000. Thus far, the verdict of investigating psychologists is that the popular-priced models are erratic or not sensitive enough to do the job. Presumably, technological advances will in time produce useful machines at a price many persons can afford.

When biofeedback machines were introduced they were quickly hailed by some behaviorists as another triumph in operant conditioning. Subjects were being conditioned to concentrate in the desired direction because of the good consequences — the reward of a green light or a bell tone — of doing so. Some early investigators even gave subjects a nickel or a free picture as a "reward" every time they succeeded in changing organic behavior in a desired direction.

But there are momentous differences between biofeedback training and much of operant conditioning. There is no comparable shaper-subject relationship. The "desired" behavior is desired by the subject in his own best interest, not something the shaper desires him to achieve. And the machine signals are as much guides as rewards. It is the will of the subject that primarily is being tested, not the skill of a controller. The power to make the change and the control over it is in the hands of the subject.

Furthermore, it is now being discovered that many of the achievements gained by will with the help of biofeedback machines can be gained in another way. And the other way certainly has nothing to do with operant conditioning. It is meditation.

Again, the pioneering work was done over centuries by religious people in Asia. In the mid-1970's Transcendental Meditation was introduced to the world on a large scale, primarily by a jolly white-bearded guru from India, Maharishi Mahesh Yogi. He trimmed out some of the more mystical, nonessential aspects of Oriental meditation to produce something that became known to millions of Western practitioners as "TM." It is now taught for a fee by some eight thousand of his teachers.

There is still some mysticism left in it. The initiate makes a symbolic offering of flowers, sweet fruit, and a white handkerchief, and is led into a private room where there is a candle and incense burning under the portrait of a great swami. After chanting a while in Sanskrit the teacher gives the initiate his soft-sounding *mantra,* or secret word. Each teacher has about seventeen words he can hand out, always in great secrecy. After that it is fairly easy. The meditator places himself in a relaxed position, shuts his eyes, and lets the mantra float around in his mind for about twenty minutes. He does this twice a day. And in each successful session he attains what TM teachers call a "higher consciousness." Others refer to the result as a kind of mental bath.

Scientific tests with groups of well-trained Transcendental Meditators show that many do indeed put their bodies through significant physical changes. The heart slows by several beats a minute. Oxygen consumption falls abruptly, often by about 15 percent. A greater density of the slow, alpha brain waves appear on the monitors. In many cases blood pressure has averaged significantly slower after a TM session.

Herbert Benson, a cardiologist at Harvard, was one of the investigators who put recording instruments on Transcendental Meditators. He tested them not only when they were in meditation but at other times, to check how long the effects lasted. His special interest is easing hypertension. He had been working with biofeedback machines to lower his patients' blood pressure when Transcendental Meditation came to his attention. His own patients on biofeedback had said they achieved drops in blood pressure by having relaxing thoughts.

It occurred to Benson to wonder why he should bother with the expensive machinery of biofeedback if meditating would work. He was unsure of the impact of the cultist aspects of Transcendental Meditation, but was impressed with the evidence of relaxation it displayed.

He began testing to see if he could skip the incense, flowers, mantra, and so on, and still come up with good results on the provable dimension of relaxation. He found he could. The four ingredients common to both

Transcendental Meditation and the new Relaxation Response approach he developed were:[5]

• Quiet surroundings; preferably, the eyes should be closed.
• A totally passive attitude.
• A comfortable position; but one must not lie down, which would encourage one to lapse into naps.
• Something to concentrate on, such as a word or a particular feeling. Benson finds that concentrating on repeating the word *one* whenever one draws a breath seems to work just about as well as a secret mantra.

Patients with high blood pressure, he reports, show a significant, *sustained* reduction of ten to fifteen points on their high reading after a couple of months of home practice.

He has found that hypnosis with suggested deep relaxation produces many of the same physical changes as meditation. Possibly meditation is a mild form of self-hypnosis.

Pure rest may also contribute somewhat to the reduction of stress. Investigators at the University of Michigan reported in 1976 that blood samples were taken of two groups. One group consisted of TM teachers. The second consisted of untrained people who just rested for twenty to thirty minutes with their eyes closed. The levels for catecholamine, a body chemical associated with stress, responded about the same for both groups.[6]

Benson contends, however, that meditation is not a form of sleep, even though meditators may occasionally doze off. The drop in oxygen consumption is more dramatic and far more abrupt than in sleep. The preponderance of alpha brain waves found with meditators is not present with sleepers.

All these various approaches to the self-direction of behavior offer encouraging evidence that despite the drift toward manipulation, the individual can still thrive on his own efforts.

Millions of people are seeking a greater measure of self-direction. And they have good reason to believe that they can achieve it.

Toward a More Robust Model of Man

26

*We are witnessing the erosion, perhaps the final
erosion, of the idea of man as something splendid.*
— Leon Kass, molecular biologist and ethicist

What we have been examining throughout most of this book are mani-
festations of a new, reductive view of Man. In the process of the examination we have found ourselves coming up against some rather large
questions about human existence in the late twentieth century. Here is
just one example: What happens to the historic concept of the sanctity of
the individual if surgeons can transplant heads, if biologists can double
the size of the head, and if physiologists can put electrical machinery
inside the head?

The technologies for behavior control and for manipulating the beginning and end of life have produced some happy results, as I have discussed on previous chapters. But on the whole they add up to regress
rather than progress. And as just noted, many people are searching in a
variety of ways to prove that the individual still counts and can be self-
directing.

When people are subject to ever-new forms of manipulation it becomes
more difficult for them to feel they are anything special, let alone splendid.
Our justification for feeling unique is under assault. Experimenters keep
devising new ways to demonstrate the malleability of Man.

To recapitulate, let us list the kinds of manipulations of Man that scientists have succeeded in making or are trying to make:

- The manipulation of behavior according to scientifically precise methods
- The manipulation of moods
- The radical manipulation of human reproduction

- The manipulation of personality
- The manipulation of the brain and its functioning
- The manipulation of genetic traits
- The manipulation of longevity and human body parts
- The manipulation of situations in which we make decisions
- The manipulations that deprive us of privacy
- The manipulation of the uniqueness of the human species

Most of this power to manipulate is being achieved because the "sheer fascination with experimentation has outstripped our moral imagination," as one ethicist put it. It has also outstripped our capacity to evaluate the long-term desirability of research before the research is launched. Social responsibility must be assumed at some point because the potentialities for the modification of Man are so far-reaching.

As I see it, the broader issues and what to do about them can be summarized in these twelve propositions:

1. Major research efforts that can result in the modification of human beings should be coordinated with social policies that seek to optimize the condition of Man for the next fifty years. The broadening of our knowledge of human nature need not be limited to facts that are pleasant or flattering to know. But in the face of the accelerating rate at which knowledge of how to alter human behavior and development is being acquired, it has become crucial that social policies be devised for the guidance and control of such research.

2. Unnatural techniques for creating human life raise extraordinarily sensitive issues. Research in these areas — from seed banks and embryo transplants to artificial wombs, gene manipulation, presexing, cloning, and chimeras — should be under the firm control of a national commission on reproductive research in each country.

3. Any scientific effort to modify an individual's behavior may well represent an undermining of his freedom unless the effort is undertaken at his request (or that of his legal guardians). It often is commendable to administer therapy that will make a person more capable of living according to his own concept of a tolerable or enjoyable life. But it usually is deplorable to condition anyone to a predictable pattern of behavior that the conditioner has thought up.

4. Dehumanization is a serious problem. A dehumanizing effect is particularly likely to result when there is laboratory control over human creation, when bodily parts involving personality are altered or replaced, when man-animal combinations are attempted, when disembodied sex organs in culture become a continuous source for the seeds of future human beings.

5. The desire of an individual to benefit from a new technology may not coincide with sound social policy. The barren wife may desire to achieve a baby by the creation of an embryo in a laboratory. On the other hand, population trends in the world raise questions about the wisdom of extraordinary efforts to add to the methods of human reproduction. And the research needed in order to help that wife fulfill her wish will help provide the means for undertaking such socially dubious undertakings as the development of artificial wombs, presexing, cloning, chimeras, and other forms of unnatural creation.

6. Many of the measures to modify disruptive behavior by conditioning, chemicals, or surgery are purportedly being undertaken to help people. Actually, the primary motivation in most cases is to make them easier to manage. An honest examination of motivation should precede all such efforts.

7. The possibility that new forms of behavior control may lead to political misuse warrants serious concern. Democracy is not a fragile flower. The fact remains, however, that the great majority of the people in the world live under dictatorial regimes eager to learn more effective and subtle control techniques. And additional countries are slipping under authoritarianism almost yearly. As natural resources become scarcer, pressures toward totalitarianism may mount throughout the world. Thus, the development and unveiling of techniques with a potential for controlling or modifying people should not be seen just as an exercise in virtuosity.

8. If Russia and China want to go ahead and develop artificial wombs in order to increase womanpower in their work forces, let them do it. They will be paying a price for the innovation that a humane, forward-looking society should be unwilling to pay.

9. Any mass program to raise IQ's by twenty points would create problems and costs outweighing the benefits. The same is true of many mass programs that would add twenty years to each of our lives.

10. The notion of trying to create ideal humans should be avoided: the concept runs counter to our incomparably greater asset of genetic diversity.

11. The behavioral sciences have given trained practitioners a capacity to control people effectively in many situations. These same behavioral sciences are discovering — in industry, education, and elsewhere — that people are most effective when they are treated as free, responsible, self-directed, reasonably unique persons.

12. Finally, life scientists are at their most reckless when they propose, as some do, undercutting the family as the basic unit of reproduction in favor of some animal-husbandry model. It may be tantalizing to test the feasibility of manufacturing babies in laboratories or hatcheries, to en-

vision magnificent male studs fathering thousands of babies via the seed banks, to arrange for females to carry other females' babies to term. What all these plans neglect to explore are the problems of getting such offspring reared.

The kind of rearing that children receive over two formative decades certainly plays a far more profound role in the kind of adults that emerge than any ingenious techniques used in the initial production of the infants. It is now often overlooked that the nurture and rearing of a fine youngster is a gratifying achievement and a difficult human goal to realize.

Virtually every society in the world has found that an intact, natural family is the best environment in which to rear children. It provides them with warmth, protection, identity. Only in a good family — or an arrangement that provides the warmth, protection, and identity of a good family — can a child be assured of automatic love. The parental and reproductive drives of the husband and wife give the child stability and continuing reassurance. No other arrangement thus far tested has proved that it can compete with the family on a large-scale basis.

Proposals for separating the recreational from the procreational aspects of human intercourse are dubious. Leon Kass asks:

"Is there possibly some wisdom in that mystery of nature which joins the pleasure of sex, the communication of love and the desire for children in the very activity by which we continue the chain of human existence? I think there is."

The family is already a threatened institution because of the current vogue for temporary relationships. Before we take too seriously the novel forms of procreation made possible by technology we should demand that the proponents broaden their wisdom by some interdisciplinary sharing of knowledge with child psychologists and family-life specialists.

GOALS FOR MAN

To cope with the increasing number of people and institutions seeking to shape our lives — for whatever benevolent, malevolent or misguided reasons — we need to assess our position and erect defenses. Our defenses will be greatly enhanced if we can achieve a rough consensus on human goals. What values should be discouraged and encouraged?

Setting goals can itself be a problem. Some of the manipulators have announced goals, too, some of them banal ones. I have noted H. J. Muller's long list of the human characteristics he thought should be promoted, many of them having little or nothing to do with genes. B. F. Skinner has from time to time felt impelled to set goals for Man. He has added goals and dropped goals. At one point he proposed: "Let men be happy, informed, skillful, well behaved and productive." His friend and ideological rival, the psychologist Carl Rogers, joshed him about two of these goals: "well behaved" and "happy." He suggested that Skinner chose them for other people, not for himself: "I would hate to see Skinner become 'well-behaved' as the term would be defined for him by behavioral scientists." As for "happy," Rogers said of Skinner, "The most awful fate I can imagine for him would be to have him constantly 'happy.' It is the fact that he is very unhappy about many things which makes me prize him."[1]

At other times Skinner has said that a good society should provide for "the pursuit of happiness." That sounds safe enough since it is in the Declaration of Independence. But the Austrian psychiatrist Viktor Frankl, one of the wisest men I ever met, would fault him even on that. Frankl suggests that the pursuit of happiness is a frivolous, frustrating preoccupation. Rather, he said, Man should immerse himself in projects that will give him good *reasons* to feel happy.

Ideally, any plan for coping with people shapers should include the development of a framework of relevant positive and negative values. Perhaps some well-endowed foundation could underwrite the research for, and dissemination of, a Hierarchy of Human Values. It could be used for assessing the consequences of all proposals to alter human behavior and development. Perhaps in each advanced country, boards of wise men and women could be established by the national science associations to make evaluations and judgments.

Some of the better ideas for a hierarchy of values probably would come out of humanistic or existentialist psychology. At any rate, it would be well if Man spent considerable time thinking about values and discussing them in connection with scientific innovations.

Such contemplation would be especially desirable in medical schools and in the biology and computer-science departments of universities.

My own thoughts on a hierarchy of values (offered below) come out of reflecting on the developments reported in this book. At the risk of sounding even less plausible than Skinner or Muller I present them here.

A low value should be placed on developments that make Man

- More predictable
- More remote from family ties
- More irresponsible
- More dehumanized
- More adulterated
- More immediacy-oriented
- More dependent
- More malleable

On the other hand, societies should place a high value on individuals who manage to achieve

- Responsible self-direction
- Individual fulfillment
- The rearing of fine children
- Clear-cut uniqueness as a person
- A spontaneous way of life
- A capacity for independent thinking

And societies themselves should be esteemed to the extent that they place a high value on:

- Esteeming individual growth more than the remodeling of people
- Cherishing the dignity, strength, and importance of each individual
- Planning predictable machines but not predictable people
- Encouraging people to strive to be pilots rather than pawns
- Providing for the right to a large degree of individual privacy
- Guaranteeing free citizens freedom from coercion
- Promoting respect for the evolutionary miracle of human life
- Demonstrating social imagination by seeking to anticipate the implications of innovations that would affect human behavior and development
- Promoting awareness as a defense against manipulation

On this last point it might be noted that possession of a "recognition reflex" in itself makes people less vulnerable to being unwittingly controlled, modified, shaped, swayed, or manipulated.

The above nine goals could be a start toward a reasonable social policy for any society confronted with revolutionary scientific developments intimately affecting the lives of people.

A MORE STURDY MODEL OF MAN

The high and low values cited might also contribute to evolving a realistically robust model of Man in the coming years.

We started out this exploration by noting six images of Man developed by twentieth-century scientists in accordance with their perceptions of Man's nature. One saw Man as captain of his fate. But most saw Man in terms of his malleability, his plasticity.

These plastic models of Man have given many scientists what they feel is their right to try to change Man mentally, emotionally, and physically in any way they think appropriate. This has been particularly true of behaviorists, psychosurgeons, reproductive biologists, molecular biologists, and biocomputer specialists. Most are persons of good-will and generous impulses. But they have often been thoughtless about the human implications of what they have been doing, and proposing.

We can accept the fact that humans are potentially malleable. We can accept the deterministic view that genes, instincts, and environment play a shaping role in our lives. We can accept that all three can be modified or manipulated by others.

The fact remains, however, that with will and skill we all can largely control our environmental influences, curb or sublimate our instincts, make the most of our genetic inheritance, and minimize the effect of manipulation.

Given the capabilities of the scientists, which cannot be denied, and given Man's major aspirations, I propose as valid and useful this model of Man for the coming decades:

Man is many things, admirable and unadmirable, but he has the potential for self-mastery and social direction, and he is at his best when he is achieving them.

To a very great extent each of us can be his own shaper.

Appendix: Man's Nature

As Viewed by Major Schools of Psychological and Philosophical Thought

The viewpoints are given as the answers to five questions about the nature of Man that are especially relevant to this book. They were developed by Natalie Terbovic Warren, a psychologist of New York, after an exchange of thoughts with the author.

1. Is human nature basically good, bad, or neutral?

NEGATIVE VIEWPOINTS

Freudians. Man is evil by nature, driven by biologically rooted instincts, particularly sex and aggression, which are merely masked by the conventions of society.

Empiricists (Hobbes). Man acts only in his own self-interest.

Utilitarians (Bentham; Mill). All human acts are only the products of self-interest.

Hedonists. Man acts only to fulfill his needs for pleasure and to escape encounters with pain.

Ethologists (Lorenz). Man is innately evil in the sense that he is born with a need to aggress against others of his own species.

Orthomolecular psychiatrists (Newbold). Substantially the same as the ethologists' view of Man's aggressiveness.

NEUTRAL VIEWPOINTS

Behaviorists (Watson; Skinner). Man is not naturally evil or good. He becomes whatever his environment makes of him.

Social learning theorists (Bandura; Mischel). People learn to be either good or bad as a result of what gains them rewards and allows them to avoid punishments.

Existentialists (Sartre). Man is not essentially good or bad but every act that every man undertakes contributes to the essence of human nature. Therefore, if all men are good, then human nature is good and vice versa.

POSITIVE VIEWPOINTS

Neo-Freudians (Fromm; Erikson). Man has a distinct potential for goodness but whether or not he attains goodness depends on the nature of the society in which he lives and the attitudes of the individuals with whom he interacts, especially during his childhood. Good deeds are not the manifestation of underlying biological needs as Freud suggested.

Humanists (Maslow; Rogers). Man has a potential for goodness that will come to the forefront unless social demands or poor decisions on the part of the individual interfere.

Romanticists (Rousseau). Man possesses a natural goodness from birth and any evil acts that are committed by men are caused by detrimental social conditions rather than anything inherent in Man.

2. **Is individual human behavior the result of free will or is it almost completely determined by environment, heredity, early childhood, or God?**

The viewpoints of the various schools of thought can best be shown graphically in a continuum from complete determinism to no determinism:

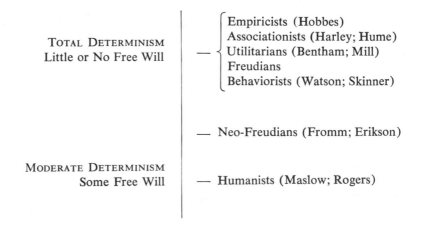

TOTAL DETERMINISM
Little or No Free Will

— Empiricists (Hobbes)
Associationists (Harley; Hume)
Utilitarians (Bentham; Mill)
Freudians
Behaviorists (Watson; Skinner)

— Neo-Freudians (Fromm; Erikson)

MODERATE DETERMINISM
Some Free Will

— Humanists (Maslow; Rogers)

$$
- \begin{cases} \text{Rationalists (Descartes)} \\ \text{Existential Humanists (May;} \\ \quad \text{Frankl)} \\ \text{Theistic Existentialists (Buber;} \\ \quad \text{Tillich; Fournier)} \end{cases}
$$

No Determinism
Complete Free Will

$$
- \begin{cases} \text{Transcendentalists (Kant)} \\ \text{Existentialists (Sartre)} \end{cases}
$$

3. What are the sources of human determinism?

INSTINCTUAL DRIVES

Freudians. Man is controlled by instinctive, biological drives (sex, hunger, elimination, and the like). All human behavior is merely a product of compromises between these instinctual needs and the conflicting demands of society. Instinctual drives often operate at the unconscious level so that man is not only controlled, but is often unaware of this control.

Ethologists (Lorenz). Much of man's behavior stems from an inborn drive to aggress against other men.

Orthomolecular psychiatrists (Newbold). Similar to the position of the ethologists.

Humanists (Rogers; Maslow). Man has an innate drive toward self-actualization.

GENETIC ENDOWMENT

"Inherited I.Q. school" (Jensen; Shockley; Herrnstein). Human intelligence is largely inherited. All behavior related to intelligence has predetermined restrictions placed on it by an individual's inheritance of "intelligence genes."

Eysenck's biological theory of personality. Much human behavior is the result of inherited tendencies to neuroticism and extroversion or introversion.

ENVIRONMENTAL FORCES

Methodological behaviorists (Watson). Behavior is determined completely by the environment.

Radical behaviorists (Skinner). Environmental factors are by far the most important determinants of behavior, though some genetic factors do play a role.

Social learning theorists (Bandura; Berkowitz). Most behavior, particularly social behavior, is the product of environmental learning rather than the expression of innate predispositions.

Neo-Freudians (Fromm; Erikson). The social and cultural environment is the primary force that shapes human behavior. Biological drives are far less important.

Marxists. The means of production or the economic system plays a major role in shaping the individual's beliefs and values and, in turn, determines much of his behavior.

Humanists (Rogers; Maslow). Behavior can be determined by certain social and other environmental factors. If the basic needs for physical well-being and self-esteem are not met, then these environmental factors can warp the individual and cause him to behave in ways which fail to enhance his personal development. Once these basic needs are met, however, the individual can proceed to realize goals of a higher order, like self-actualization, which are not under external environmental control.

SPIRITUAL FORCES

These are widely ignored by scientists because, if they do exist, they can't be measured or explained. Theologians of most of the major religions credit God or gods with influencing much of human behavior. Throughout history, until recent decades, most cultures have emphasized the spiritual over the material.

4. Do human beings and animals function according to the same principles of behavior or are there innate tendencies in man that go beyond simple animal needs?

Associationists (Hume; Hartley). All human behavior, like all other animal behavior, is simply a product of mechanically occurring associations between sensations.

Empiricists (Hobbes). Human nature is an entirely mechanical phenomenon subject only to the laws of motion. There is no higher quality such as a soul present in Man.

Freudians. Man, like all other animals, is motivated entirely by instinctually based wishes for the reduction of tensions that stem from unfulfilled biological needs. Human behavior is governed by a pleasure-seeking, discomfort-avoiding principle. Even behavior that would appear to be motivated by higher goals really only represents the sublimation of baser desires.

Behaviorists (Skinner). All behavior, human and otherwise, is governed by the principles of operant conditioning. Human consciousness is simply a by-product of behavior and not something that makes human behavior different from animal behavior. Such concepts as free will, "inner motivation," and "autonomous man" are inaccurate, useless, and even dangerous in that they delude Man into believing that he is something special when he is not.

Rationalists (Descartes). The lower animals are like machines in that their behavior is totally controlled by physical laws. Man has, in addition to an animal nature, a rational nature, which allows for judgment, choice, and free will.

Neo-Freudians (Fromm, Erikson). Man has capacities that extend beyond the fulfillment of simple, biological needs. Humans have a special potential for achieving goodness. Social factors determine whether this potential will be fulfilled. Man's good works can come from motives of a higher order; they are not simply distortions of baser goals.

Humanists (Maslow, Rogers). Human nature is distinct from and, in some ways, superior to animal nature. Every human being has the capacity to grow toward an ideal form, to become self-actualized. The drive for self-actualization can be warped or even destroyed by adverse environmental conditions, including a poor social environment. Examples of needs that are important to man but not to animals are those for love, self-esteem, recognition, integrity, self-respect, and self-understanding.

Existentialists (Sartre). Man differs from other animals in that he has the capacity to recognize that he alone is responsible for his actions. This recognition leads man to a state of despair and aloneness which is distinctly human.

5. What should be done to enhance man's future? What are the ultimate goals of human development?

Freudians. The outlook is pessimistic. Man's basically selfish nature is the root of his problems and the biological origin of this nature makes it difficult to modify except through evolution. Freud suggested that encouragement of the life instincts (like sex) and de-emphasis of the destructive instincts (like aggression) might improve Man's lot but he was not too hopeful about the outcome of such measures.

Neo-Freudians (Fromm; Erikson). Man's faults stem from the negative influences of society. Therefore, if society can be changed to enhance Man's good points and discourage his weaknesses, then his lot will improve. The ultimate goal of human development is the creation of a society that allows all men to fulfill their potential for creativeness and good works.

Marxists (Marx; Fromm). The development of a socialist society in which all citizens share in the means of production and in what is produced is the solution to human problems and the ultimate goal of human progress. Individuals fulfill themselves most fully as they contribute to group objectives.

Behaviorists (Skinner). The ultimate goal of human development is the *survival* of the human species and anything that facilitates this goal is by definition desirable. The arrangement of environments is the key to encouraging survival. Properly arranged environments can be used to encourage behaviors that increase the likelihood of survival (productivity, peace, population control, and so on) and discourage behaviors that decrease the likelihood of survival (violence, aggressiveness, overpopulation, and so on).

Empiricists (Hobbes). The use of empirical data to predict and control human behavior is a desirable way to enhance man's development.

Utilitarians (Bentham; Mill). Society should control behavior so that the greatest good will come to the greatest number of people.

Humanists (Maslow; Rogers). Maslow: Every individual has within himself an innate will to grow and seek self-actualization but this inner nature is delicate and is easily set aside or ignored in the face of contradictory environmental pressures. Therefore, the key to improving the quality of human life lies in society's recognition and encouragement of the expression of this inner nature. Rogers: Individuals need unconditional approval from others in order to accept themselves as persons and then to begin to grow and develop to their full potential. Therefore, an increase in the amount of approval is seen as the key to improving the human condition. Self-actualization for all is the goal of human development.

Existential humanists (May; Frankl). May: Modern man must recognize that he still has a will of his own and learn to recognize those occasions when he exercises it. By regaining the sense of will, man can improve his condition. Frankl: Every man needs to find meaning in his life by discovering something or someone to live for. This source of meaning provides a basis for making decisions and removes him from a state of aloneness and despair. The meaningfulness of one's life is prerequisite to human development.

Theistic existentialists (Buber; Tillich; Tournier). Open, honest interaction with God and other human beings is a source of guidance in the exercise of free will and the enhancement of human development.

Chapter References

1. *The Emerging Plastic Image of Man*

1. Lloyd Homme, Polo C'de Baca, Lon Cottingham, and Angela Home, "What Behavioral Engineering Is," *Psychological Record* 18 (1968): 431.
2. For a book-length study of Man's images, see *Changing Ideas of Man,* Policy Research Report 4, prepared by the Center for the Study of Social Policy, Stanford Research Institute, Menlo Park, Calif., May 1974.
3. H. L. Newbold, *The Psychiatric Programming of People* (New York: Pergamon Press, 1972), p. 59.
4. *Physical Manipulation of the Brain,* special supplement to *Hastings Center Report* (May 1973), p. 3. A report of a discussion held by the Research Group on Behavior Control of the Institute of Society, Ethics and the Life Sciences, Hastings, N.Y.

2. *Pioneers in Programming Pigeons and People*

1. Vincent Bugliosi, with Curt Gentry, *Helter Skelter* (New York: W. W. Norton & Company, 1974). See especially pp. 232–238.
2. I. E. Farber, Harry F. Harlow, and Louis Jolyon West, "Brainwashing, Conditioning and DDD (Debility, Dependency and Dread)," in Roger Ulrich, Thomas Stachnik, and John Mabry, eds., *Control of Human Behavior* (Glenview, Ill.: Scott Foresman and Company, 1966), pp. 322–330; reprinted from *Sociometry* 20 (1957): 271–283. See also Selma Fraiberg, "The Science of Thought Control," *Commentary* (May 1962), pp. 220–229, and Albert D. Biderman and Herbert Zimmer, eds., *The Manipulation of Human Behavior* (New York: John Wiley & Sons, 1961).
3. Harvey Wheeler, ed., *Beyond the Punitive Society* (San Francisco: W. H. Freeman and Co., 1973), pp. 8–9.

4. Elizabeth Hall, "Will Sucess Spoil B. F. Skinner?" *Psychology Today* (Nov. 1972), p. 65.

5. Kenneth Goodall, "Shapers at Work," *Psychology Today* (Nov. 1972), pp. 61, 62.

3. *The Behavior Shapers Take On the Public*

1. Perry London, *Behavior Control* (New York: Harper & Row, 1969), p. 121.

2. David H. Barlow, Harold Leitenberg, and W. Stewart Agras, "Experimental Control of Sexual Deviation Through Manipulation of the Noxious Scene in Covert Sensitization," *Journal of Abnormal Psychology* 74, no. 5 (1969): 596–601.

3. Robert P. Hawkins, Robert F. Peterson, Edda Schweid, and Sidney W. Bijou, "Behavior Therapy in the Home: Amelioration of Problem Parent-Child Relations with Parent in a Therapeutic Role," *Journal of Experimental Child Psychology* 4 (Sept. 1966): 99–107.

4. Martha E. Bernal, John S. Duryee, Harold L. Pruett, and Beverlee J. Burns, "Behavior Modification and the Brat Syndrome," *Journal of Consulting and Clinical Psychology* 32 (1968): 447–455.

5. Kenneth Goodall, "Shapers at Work," *Psychology Today* (Nov. 1972), pp. 132, 133.

6. C. Quarti and J. Renaud, "A New Treatment of Constipation by Conditioning: A Preliminary Report," in C. M. Franks, *Conditioning Techniques in Clinical Practice and Research* (New York: Springer Publishing Co., 1964), pp. 219–227.

7. "Connubial Bliss Through Tokens," *Human Behavior* (Aug. 1974), p. 31, and Richard B. Stuart, "Operant-Interpersonal Treatment for Marital Discord," *Journal of Consulting and Clinical Psychology* 33, no. 6 (1969): 675–682.

8. Goodall, p. 134.

9. Marvin Karlins and Lewis M. Andrews, *Biofeedback* (New York: Warner Press, 1972), p. 138.

10. Richard A. Winett and Robin C. Winkler, "Current Behavior Modification in the Classroom: Be Still, Be Quiet, Be Docile," *Journal of Applied Behavior Analysis* 5 (1972): 499–504, and Daniel K. O'Leary, "Behavior Modification in the Classroom: A Rejoinder to Winett and Winkler," ibid., 505–511.

11. Chris Argyris, review of *Beyond Freedom and Dignity* by B. F. Skinner, in *Harvard Educational Review* (Nov. 1971), pp. 550–567. For examples of relapse, see reports dealing with schools and delinquency in Harold Leitenberg, ed., *Handbook on Behavior Modification and Behavior Therapy* (Englewood Cliffs, N.J.: Prentice-Hall, 1976), pp. 405–452 and 475–515.

12. Karl H. Pribram, "Operant Behaviorism: Fad, Fact-ory and Fantasy," in Harvey Wheeler, ed., *Beyond the Punitive Society* (San Francisco: W. H. Freeman & Co., 1973), p. 108.

4. *Mood Management*

1. Roger W. Sperry, "Left-Brain, Right-Brain," *Saturday Review* (Aug. 9 1975), pp. 30–33.
2. Jim Warren, "Peace Pills for Presidents?" *Psychology Today* (Oct. 1973), p. 59.
3. Gerald Jonas, "Into the Brain," *New Yorker* (July 1, 1974), p. 52.
4. Albert Rosenfeld, *The Second Genesis: The Coming Control of Life* (New York: Pyramid Communications, 1972), p. 242.
5. Douglas E. Smith, Melvyn B. King, and Bartley G. Hoebel, "Lateral Hypothalamic Control of Killing: Evidence for a Cholinoceptive Mechanism," *Science* (Feb. 6, 1970), pp. 900, 901.
6. James P. Miller, "The Brain Machines Are Here," *Human Behavior* (Aug. 1974), pp. 17–22.
7. Leo E. Hollister, "Drug Therapy," *New England Journal of Medicine* 286, no. 22 (1972): 1195.
8. Laurie Jacobson, "Feedback on Biofeedback," *Human Behavior* (July 1974), p. 50.
9. Gilbert Honigfeld and Alfreda Howard, *Psychiatric Drugs* (New York: Academic Press, 1972), pp. 48–51.
10. Marilyn Ferguson, *The Brain Revolution* (New York: Taplinger Publishing Co., 1973), p. 189.
11. José M. R. Delgado, *Physical Control of the Mind: Toward a Psychocivilized Society* (New York: Harper & Row, Colophon Books, 1971), p. 147.
12. Robert G. Heath, "Electrical Self-Stimulation of the Brain in Man," *American Journal of Psychiatry* 120 (1963): 553–554.
13. Elliot S. Valenstein, *Brain Control: A Critical Examination of Brain Stimulation and Psychosurgery* (New York: John Wiley & Sons, 1973), p. 105.
14. Ibid., p. 122.
15. Rosenfeld, p. 205.
16. Delgado, p. 166.
17. Ibid., p. 116.
18. Ibid., p. 108.
19. Robert Coughlan, "The Chemical Mind-Changers," *Life* (Mar. 14, 1963), p. 81.
20. Nathan S. Kline, "The Future of Drugs and Drugs of the Future," *Journal of Social Issues* 27, no. 3 (1971): 81, 82.
21. Robert Coughlan, "Behavior by Electronics," *Life* (Mar. 7, 1963), p. 92.

5. *New Personalities for Old*

1. Maya Pines, *The Brain Changers* (New York: Harcourt Brace Jovanovich, 1973), pp. 123–124.
2. Ibid., p. 127.
3. P. H. Gray, "Theory and Evidence of Imprinting in Human Infants," *Journal of Psychology* 46 (1958): 155–156.

4. John Money, Jean G. Hampson, and John L. Hampson, "Imprinting and the Establishment of Gender Role," *Archives of Neurology and Psychiatry* 77 (Mar. 1957): 333–334.
5. Maya Pines, "Head Head Start," *New York Times Magazine* (Oct. 26, 1975), p. 60.
6. *Individual Rights and the Federal Role in Behavior Modification,* a study prepared by the staff of the Subcommittee on Constitutional Rights, Committee on the Judiciary, U.S. Senate, 93d Cong., 2d sess. (Washington, D.C.: Government Printing Office, Nov. 1974), Appendix G, p. 190.
7. Elliot S. Valenstein, *Brain Control: A Critical Examination of Brain Stimulation and Psychosurgery* (New York: John Wiley & Sons, 1973), pp. 174–177.
8. Vernon H. Mark and Frank R. Ervin, *Violence and the Brain* (New York: Harper & Row, 1970), p. 86.
9. Lee Edson, "The Psyche and the Surgeon," *New York Times Magazine* (Sept. 30, 1973), p. 14.
10. Stephan L. Chorover, "Psychosurgery: A Neuropsychological Perspective," *Boston University Law Review* (1974), pp. 244–245.
11. Ibid., pp. 238–240.

6. *On Making Man More Tractable*

1. H. L. Newbold, *The Psychiatric Programming of People* (New York: Pergamon Press, 1972), pp. 23–24.
2. Elliot S. Valenstein, *Brain Control: A Critical Examination of Brain Stimulation and Psychosurgery* (New York: John Wiley & Sons, 1973), pp. 254, 255.
3. Maya Pines, *The Brain Changers* (New York: Harcourt Brace Jovanovich, 1973), p. 208.
4. Peter R. Breggin, "The Return of Lobotomy and Psychosurgery," *Congressional Record,* Feb. 24, 1972, p. 5572.
5. Valenstein, p. 223.
6. Lee Edson, "The Psyche and the Surgeon," *New York Times Magazine* (Sept. 30, 1973), p. 14.
7. Peter R. Breggin, "Psychosurgery for the Control of Violence — Including a Critical Examination of the Work of Vernon Mark and Frank Ervin," *Congressional Record,* Mar. 20, 1972, pp. E3380–E3386.
8. Stephan L. Chorover, "Big Brother and Psychotechnology," *Psychology Today* (Oct. 1973), p. 43.
9. Pines, pp. 205, 206.
10. Robert A. Burt, "Why We Should Keep Prisoners from the Doctors," *Hastings Center Report* (Feb. 1975). Published by the Institute of Society, Ethics and the Life Sciences, Hastings, N.Y.
11. Gordon Rattray Taylor, *The Biological Time Bomb* (New York: New American Library, Mentor Books, 1969), p. 134.

12. Michael Sheard, "Aggression and Lithium," *Science News* (Apr. 3, 1971), p. 240.
13. Nick DiSpoldo, "Arizona's 'Clockwork Orange' Bill," *New York Times,* June 20, 1974.
14. Jessica Mitford, *Kind and Usual Punishment: The Prison Business* (New York: Alfred A. Knopf, 1973), pp. 120–124. See also " 'Behavior Mod' Behind the Walls," *Time* (Mar. 11, 1974), pp. 74, 75.
15. Barbara Yuncker, "What Is 'B-Mod'?" *New York Post,* Mar. 9, 1974.
16. Wayne Sage, "Crime and the Clockwork Lemon," *Human Behavior* (Sept. 1974), pp. 19, 20.
17. " 'Behavior Mod' Behind the Walls," as cited in n. 14.
18. Breggin, *Congressional Record,* Feb. 24, 1972, p. 5569.
19. Statement by William R. Hutton, executive director of the National Council of Senior Citizens, before the U.S. Senate Special Committee on Aging, Chicago, Sept. 14, 1971, in "Hearings on Long-Term Care of the Special Committee on Aging," pp. 1424–1448.
20. Gardner C. Quarton, "Deliberate Efforts to Control Human Behavior and Modify Personality," *Daedalus* (Summer 1967), p. 848.
21. Stanley S. Robin and James J. Bosco, "Ritalin for School Children: The Teachers' Perspective," *Journal of School Health* 43, no. 10 (Dec. 1973): 624–628. (Version of paper presented in Feb. 1973 at the American Educational Research Association meeting, New Orleans.)
22. Peter Schrag and Diane Divoky, *The Myth of the Hyperactive Child* (New York: Pantheon Books, 1975), p. 75.
23. Connie Bruck, "Battle Lines in the Ritalin War," *Human Behavior* (Aug. 1976), pp. 25–33.
24. Ibid., p. 32.
25. Edward T. Ladd, "Pills for Classroom Peace?" *Saturday Review* (Nov. 21, 1970), p. 68.
26. Robert J. Bazell, "Panel Sanctions Amphetamines for Hyperkinetic Children," *Science* (Mar. 26, 1971), p. 1223.
27. T. Kenny, R. Clemmens, B. Hudson, G. Lents, R. Cicci, and P. Nair, "Characteristics of Children Referred Because of Hyperactivity," *Journal of Pediatrics* 79, no. 4 (1971): 618–623.
28. Gerald Solomons, a commentary on drugs in managing hyperkinetic children, *Medical Opinion* (May 1972), p. 59.
29. Sydney Walker III, "We're Too Cavalier about Hyperactivity," *Psychology Today* (Dec. 1974), p. 43.
30. Nicholas Von Hoffman, "A Crime-causing Diet," *New York Post,* Sept. 12, 1974.
31. Herbert E. Rie, "Ritalin Controversy Continues," *New York Times,* June 26, 1974.
32. Daniel Safer, Richard Allen, and Evelyn Barr, "Depression of Growth in Hyperactive Children on Stimulant Drugs," *New England Journal of Medicine* (Aug. 3, 1972), p. 218.

7. *Building Brighter — or Duller — People*

1. Albert Rosenfeld, *The Second Genesis: The Coming Control of Life* (New York: Pyramid Communications, 1972), p. 248.
2. Ibid., p. 252.
3. Marilyn Ferguson, *The Brain Revolution* (New York: Taplinger Publishing Co., 1973), p. 255.
4. Roger Lewin, "Starved Brains," *Psychology Today* (Sept. 1975), pp. 32, 33.
5. Gerald Leach, *The Biocrats* (London and Baltimore: Penguin Books, 1972). See Chapter 8, "Brains," p. 220.
6. "Chemicals to Aid Learning Tested on Monkeys," *Behavior Today* (Mar. 11, 1974), p. 69.
7. Joan Arehart-Treichel, "The Role of Hormones in Learning, Memory and Behavior," *Science News* (May 20, 1972), pp. 334, 335.
8. Robert T. Francoeur, *Utopian Motherhood: New Trends in Human Reproduction* (South Brunswick & New York: A. S. Barnes & Co., Perpetua Books, 1973), pp. 163–170.
9. Philip W. Landfield, Ronald J. Tusa, and James L. McGaugh, "Effects of Posttrial Hippocampal Stimulation on Memory Storage and EEG Activity," *Behavioral Biology* (Apr. 1973), p. 485.
10. G. Ungar, D. M. Desidero, and W. Parr, "Isolation, Identification and Synthesis of a Specific Behavior Inducing Brain Peptide," *Nature* 238 (1972): 198–202.
11. Gerald Jonas, "Into the Brain," *New Yorker* (July 1, 1974), p. 57.
12. Lewin, p. 33.

8. *Keeping Track of — and Controlling — the Populace*

1. Jessica Mitford, *Kind and Usual Punishment: The Prison Business* (New York: Alfred A. Knopf, 1973), pp. 225–226.
2. J. A. Meyer, "Crime Deterrent Transponder System," Institute of Electrical and Electronic Engineers' *Transactions on Aerospace and Electronic Systems,* vol. AES-7, no. 1 (Jan. 1971) pp. 2–20.
3. Ralph K. Schwitzgebel, "Electronic Innovation in the Behavioral Sciences: A Call to Responsibility," *American Psychologist* 22 (1967): 364–370.
4. Mitford, p. 227.
5. Alan F. Westin, *Privacy and Freedom* (New York: Atheneum Publishers, 1967), p. 88.
6. *U.S.* v. *King,* 335 F.Supp. 523, 542–543, S.D. Calif., 1971.
7. Westin, p. 309.
8. Text of editorials read over WFSB-TV, Hartford, Connecticut, on May 9, 10, 22, and 23, 1974.
9. Paul S. Entmacher, "Computerized Insurance Records," *Hastings Center Report* (Nov. 1973), pp. 8, 9.
10. Aryeh Neier, *Dossier* (New York: Stein & Day, 1974), pp. 200, 210.

11. "Talk of the Town," *New Yorker* (Apr. 8, 1974), p. 31.
12. David E. Rosenbaum, "Threats by Nixon Reported on Tape Heard by Inquiry," *New York Times,* May 16, 1974.

9. *Molding Super Consumers, Super Athletes, Super Employees*

1. Vance Packard, *The Hidden Persuaders* (New York: David McKay Co., 1957); *The Status Seekers* (New York: David McKay Co., 1959); *The Waste Makers* (New York: David McKay Co., 1960).
2. "The New Six," *Media Decisions* (May 1973), pp. 68–76.
3. Michel Pierre Janisse and W. Scott Peavler, "Pupillary Research Today: Emotion in the Eye," *Psychology Today* (Feb. 1974), p. 63.
4. Marilyn Elias, "How to Win Friends and Influence Kids on Television," *Human Behavior* (Apr. 1974), pp. 16, 17.
5. Alan F. Westin, *Privacy and Freedom* (New York: Atheneum Publishers, 1967), pp. 295–296.
6. Del Hawkins, "The Effects of Subliminal Stimulation on Drive Level and Brand Preference," *Journal of Marketing Research* (Aug. 1970), p. 322.
7. Arnold J. Mandell, "A Psychiatric Study of Professional Football," *Saturday Review WORLD* (Oct. 5, 1974), pp. 12–16.
8. Arnold J. Mandell, "Pro Football Fumbles the Drug Scandal," *Psychology Today* (June 1975), p. 39.
9. Gwilym S. Brown, "Winning One for the Ripper," *Sports Illustrated* (Nov. 26, 1973), p. 146.

10. *Engineering Voter Approval*

1. Carl Behrens, "Bringing Biology into the Political Picture," *Science News* 98 (Nov. 28, 1970): 434.
2. Leonard V. Gordon, "The Image of Political Candidates: Values and Voter Preference," *Journal of Applied Psychology* 56, no. 5 (1972): 382–387.
3. Alan L. Otten, "Computing Democratic Winners in '72," *Wall Street Journal,* Dec. 11, 1970.
4. James Kahn, "Social Scientists' Role in Selection of Juries Sparks Legal Debate," *Wall Street Journal,* Aug. 12, 1974.
5. Edward Tivnan, "Jury by Trial," *New York Times Magazine* (Nov. 30, 1975), pp. 54–56.
6. Marcia Chambers, "A Jury Watcher Advises Mitchell's Defense on the Reactions of Panel Members," *New York Times,* Apr. 1, 1974.
7. Martin Arnold, "How Mitchell-Stans Jury Reached Acquittal Verdict," *New York Times,* May 5, 1974.
8. Wallace Turner, "California Sued over Practice of Investigating Background of Prospective Jurors," *New York Times,* Sept. 14, 1976.

11. Behavior Control by the New Hypnotechnicians

1. Ernest R. Hilgard, "Hypnosis Is No Mirage," *Psychology Today* (Nov. 1974), p. 121.
2. James H. Austin, "Eyes Left! Eyes Right!" *Saturday Review* (Aug. 9, 1975), p. 32.
3. Hilgard, "Hypnosis Is No Mirage," p. 122.
4. "Remote-Control Hypnosis," *Time* (July 2, 1965), p. 37.
5. Harry Arons, *Hypnosis in Criminal Investigation* (Springfield, Ill.: Charles C. Thomas, Publisher, 1973), p. 176.
6. "Hypnotism: Culturally Unacceptable," *Science News* (Dec. 13, 1975), p. 375.
7. G. H. Estabrooks, "Hypnosis Comes of Age," *Science Digest* (Apr. 1973), p. 44.
8. Marilyn Ferguson, *The Brain Revolution* (New York: Taplinger Publishing Co., 1973), p. 105.
9. Henry S. Tugender and William E. Ferinden, *A Handbook of Hypnooperant Therapy* (South Orange, N.J.: Power Publishers, 1972), pp. 7–17.
10. Arons, p. 31.
11. Estabrooks, "Hypnosis Comes of Age."

12. Selling and Storing the Seed of Man

1. Gordon Rattray Taylor, *The Biological Time Bomb* (New York: New American Library, Mentor Books, 1969), p. 36.
2. Robert T. Francoeur, *Utopian Motherhood: New Trends in Human Reproduction* (South Brunswick and New York: A. S. Barnes & Co., Perpetua Books, 1973), pp. 22, 23.
3. Gerda Gerstel, "A Psychoanalytic View of Artificial Donor Insemination," *American Journal of Psychotherapy* 17 (1963): 64–77.
4. L. H. Levie, "An Inquiry into the Psychological Effects on Parents of Artificial Insemination with Donor Semen," *Eugenics Review* 59, no. 97 (1967).
5. R. G. Edwards, "Fertilization of Human Eggs in Vitro: Morals, Ethics and the Law," *Quarterly Review of Biology* (Mar. 1974).

13. Starting Man in a Test Tube

1. Gebhard F. B. Schumacher, B. G. Brackett, Joseph Fletcher, J. J. Marik, L. Mastroianni, Jr., L. B. Shettles, R. Tejada, and Edward T. Tyler, "In Vitro Fertilization of Human Ova and Blastocyst Transfer: An Invitational Symposium," *Journal of Reproductive Medicine* (Nov. 1973), p. 196.
2. John Arehart-Treichel, "Test-Tube Babies in the Making," *Science News* (Feb. 24, 1973), p. 24.
3. Edward Grossman, "The Obsolescent Mother," *Atlantic* (May 1971), p. 39.
4. Robert T. Francoeur, *Utopian Motherhood: New Trends in Human Reproduction* (South Brunswick and New York: A. S. Barnes & Co., Perpetua Books, 1973), pp. 57–59.

5. David Rorvik, "The Embryo Sweepstakes," *New York Times Magazine* (Sept. 15, 1974), p. 17.
6. Ibid.
7. Francoeur, p. 286, summarizing a report by David Rorvik in *Look* (May 18, 1971).
8. *Ob.-Gyn. News* (Oct. 1, 1976).
9. D. de Dretzer et al., "Transfer of a Human Zygote," *Lancet* (Sept. 29, 1973), pp. 728, 729.
10. Francoeur, pp. 93–94.
11. R. A. S. Lawson, L. E. A. Rowson, and C. E. Adams, "The Development of Cow Eggs in the Rabbit Oviduct and Their Viability after Retransfer to Heifers," *Journal of Reproductive Fertility* 28 (1972): 313–315.
12. R. G. Edwards, "Fertilization of Human Eggs in Vitro: Morals, Ethics and the Law," *Quarterly Review of Biology* (Mar. 1974), p. 3.
13. Mariel Revillard, "Legal Aspects of Artificial Insemination and Embryo Transfer in French Domestic Law and Private International Law," Ciba Foundation Symposium (New York: Associated Scientific Publishers, 1973), pp. 77–90.
14. "Protection of Human Subjects," *Federal Register* 38, no. 221 (Nov. 16, 1973): 31738–31749.
15. Shyoso Ogawa, Kahei Satoh, Mitsuma Hamada, and Hajime Hashimoto, "In Vitro Culture of Rabbit Ova Fertilized by Epididymal Sperms in Chemically Defined Media," *Nature* (Aug. 4, 1972).

14. *Leased Wombs, Artificial Wombs, Nonhuman Wombs*

1. David Rorvik, "The Embryo Sweepstakes," *New York Times Magazine* (Sept. 15, 1974), p. 17.
2. Sheila K. Johnson, "The Business in Babies," *New York Times Magazine* (Aug. 17, 1975), p. 11.
3. "People," *Time* (Nov. 25, 1974), p. 67.
4. Robert T. Francoeur, *Utopian Motherhood: New Trends in Human Reproduction* (South Brunswick and New York: A. S. Barnes & Co., Perpetua Books, 1973), p. 106.
5. Edward Grossman, "The Obsolescent Mother," *Atlantic* (May 1971), p. 43.
6. Francoeur, p. 58.
7. Yu-Chih Hsu, "Post-Blastocyst Differentiation in Vitro," *Nature* (May 14, 1971), pp. 100–101.
8. Francoeur, pp. 54, 55.

15. *Males or Females to Order*

1. Lawrence Galton, "Parent and Child: Choosing the Sex of the Child," *New York Times Magazine* (June 30, 1974), p. 22.
2. Albert Rosenfeld, *The Second Genesis: The Coming Control of Life* (New York: Pyramid Communications, 1972), p. 124.

3. "Choice of Sex," *Intellectual Digest* (Nov. 1973), p. 56.
4. R. J. Ericsson, C. N. Langevin, and M. Nishino, "Isolation of Fractions Rich in Human Y Sperm," *Nature* (Dec. 14, 1973), p. 422.
5. David Rorvik, *Brave New Baby* (New York: Doubleday & Company, 1969), p. 49.
6. Dorothea Bennett, "Sex Ratio in Progeny of Mice Inseminated with Sperm Treated with Hy-Y Antiserum," *Nature* (Nov. 30, 1973), pp. 308, 309.
7. K. V. Chachava, P. Ya. Kintraya, T. G. Zhgenti, and K. L. Keburiya, "Effect of Fetal Sex on the Sign of the Electric Charge on Spermatozoa," *Bulletin of Experimental Biology and Medicine* 68 (1970): 572–574.
8. John Fletcher, "Moral Problems in Genetic Counseling," *Pastoral Psychology* (Apr. 1972), p. 59.
9. "The Second Sex — Even to Mom and Dad," *Behavior Today* (June 3, 1974), pp. 156–157.
10. Ibid.
11. Colin Campbell, "The Manchild Pill?" *Psychology Today* (Aug. 1976), p. 86.
12. "Wider Gap Forecast between the Number of Elderly Women and Men," *Behavior Today* (June 14, 1976), p. 7.

16. Modifying Our Genetic Blueprints

1. Joan Arehart-Treichel, "Putting Human Genes on the Map," *Science News* (Oct. 11, 1975), pp. 234–235.
2. "Can Genetic Defects Be Corrected in Cells?" *Nature New Biology* (Mar. 3, 1971), pp. 1–2.
3. Harold M. Schemeck, Jr., "Researchers Discover Method of Transferring Cell Nuclei in Living Tissue," *New York Times,* June 17, 1974.
4. James J. Nagle, "Genetic Engineering," *Bulletin of the Atomic Scientists* (Dec. 1971), pp. 43–44.
5. See Stanley N. Cohen, "The Manipulation of Genes," *Scientific American* (July 1975), p. 25; and "The Gene Transplanters," *Newsweek* (July 17, 1974), p. 54.
6. Tabitha M. Powledge, "The Genetic Engineers Still Await Guidelines," *New York Times,* Feb. 15, 1976.

17. The Quality Control of New Humans

1. Arthur J. Snider, "The Genetic Control of Man," *Science Digest* (Apr. 1971), p. 56.
2. Richard M. Restak, *Premeditated Man* (New York: Viking Press, 1975), pp. 81, 82.
3. Barbara J. Culliton, "Amniocentesis: HEW Backs Test for Prenatal Diagnosis of Disease," *Science* (Nov. 7, 1975), pp. 538–539.

4. "Abortion: The Edelin Shock Wave," *Time* (Mar. 3, 1975), pp. 36–37.
5. John A. Robertson, "Medical Ethics in the Courtroom," *Hastings Center Report* (Sept. 1974), pp. 1–3.
6. Leroy Augenstein, *Come, Let Us Play God* (New York: Harper & Row, Publishers, 1969), p. 116.
7. Carl Jay Bajema, "The Genetic Implications of Population Control," *Bio-Science* (Jan. 15, 1971), p. 72.
8. Barbara J. Culliton, "Genetic Screening: States May Be Writing the Wrong Kinds of Laws," *Science* (Mar. 5, 1976), pp. 927–928.
9. Ibid.
10. Mary A. Telfer, Gerald R. Clark, David Baker, and Claude E. Richardson, "Incidence of Gross Chromosomal Errors among Tall, Criminal American Males," *Science* 159 (1968): 1249–1250.
11. Frederick Ausubel, Jon Beckwith, and Kaaren Janssen, "The Politics of Genetic Engineering: Who Decides Who Is Defective?" *Psychology Today* (June 1974), p. 38.
12. Robert F. Murray, Jr., "Problems Behind the Promise: Ethical Issues in Mass Genetic Screening," *Hastings Center Report* (Apr. 1972), pp. 10–11.

18. *Packaging Superior People*

1. Bentley Glass, "Human Heredity and Ethical Problems," in Jay Katz, *Experimentation with Human Beings* (New York: Russell Sage Foundation, 1972), pp. 451–456. Originally delivered as the first annual address to the Society for Health and Human Values, Los Angeles, Calif., Oct. 29, 1970.
2. Carl Jay Bajema, "The Genetic Implications of Population Control," *Bio-Science* (Jan. 15, 1971).
3. E. Fuller Torrey, *Ethical Issues in Medicine* (Boston: Little, Brown and Company, 1968), p. 386.
4. Angela Haines, "Controlling Height," *New York Times Magazine* (Apr. 4, 1976), p. 74.
5. H. Tristram Engelhardt, Jr., "The Philosophy of Medicine: A New Endeavor," *Texas Reports on Biology and Medicine* 31, no. 3 (Fall 1973), pp. 448–449, and *Mind-Body: A Categorical Relation* (The Hague: Martinus Nijhoff, 1973), pp. 164–165; A. Segaloff, "Progress in the Treatment of Breast Cancer," proceedings of a symposium held at Gonville and Caius College, University of Cambridge, England, Sept. 9, 1967, p. 8.
6. Theodosius Dobzhansky, *Mankind Evolving* (New York: Bantam Books, Matrix Editions, 1970), p. 328.
7. "Families and Intellect: Scores to Increase," *Science News* (Apr. 17, 1976), pp. 245–246.

19. *Duplicating Humans of a Desired Model*

1. Leon R. Kass, "Making Babies: The New Biology and the 'Old' Morality," *Public Interest* (Winter 1972), pp. 18–56.

2. Albert Rosenfeld, *The Second Genesis: The Coming Control of Life* (New York: Pyramid Communications, 1972), p. 119.

3. Anna Witkowska, "Parthenogenetic Development of Mouse Embryos in Vivo," *Journal of Embryology and Experimental Morphology* 30, no. 3 (1973): 547–560.

20. *The Work on Man-Animals and Man-Computers*

1. "A Computer Under Your Hat," *Science* (Feb. 28, 1976), p. 133.

2. G. Poste and P. Reeve, "Formation of Hybrid Cells and Heterokaryons by Fusion of Enucleated and Nucleated Cells," *Nature New Biology* 229 (Jan. 27, 1971): 123–124.

3. "Genetic Engineering: Clashing Views," *Science News* (Nov. 2, 1974), p. 277.

4. G. H. Zeilmaker, "Fusion of Rat and Mouse Morulae and Formation of Chimaeric Blastocysts," *Nature* (Mar. 9, 1973), pp. 115–116.

5. Carl O. Povlsen, Niels E. Skakkebaek, Jørgen Rygaard, and Gerd Jensen, "Heterotransplantation of Human Foetal Organs to the Mouse Mutant Nude," *Nature* (Mar. 15, 1974), p. 248.

6. Jean Rostand, *Humanly Possible* (New York: Saturday Review Press, 1973), p. 72.

7. Maya Pines, *The Brain Changers* (New York: Harcourt Brace Jovanovich, 1973), pp. 24–26.

8. Ibid.

9. Lawrence R. Pinneo, "Persistent EEG Patterns Associated With Overt and Covert Speech" (mimeographed). For a copy write to the Neurophysiology Program, Stanford Research Institute, Menlo Park, California 94025.

21. *Resetting the Clocks in Our Bodies*

1. Vance Packard, *The Waste Makers* (New York: David McKay Co., 1960). First twelve chapters.

2. Paul Ferris, "The Fountain of Youth Updated," *New York Times Magazine* (Dec. 2, 1973), p. 38.

3. Walter Sullivan, "Science Seeks to End the Miseries of Aging," *New York Times,* Nov. 6, 1975.

4. Barnett Rosenberg, Gabor Kemeny, Lawrence G. Smith, Ira D. Skurnick, and Mary J. Bandurski, "The Kinetics and Thermodynamics of Death in Multicellular Organisms," *Mechanisms of Ageing and Development* 2 (1975): 290–291.

5. Gene Bylinsky, "Science on the Trail of the Fountain of Youth," *Fortune* (July 1976), p. 139.

6. Rona Cherry and Laurence Cherry, "Slowing the Clock of Age," *New York Times Magazine* (May 12, 1974), p. 20.

7. Gene Bylinsky, "Science on the Trail," pp. 136–138.

8. Jean L. Marx, "Aging Research (1): Cellular Theories of Senescence," *Science* (Dec. 20, 1974), p. 1107.

9. Cherry and Cherry, "Slowing the Clock of Age."
10. Albert Rosenfeld, "The Longevity Seekers," *Saturday Review* (Mar. 1973), p. 46.
11. "Treating Senility," *Science Digest* (June 1976), p. 20.

22. *The Human of Totally Replaceable Parts*

1. Joan Arehart-Treichel, "Organ Transplants: What Hope for Patients?" *Science News* (Nov. 16, 1974), pp. 314–315.
2. Gerald Leach, *The Biocrats* (Baltimore: Penguin Books, 1972), pp. 323–324, and correspondence with Yukihiko Nosé of the Cleveland Clinic.
3. Willard Gaylin, "Harvesting the Dead: The Potential for Recycling Human Bodies," *Harper's* (Sept. 1974), pp. 23–30.
4. "Kidney Patients' Program in Trouble," *New York Times,* Sept. 25, 1975.
5. "The Totally Implanted Heart," a report by the Artificial Heart Assessment Panel of the National Heart and Lung Institute, June 1973, p. 115.
6. Robert J. White, L. R. Wolin, L. C. Massopust, Jr., N. Taslitz, and J. Verdura, "Primate Cephalic Transplantation: Neurogenic Separation, Vascular Association," *Transplantation Proceedings* 3, no. 1 (Mar. 1971). See also White et al., "Cephalic Exchange Transplantation in the Monkey," *Surgery* 70, no. 1 (July 1971): 602–603.
7. Oriana Fallaci, "The Dead Body and the Living Brain," *Look* (Nov. 28, 1967), p. 99.
8. Joan Arehart-Treichel, "Nerve Regeneration," *Science News* (July 17, 1976), p. 42.

23. *Second Thoughts by the Human Engineers*

1. David Rorvik, *Brave New Baby* (New York: Doubleday & Co., 1969), p. 106.
2. Horace Freeland Judson, "Fearful of Science," *Harper's* (Mar. 1975), p. 74.
3. "The Ethics of Human Experimentation," *Science News* (Mar. 1, 1975), p. 134.
4. Richard M. Restak, *Premeditated Man* (New York: Viking Press, 1975), pp. 111–114.
5. Herrman L. Blumgart, "The Medical Framework," *Daedalus* (Spring, 1969), p. 257.
6. Restak, pp. 114–116.
7. American Psychological Association, Inc., *Ethical Principles in the Conduct of Research with Human Participants* (Washington, D.C., 1972), p. 74.

24. *On Controlling the Controllers*

1. "Politics and Genes," *Newsweek* (Jan. 12, 1976), pp. 50–52.
2. Richard M. Restak, *Premeditated Man* (New York: Viking Press, 1975), pp. 157–158.

25. *New Trends That Can Enhance Self-direction*

1. Albert Bandura, "Behavior Theory and the Models of Man," *American Psychologist* (Dec. 1974), pp. 859–869.
2. Robert J. Trotter, "Peter Breggin's Private War," *Human Behavior* (Nov. 1973), p. 56.
3. Gerald Jonas, *Visceral Learning* (New York: Viking Press, 1972). See especially pp. 99–106.
4. Ibid., pp. 120–122.
5. Herbert Benson, *The Relaxation Response* (New York: William Morrow and Co., Inc., 1975), pp. 78–79.
6. "TM: Understanding the Rest of It," *Science News* (June 19, 1976), p. 390.

26. *Toward a More Robust Model of Man*

1. Carl R. Rogers and B. F. Skinner, "Some Issues Concerning the Control of Human Behavior: A Symposium," *Science* (Nov. 30, 1956), p. 1057.

Acknowledgments

During the four and a half years in which this book was in preparation I became indebted to a great many people for information, counsel, or assistance. A few dozen were friends. A number of the scientists and other experts who were helpful are indicated in the text. However, I want to single out several by name because of their repeated acts of helpfulness:

James Bosco, educational researcher; Vincent Collins, anesthesiologist: José Delgado, physiologist; Joseph Feldschuh, sperm-bank director; Denham Harman, biochemist; David Krech, psychologist; James McConnell, psychologist; James Nagle, geneticist; Yukihiko Nosé, artificial-organ designer; Stanley Robin, sociologist; Barnett Rosenberg, biophysicist; Landrum Shettles, obstetrician; Natalie Terbovic Warren, social psychologist; Robert White, brain surgeon.

Others who were repeatedly helpful were John and Kay Tebbel, Kennett and Eleanor Rawson, Nancy Mitchell, Kathy Hammell, Charles Concino, Carol Colman, David Reif, Sandra Clark, Sara Crafts, Tom Clark, Sandra Wood, and all the members of my family.

I am indebted to assistance received from officials of the Institute of Society, Ethics and the Life Sciences, Hastings, New York, and personnel of the American Psychiatric Association and the American Psychological Association.

Index